PROTEIN SYNTHESIS

A SERIES OF ADVANCES

VOLUME 2

Edited by Edwin H. McConkey

Department of Molecular, Cellular and
*   Developmental Biology*
University of Colorado
College of Arts and Sciences
Boulder, Colorado

MARCEL DEKKER, INC.,   New York and Basel

MARCEL DEKKER, INC.

270 Madison Avenue, New York, New York 10016

LIBRARY OF CONGRESS CATALOG CARD NUMBER: 75-155743

ISBN: 0-8247-1464-4

Current printing (last digit):
10 9 8 7 6 5 4 3 2 1

PRINTED IN THE UNITED STATES OF AMERICA

ABOUT THE SERIES

These volumes will present reviews on the structure, functions, bio-synthesis, and regulation of the protein-synthesizing apparatus in prokaryotic and eukaryotic cells. Articles on various RNA's will be included regularly because all known functions of RNA are related directly or indirectly to protein synthesis or the regulation of protein synthesis. Although it could be argued that *all* of molecular biology and biochemistry is indirectly related to protein synthesis, this series will maintain a relatively restricted focus. The molecular biology of DNA or the mechanisms of enzyme action, for example, will not be considered as general subjects, although specific genes, such as those that code for ribosomal components, and specific enzymes, such as peptidyl transferase and amino acid: tRNA ligases may be treated in detail.

A unified series of reviews covering all aspects of protein synthesis is much needed. Although the principal reactions of protein synthesis have been identified, the subject is far from exhausted. The number of researchers active in the field is very large, and the rate of publication appears to increase continually. Consider, for example, the *Proceedings of the National Academy of Sciences* and the *Journal of Molecular Biology,* two major sources of literature on protein synthesis. Together these journals published approximately 10,000 pages in 1970; by my interpretation, the *Proceedings* contained nearly 1000 pages and the *Journal of Molecular Biology* contained nearly 2000 pages of articles directly pertinent to protein synthesis. This is only a minor part of the total current literature on the subject, of course.

There are few persons fortunate enough to be able to keep abreast of this deluge of new information through personal contacts. Most of those who are either active or interested in the field must depend almost entirely on published articles and many find that their attempts to maintain familiarity with current research are becoming less and less adequate. There is no shortage of reviews; indeed, one can scarcely do a thorough job of reading reviews, even if one almost abandons original reports.

Unfortunately, the intentions and achievements of reviewers vary as widely as the quality of the research that they review. Excellence appears as rarely in scientific work as in any other profession, and a reviewer who uncritically catalogs the contents of abstracts does his colleagues no service. *Protein Synthesis: A Series of Advances* will provide critical evaluations, rather than mere summaries of recent data. Attempts to reconcile conflicting observations will be featured. In order to maximize the usefulness of these reviews as reference sources, full literature citations will be given, alphabetically arranged.

Professional scientists and graduate students should find these volumes helpful in following the development of this complex subject. When possible, several closely related articles will be published in the same volume. The basic mechanisms of protein synthesis, which were exhaustively documented in the 1969 *Cold Spring Harbor Symposium,* are also the subject of all the articles of Volume 1 of *Protein Synthesis*. This subject will undoubtedly generate many future reviews. The intricate molecular anatomy of ribosomes is being investigated vigorously, and the details of intermolecular associations between ribosomal proteins and supernatant factors or between tRNA's and the ligases have barely begun to be elucidated. Soon the sequencers will become dominant, and in due time, we shall almost certainly know every amino acid and every nucleotide in the roughly 150 macromolecules that constitute the protein-synthesizing machinery of *Escherichia coli.*

All manifestations of protein synthesis are not found in bacteria, although the basic mechanisms are remarkably similar throughout the living world. Eukaryotes, however, may possess unique control mechanisms, such as initiation factors specific for certain types of mRNA or repressors that function at the polysome level. The process by which mRNA's reach the cytoplasm is still mysterious, and the eukaryote nucleus has become a jungle of orphan RNA's of all imaginable sizes and properties, on which speculation grows luxuriantly. Interest in these problems has increased tremendously in the past year or two, partly because some of the excitement of exploring bacterial systems has passed, and partly because the United States government has begun to define health-related research more narrowly.

Edwin H. McConkey

PREFACE

The emphasis in this volume is on eukaryotes. Brandhorst reviews the perplexing problem of the origins of messenger RNA, a decade-old dilemma that remains unresolved in spite of the intensive and ingenious efforts that have been lavished upon it. Pederson describes the biosynthesis of histones, which serves simultaneously as a model system for some general features of eukaryotic translation and at the same time displays several intriguing features that may be unique. Grollman and Huang summarize the increasingly complex array of inhibitors that affect protein synthesis in eukaryotes. Hoober and O'Brien describe the unique features of chloroplast and mitochondrial protein synthesis, respectively, comparing these systems with their counterparts in both eukaryotic and prokaryotic cytoplasms. Finally, Gallant and Lazzarini represent *Escherichia coli* and its relatives with the story of magic spots and the control of ribosomal RNA synthesis.

A number of other topics might have been included in this volume, such as a critical appraisal of the multitudinous claims and counterclaims about translational control mechanisms in higher eukaryotes, or a review of the interactions between ribosomes and membranes. Lack of space and the fact that some subjects are in a state of rapid development prevented an attempt to cover them. It is obvious that the story of protein synthesis will continue to unfold for many years to come.

Edwin H. McConkey

CONTRIBUTORS TO VOLUME 2

Bruce P. Brandhorst, Department of Biology, McGill University,
    Montreal, Quebec, Canada

Jonathan Gallant, Department of Genetics, University of Washington,
    Seattle, Washington

A. P. Grollman, Department of Pharmacological Sciences, State
    University of New York at Stony Brook, Stony Brook, New York

J. Kenneth Hoober, Department of Biochemistry, Temple University
    School of Medicine, Philadelphia, Pennsylvania

M.-T. Huang,* Department of Pharmacology, Albert Einstein College of
    Medicine, Bronx, New York

Robert A. Lazzarini, Laboratory of Molecular Biology, NINCDS,
    National Institutes of Health, Bethesda, Maryland

Thomas W. O'Brien, Department of Biochemistry, University of
    Florida, Gainesville, Florida

Thoru Pederson, The Worcester Foundation for Experimental Biology,
    Shrewsbury, Massachusetts

*Present address: Department of Biochemistry and Drug
Metabolism, Hoffmann-La Roche Inc., Nutley, New Jersey.

CONTENTS

CONTENTS OF VOLUME I

Chapter 1
HETEROGENEOUS NUCLEAR RNA OF ANIMAL CELLS
AND ITS RELATIONSHIP TO MESSENGER RNA

Bruce P. Brandhorst

Department of Biology
McGill University
Montreal, Quebec
Canada

I. INTRODUCTION

Protein synthesis in animal cells is directed by messenger RNA (mRNA) on polyribosomes (polysomes). Messenger RNA is heterogeneous in size and, unlike ribosomal RNA (rRNA) or transfer RNA, as a class has a base composition similar to that of DNA. Eukaryotic cells synthesize in the nucleus a class of RNA molecules that is heterogeneous in size and has a DNA-like base composition. It has long been assumed that at least some of this heterogeneous nuclear RNA (hnRNA) is a precursor of the mRNA translated on cytoplasmic polysomes. The purpose of this review is to evaluate recent literature concerning the properties of and relationship between hnRNA and mRNA. It has been established in numerous investigations, some of which are reviewed below, that most of the hnRNA synthesized in higher animal cells decays rapidly without serving as a precursor of the much more stable mRNA. It is virtually certain, however, that some hnRNA serves as a precursor of mRNA. The principal question of current concern is whether mRNA is derived from a much larger nuclear precursor (giant hnRNA) which is processed to yield the smaller mRNA molecule or whether mRNA is derived from a similarly sized nuclear precursor. There is much evidence favoring the giant nuclear precursor hypothesis for at least some mRNA molecules, but the evidence is not compelling. Until there is conclusive evidence for this hypothesis, it cannot be ruled out that all mRNA is derived from a similarly sized nuclear precursor. It does appear almost certain that in mammalian cells not all hnRNA molecules contain sequences that become functional mRNA (directing protein synthesis on cytoplasmic polysomes).

The scope of this review encompasses the heterogeneous RNA of animal cells in general, but the majority of the literature cited concerns cultured mammalian cells. Work on nonmammalian cells is included when it is particularly useful. For instance, the insignificant rate of rRNA synthesis in sea urchin embryos makes them particularly useful for some kinds of investigations that are impossible without the use of drugs in mammalian systems. On the other hand, there are significant differences in the details of metabolism and

other characteristics of hnRNA in different phyla; when appropriate,
these differences are pointed out.

This review is selective: It was not intended to cover all the
literature, or even all the best literature. Readers are referred
to references in the reports cited and to earlier reviews by Darnell
(1968), Weinberg (1973), Darnell et al. (1973), Mathews (1973), and
Davidson and Britten (1973), which cover in more detail the earlier
work on some of the topics dealt with in this review.

I consider first the techniques used for identification and isolation of hnRNA and mRNA. I then record and evaluate the evidence
that mRNA is derived from a much larger nuclear precursor. Finally,
I review recent investigations of the physical properties of mRNA and
hnRNA which establish that there are extensive similarities and differences between these two classes of heterogeneous RNA. Hypotheses
are discussed concerning the possible role of the rapidly decaying
hnRNA that does not become mRNA.

Reviews of the literature are appropriate when a new level of understanding has been achieved, or when intense activity leads to an
overwhelming amount of new but incomplete or conflicting information.
The understanding of the relationship between hnRNA and mRNA is at
the latter point. It is hoped that this review will focus attention
on the few remaining problems in the interpretation of the rapidly
accumulating new information. It is quite possible that by the time
this review appears it will have been conclusively shown that some
mRNA molecules are derived from giant hnRNA molecules. To prove that
all mRNA is derived from similarly sized hnRNA molecules would be more
difficult. Although we appear to be quite close to an understanding
of what happens in the metabolism of heterogeneous RNA, we are probably far from understanding how and why.

II. IDENTIFICATION AND ISOLATION OF HETEROGENEOUS RNA

Some of the procedures used for comparing and characterizing heterogeneous RNA molecules are extremely sensitive and are thus highly susceptible to the influence of contaminants. The conclusions drawn

often depend on the purity of the RNA fraction. Therefore, it is always necessary to consider carefully the criteria used to estimate
this purity.

A. hnRNA

Heterogeneous nuclear RNA is defined as all nuclear RNA having
sedimentation coefficients greater than about 8S, excluding rRNA;
transfer RNA and low molecular weight heterogeneous RNA are excluded
(Weinberg, 1973). Growing cells synthesize rRNA at a substantial rate.
Although rRNA is synthesized as a precursor molecule of discrete size
and is processed via a series of intermediates of discrete size (see
Darnell, 1968; Weinberg, 1973), many hnRNA molecules sediment at the
same rate as these ribosomal precursors and rRNA molecules.

A number of approaches have been used to resolve hnRNA from rRNA.
In some investigations only those nuclear RNA molecules having sedimentation coefficients greater than that of ribosomal precursor (45S
in mammalian cells) are isolated for analysis; since a substantial
portion of hnRNA is of lower molecular weight than ribosomal precursor, the latter is not included in these investigations of hnRNA.
Considering the evidence that some physical properties of hnRNA vary
according to its size (see Sec. IV,B), this is not a satisfactory approach in many investigations.

Ribosomal RNA in mammalian cells has a high G + C content. While
the base composition of hnRNA varies somewhat according to size (Attardi et al., 1966; Emerson and Humphreys, 1970), it is of considerably lower G + C content than rRNA. The nearest neighbor base composition of hnRNA sedimenting more rapidly than ribosomal precursor RNA
can be determined and assumed to represent the average base composition of all hnRNA. The intermediate base composition of hnRNA contaminated with rRNA can be used to estimate the relative amounts of
hnRNA and rRNA. This method has been used effectively in quantitative
kinetic analyses of hnRNA metabolism (Soeiro et al., 1968), but does
not allow for the investigation of the physical properties of hnRNA.

Ribosomal RNA synthesis is especially sensitive to certain drugs,
particularly low concentrations of actinomycin D (Penman et al., 1968;

Perry and Kelley, 1970). The synthesis of rRNA can thus be selectively inhibited. A drawback of this approach is that some hnRNA synthesis is also inhibited, the extent of inhibition increasing with time
(Penman et al., 1968; Brandhorst, 1972). The use of drugs is thus
not always desirable in investigations of the metabolism of hnRNA,
but many useful investigations of the properties of hnRNA have relied
on the use of drugs to selectively inhibit synthesis of rRNA.

   Some nongrowing cells exhibit a low or negligible rate of rRNA
synthesis and accumulation relative to hnRNA synthesis and accumulation. Frequently investigated cells are cultured avian erythroblasts
(Attardi et al., 1966; Scherrer et al., 1966) and sea urchin embryos
(Gross et al., 1964; Emerson and Humphreys, 1970; Brandhorst and Humphreys, 1972).

   Nuclei can be fractionated into nucleoli containing most of the
rRNA and nucleoplasm with which most hnRNA is associated (Penman et
al., 1966). The methods are tedious and cross contamination occurs.

   A large fraction (20-40%) of hnRNA molecules are polyadenylated
in mammalian cells (Edmonds and Caramela, 1969; Sheldon et al., 1972a;
Jelinek et al., 1973; Greenberg and Perry, 1972) and can be selected
for by the affinity of poly(A) for poly(U) or oligo(dT), for which
rRNA has little or no affinity. This polyadenylated fraction of RNA
is not likely to be a random sample of hnRNA. Thus there is no universally applicable method for the detection, quantitation, and isolation of hnRNA. The use of a variety of criteria and methods often
leads to consistent results.

   While some investigators have been content to define hnRNA as any
rapidly sedimenting fraction of cellular RNA, most begin by isolating
nuclei. A wide variety of nuclear isolation procedures have been described, and it is clear that different conditions are required for
different types of cells. In many cases the extent of contamination
of nuclei by cytoplasm is not evaluated and may be quite high. Often
fixed nuclei are examined with the electron microscope for "cytoplasmic tags," or associated ribosomes, a procedure having little quantitative value. In mammalian cells, at least, when the 18S rRNA is completed it rapidly passes to the cytoplasm; the presence of a signifi-

cant amount of 18S rRNA in an extract of nuclear RNA is an indication of contamination by cytoplasmic ribosomes (Penman, 1966). Typically this contamination can be reduced to less than 1%. Even so, this level of contamination could seriously hinder the interpretation of some types of experiments. A further problem with nuclear isolation procedures is that hnRNA is often lost to the cytoplasm (Perry and Kelley, 1968; Fromson and Nemer, 1970; Brandhorst and McConkey, 1974).

Most investigations of hnRNA use methods for the preparation and identification of hnRNA that are adequate for the experiments described, but the question of adequacy should always be considered and evaluated, particularly in recent investigations (see Sect. III,D,4).

B. mRNA

Messenger RNA is generally defined by its function: the direction of protein synthesis on polysomes. In practice, however, heterogenous RNA associated with polysomes is usually defined as mRNA. In HeLa and L cells most mRNA molecules, with the notable exception of histone mRNA, are covalently linked to a sequence of poly(A) at their 3' ends (Adesnik et al., 1972; Greenberg and Perry, 1972). It has not yet been established that most mRNA molecules in other cells are polyadenylated; in fact, a large fraction (greater than half) of mRNA molecules in sea urchin embryos are almost certainly not polyadenylated (Fromson and Duchastel, 1975) and are different in nucleotide sequence from polyadenylated mRNA (Nemer et al., 1974). As much as 30% of the mRNA of HeLa cells may not be polyadenylated and is not histone mRNA (Milcarek et al., 1974). A cDNA copy of polyadenylated mRNA does not hybridize significantly to the nonpolyadenylated mRNA, providing evidence that the two classes of mRNA are really different. However, since it was not established that the cDNA copies represent sequences other than those immediately adjacent to the poly(A) segment, the unlikely possibility cannot be eliminated that nonpolyadenylated molecules are all derived by spurious endonucleolytic cleavage of polyadenylated mRNA. Nonpolyadenylated mRNA of sea urchin embryos is translated in vitro into proteins having physical properties different from those coded for by polyadenylated mRNA (D. Fromson and D. P. Verma, 1976).

Messenger RNA molecules that are polyadenylated can be resolved from ribosomal RNA coextracted from polysomes by virtue of the affinity of poly(A) for poly(U) or oligo(dT). Affinity chromatography is employed, often using poly(U)-sepharose or poly(U)-cellulose or oligo(dT)-cellulose. The latter appears to be more reproducible for most applications, probably because the oligo(dT) remains totally bound to the cellulose (Wu and Wilt, 1973). Although affinity chromatography has a high degree of specificity, it is not absolutely specific: Some ribosomal RNA sometimes binds and elutes with polyadenylated RNA (Singer and Penman, 1973). The preliminary hybridization of poly(U) to the poly(A) segment of mRNA followed by hydroxyapatite chromatography may be the best method at present for preparing uncontaminated mRNA (Greenberg and Perry, 1972). The convenient use of filters on which poly(U) has been immobilized (Sheldon et al., 1972a) to bind polyadenylated RNA has had varying degrees of success in different laboratories. Polyadenylated RNA has an affinity for nitrocellulose filters or cellulose under certain ionic conditions (Brawerman et al., 1971; Schutz et al., 1972). These media sometimes bind rRNA as well, and the lack of understanding of their mode of action makes them less desirable for affinity chromatography. The yields of polyadenylated RNA and the extent of contamination by rRNA should always be evaluated, but in many cases are not. However, for many types of investigations a low level of contamination by rRNA or yields that are less than quantitative are tolerable.

A serious difficulty in the preparation of mRNA is the cosedimentation with polysomes of heterogeneous RNA in ribonucleoprotein particles (RNP's) (Penman et al., 1968). Many of the RNP's appear to contain hnRNA released from nuclei during cellular fractionation, because their content can be reduced when care is taken to minimize damage to nuclei (Perry and Kelley, 1968; Fromson and Nemer, 1970). Messenger RNA in polysomes can be released as RNP's by exposure to ethylenediaminetetraacetate (EDTA) or to puromycin. These mRNP's

sediment more slowly than polysomes, while the sedimentation of large
RNP contaminants is not altered by EDTA or puromycin (Penman et al.,
1968). Messenger RNA can thus be isolated nearly free of these contaminants by sucrose gradient centrifugation, isolation of the polysome fraction, release of mRNP's from ribosomes by EDTA or puromycin-
high salt (Blobel, 1971), and collection of mRNP's. A drawback of
this approach is the long period of time and extensive handling required, which increase the likelihood of endonucleolytic decay or
breakage of the mRNA. The extent of contamination of polysomal RNA
by RNA of RNP's can be estimated by formaldehyde fixation of the polysomal region of the sucrose gradient followed by isopycnic centrifugation in CsCl: RNP's derived from the nucleus and from polysomes have
different densities (Perry and Kelley, 1968; Schochetman and Perry,
1972a). The use of affinity chromatography does not resolve mRNA from
all RNA of RNP's (whether of nuclear or cytoplasmic origin), because
the latter RNA is also polyadenylated (Perry et al., 1974; Nakazato
and Edmonds, 1972). Mathews (1973) has reviewed some methods for the
preparation of mRNA.

Preparations of mRNA have been enriched for one of a wide variety
of different specific mRNA's (see Mathews, 1973). Some of these "pure"
mRNA molecules approach a high degree of homogeneity, particularly
globin mRNA (e.g., Leder et al., 1973), one of the histone mRNA's
(Grunstein et al., 1973), and silk gland fibroin mRNA, for which nucleotide fingerprints agree with amino acid sequence information (Suzuki and Brown, 1972). All "purified" mRNA's are probably slightly
contaminated with other classes of RNA, and many are only slightly enriched. The enrichment for a specific mRNA usually involves the preparation and deproteinization of polysomes, affinity chromatography to
isolate polyadenylated mRNA, and fractionation by size for very large
(e.g., myosin, collagen, fibroin) or small (histone, globin, immuno-
globin light chain) mRNA. In some cases immunological precipitation
can be achieved with specific antisera to the polypeptide coded for
by the mRNA (e.g., Palacios et al., 1973). Although this method has
great potential, there is often a contamination of the specific polysomes by other polysomes, and the antibody preparations used usually

contain a ribonuclease activity. These complications have apparently
been overcome by the use of the pepsin antibody fragment $F(ab')_2$ in
the preparation of light-chain mRNA from myeloma polysomes (Delovitch
et al., 1972). A particularly useful situation exists in myeloma
cells: Completed myeloma immunoglobulin appears to bind specifically
heavy-chain mRNA, which can then be precipitated by antimyeloma protein antibodies (Stevens and Williamson, 1973a).

To date, preparations enriched for specific mRNA molecules have
been derived from cells in which a large fraction of the protein synthesis is devoted to the proteins coded for by the specific mRNA. The
results from the use of immunoprecipitation procedures and/or affinity
chromatography imply that preparations of highly enriched mRNA for non-
specialized proteins will ultimately be obtainable. In all these procedures, extensive handling is required, which introduces the possibility of degradation of mRNA molecules by endonucleolytic attack or physical breakage. Although most of the "purified" mRNA fractions are not
highly homogeneous, they have been useful for some types of physical
characterization, as described below. The use of reverse transciptase
to make DNA copies of mRNA molecules for use as hybridizing probes is
an extremely valuable experimental tool, only beginning to be exploited
(Verma et al., 1972; Kacian et al., 1972; Ross et al., 1972).

In conclusion, I have reviewed methods commonly used to prepare
RNA from animal cells, emphasizing their possible drawbacks. In any
investigation of the properties of hnRNA or mRNA, much consideration
must be given to adequately defining the characteristics and purity
of the RNA fraction for the questions posed. The RNA preparations used
are often not adequately evaluated and documented. In the following
sections I have often attempted to make judgments about the adequacy
of the RNA preparations, although there is often insufficient information for such judgments.

III. EVIDENCE THAT mRNA IS DERIVED FROM A
      LARGER NUCLEAR PRECURSOR

A. Historical and Circumstantial Evidence

From the time of its discovery, hnRNA has been proposed as a precursor of mRNA (see Darnell, 1968). This proposal was developed originally because both mRNA and hnRNA have a similar base composition re-
flecting that of DNA, and both are broadly heterogeneous in size. When
it was found that hnRNA sediments very rapidly on aqueous sucrose gradients (Attardi et al., 1966; Houssais and Attardi, 1966; Scherrer et
al., 1966), while mRNA extracted from polysomes sediments more slowly
(Latham and Darnell, 1965), it was proposed that giant hnRNA molecules
(i.e., those sedimenting more rapidly than the largest mRNA) are precursors of mRNA molecules which are processed to a smaller size in the
nucleus before entering the cytoplasm. This notion of a processing by
size was particularly appealing in light of the established stepwise
reduction in size of the precursor of rRNA (see Darnell, 1968; Weinberg, 1973). Just as approximately half of the ribosomal precursor
RNA decays in the nucleus, it was found that far more heterogeneous
RNA is synthesized in the nucleus than can be accounted for by heterogeneous RNA accumulating in the cytoplasm; i.e., much of the hnRNA decays in the nucleus (Harris, 1959; Scherrer et al., 1966; Houssais and
Attardi, 1966; Attardi et al., 1966; Soeiro et al., 1968; Penman et
al., 1968; Brandhorst and McConkey, 1974). A maximum rate of accumulation of radioactive hnRNA in nuclei is seen immediately upon addition of radioactive nucleoside precursors to the cell medium, but the
maximum rate of accumulation of mRNA in polysomes begins only after a
lag of 15-30 min (e.g., Penman et al., 1968; Brandhorst and Humphreys,
1972). These kinetics are consistent with, but do not prove, the hypothesis that some hnRNA is a precursor of mRNA (Britten and McCarthy,
1963). The heterogeneity in size of hnRNA as a population makes it
impossible to follow kinetically the accumulation of nuclear intermediates, as was possible for the discrete nuclear precursors of rRNA.
Moreover, pulse-chase experiments are essentially impossible because
of the slow decrease in the specific radioactivities of the RNA precursor pools in physiological chase conditions.

The proposal that part of a giant hnRNA molecule is a precursor of mRNA was enhanced by the observation that in RNA-DNA hybridization tests nuclear RNA could compete against all polysomal RNA, while polysomal RNA could compete against only a fraction of the hnRNA (Shearer and McCarthy, 1967; Church and McCarthy, 1967b; Georgiev, 1972). These competition experiments were done under conditions in which only RNA transcribed from repetitive DNA sequences hybridizes (see Sect. IV,C). They indicate that many nucleotides in the cytoplasmic fraction are similar to, if not identical to, many but not all nuclear RNA sequences transcribed from repetitions of similar if not identical DNA sequences (Britten and Kohne, 1968). The level of hybridization, though, was so low that only a small fraction of the RNA nucleotide sequences was being compared. Moreover, the extent of contamination of the polysomal fraction by hnRNA was not critically evaluated in these experiments. At high ratios of mRNA competitor to hnRNA, this contamination may have been substantial.

The hybridizability of polysomal and nuclear RNA to nuclear DNA and the kinetics of accumulation of polysomal RNA combine to provide overwhelming evidence that most mRNA is derived from nuclear RNA. What is not clear is whether mRNA is derived from hnRNA of similar size or from considerably larger molecules by intranuclear processing. In the following discussion "giant" hnRNA will refer to hnRNA molecules having sedimentation coefficients greater than most or all mRNA molecules; the specific definition varies in different investigations (from greater than 28S to greater than 50S). Of importance is that giant hnRNA, if really as large as its sedimentation coefficients in aqueous sucrose gradients or polyacrylamide gels indicate, is too large to serve as a direct precursor of mRNA without reduction in size.

Recent investigations have led to the accumulation of more direct evidence favoring the concept that at least some giant hnRNA is processed into smaller mRNA. The new evidence, which is by no means conclusive, is that, on the average, hnRNA molecules really are longer than mRNA molecules, that viral-specific RNA sequences in transformed cells are included in giant hnRNA molecules but accumulate in poly-

somes as smaller molecules, and that sequences coding for specific
cellular proteins are identifiable in fractions defined as giant hnRNA.

B. The Size Distribution of RNA

In 1964 Penman et al. released viral RNA from polysomes of HeLa
cells infected with poliovirus and observed that it sediments on sucrose gradients with the same size distribution as RNA from intact virus. Using the same procedure, Latham and Darnell (1965) showed that
pulse-labeled RNA released from polysomes of uninfected HeLa cells
sediments predominantly between 4 and 30S. Other investigations have
shown that heterogeneous RNA extracted from HeLa polysomes sediments
more slowly than 30S, with a peak at about 18S (Murphy and Attardi,
1973; Nakazato and Edmonds, 1972). On the other hand, hnRNA from
mammalian cells sediments predominantly between 30 and 70S or more on
aqueous sucrose gradients, although some overlaps the size distribution of mRNA (Houssais and Attardi, 1966; Soeiro et al., 1968; Greenberg and Perry, 1972). Messenger RNA extracted from polysomes of
other cell types sediments more rapidly than HeLa mRNA, and the average sedimentation coefficient increases as more careful extraction
procedures are used. For instance, by 1971 heterogeneous RNA from L
cell polysomes labeled in the presence of a low dose of actinomycin D
peaked at 28S in aqueous conditions; nuclear RNA sedimented considerably faster on the average (Greenberg and Perry, 1971). Thus the
sedimentation coefficients of a population of RNA molecules should
represent a minimum because the possibility that its size has been
reduced in preparation usually cannot be eliminated. Recently, it
has been possible to prepare heterogeneous RNA from polysomes of HeLa
and L cells which is nearly all polyadenylated (Adesnik et al., 1972;
Greenberg and Perry, 1972). Since poly(A) is attached to the 3' termini of the mRNA molecules (Mendecki et al., 1972; Molloy et al.,
1972a; Sheldon et al., 1972b; Nakazato et al., 1973), a single, randomly placed endonucleolytic nick would reduce the binding of radioactive RNA to poly(U) or oligo(dT) by about 50%. It can be concluded
that most mRNA of HeLa and L cells is polyadenylated and that the size

distribution observed for this RNA is that of intact mRNA. Polyadenylated HeLa mRNA peaks about 18S (based on rRNA, having a molecular
weight of 7×10^5 or 2000 nucleotides in length) (Murphy and Attardi,
1973; Nakazato and Edmonds, 1972), and L cell mRNA peaks about 28S
(1.4×10^6 daltons), although the mass average sedimentation coefficient is somewhat less. Nearly all the mRNA has an estimated molecular weight of less than 2.5×10^6. This is considerably smaller than
estimates of the size distribution of hnRNA molecules (including polyadenylated hnRNA) in these cells, which range up to 2×10^7 daltons
or more (e.g., Greenberg and Perry, 1972). Thus it is clear that in
aqueous sucrose gradients (and by aqueous polyacrylamide gel electrophoresis) hnRNA has a much larger average size than mRNA, though some
hnRNA and mRNA overlap in size distribution.

Several recent investigations have cast doubt on the size distribution of mammalian and avian hnRNA. When sedimented in aqueous sucrose gradients after formaldehyde denaturation, or in sucrose gradients made up in high concentrations (98-100%) of dimethyl sulfoxide
or formamide which denature base-paired RNA, eliminating secondary
structure, hnRNA was observed to sediment more slowly, though to a
varying degree (Mayo and de Kloet, 1971; Bramwell, 1972; Imaizumi
et al., 1973; MacNaughton et al., 1974). These investigators interpret this reduction in apparent size after denaturation as an indication that RNA on aqueous gradients forms aggregates of smaller molecules. They could not rule out the possibility that breaks in the
RNA are introduced under their conditions, which are not introduced
in aqueous media. Holmes and Bonner (1973) have described an extraction procedure in which the size distribution of giant hnRNA (selected
for by molecular sieving) is the same in aqueous, formaldehyde, and
dimethyl sulfoxide sucrose gradients. They conclude that, even under
denaturing conditions, most hnRNA molecules from rat ascites cells,
after a 30-min pulse, have molecular weights of $5-10 \times 10^6$ based on
sedimentation in sucrose gradients. Moreover, in the electron microscope, most giant hnRNA prepared under denaturing conditions has a
contour length of 4-9 μm, corresponding to molecular weights of 5-10 x

10^6 daltons. Thus it is considerably larger than mRNA under any conditions. Peltz (1973) and Kung (1974) have prepared nuclear RNA from sea urchin embryos in which there is little change in size distribution after denaturation. While it is conceivable that even under these presumably complete denaturing conditions smaller RNA molecules are aggregated, perhaps by a hydrophilic interaction with residual protein contaminants, such a change in size is unlikely. Nevertheless, the use of denaturing gradients appears to be an art rather than a science. To date, the clear indication of the investigations is that many hnRNA molecules are much larger than mRNA in mammalian and avian cells.

The situation is somewhat different in other cells. Differentially labeled nuclear and cytoplasmic RNA extracted from whole sea urchin blastulae and gastrulae sediment with similar size distributions in aqueous sucrose gradients, peaking at about 32S, corresponding to about 2 million daltons (Brandhorst and Humphreys, 1972). Polyadenylated nuclear and polysomal RNA of sea urchin embryos have similar size distributions when compared under denaturing conditions, peaking around 30S (Kung, 1974). Giant hnRNA cannot be detected in a variety of unicellular organisms (Gamow and Prescott, 1972). In the cellular slime mold *Dictyostelium discoideum,* hnRNA has an average molecular weight of 500,000, only 20% larger than cytoplasmic mRNA; the size distributions are unaltered in dimethyl sulfoxide gradients (Firtel and Lodish, 1973). Thus it appears that the size of hnRNA is larger in organisms higher on the evolutionary scale. Kung (1974) has pointed out that there appears to be a larger size for the rRNA precursor in more complex organisms as well. This is consistent with the observation that a much smaller fraction of the nucleotides in hnRNA appears to enter the cytoplasm as mRNA in more complex organisms: Most hnRNA appears to become mRNA in *Dictyolstelium* (Firtel and Lodish, 1972), about 10% in sea urchin blastulae (Brandhorst and Humphreys, 1972), and about 2% in mouse L cells (Brandhorst and McConkey, 1974). Perhaps the increased size of rRNA and hnRNA in more complex organisms is related to more complex processing and regulation thereof.

A particularly interesting situation has been reported in detail for dipteran larval salivary glands as reviewed recently by Daneholt

(1973) and Daneholt and Hossick (1973b). When salivary glands of
Chironomus are exposed to radioactive precursors of RNA, the RNA in
Balbiani ring 2 (BR-2), a puff on the polytene chromosome IV, is heavily labeled. Daneholt and co-workers can isolate BR-2 by microdissec-
tion. The RNA extracted from this isolated puff hybridizes in situ
to BR-2 (but not to other chromosome sites), demonstrating that it
is transcribed from BR-2 DNA (Lambert, 1973a). An electrophoretic
analysis of BR-2 RNA indicates that it has a main peak at 75S (Daneholt, 1972). Nuclear sap also has a prominent 75S peak. After a
lag of over 90 min, a uridine-labeled peak becomes detectable in the
cytoplasm, and comprises about 1.5% of the optical absorbance of cytoplasmic RNA (Daneholt and Hosick, 1973a). RNA from nuclear sap
and cytoplasm hybridize in situ to BR-2 DNA (Lambert, 1973b). Thus
it appears that BR-2 RNA is synthesized, passes through the nucleus,
and enters the cytoplasm as a 75S molecule without substantial change
in size (aggregation of smaller RNA molecules has not been ruled out).
The cytoplasmic molecule appears to be quite stable (Daneholt and
Hosick, 1973a). It has not been shown that the 75S cytoplasmic RNA
enters polysomes to serve as mRNA. Since more than 80% of the salivary gland protein synthesis is of specific salivary proteins (Doyle
and Laufer, 1969; Grossbach, 1973) a prominent heterogeneous RNA species in the cytoplasm is likely to correspond to a special mRNA. The
75S RNA has been estimated to have a molecular weight of $15-35 \times 10^6$
(Daneholt, 1972). The estimate is very crude because of the lack of
suitable markers for comparison. While this size far exceeds the sequence length required for most proteins, Grossbach (1973) has reported the synthesis of a major polypeptide of 5×10^5 daltons; the
75S RNA molecule may be only 3-8 times larger than required to code
for this molecule. The silk fibroin message of Bombyx mori is also
very large (45-65S) (Suzuki and Brown, 1972). It can be concluded
that in insects giant mRNA molecules are translated, and that in the
dipteran salivary gland, a giant probable mRNA is derived from a similarly sized nuclear RNA molecule.

Thus processing of RNA by extensive change in size does not appear
to be necessary in the cells of some organisms, if any. Edstrom and

Tanguay (1973) have proposed that there are 2 types of mRNA in insect
cells having polytene chromosomes: that derived from puffs including
BR-2 RNA which enters the cytoplasm after a prolonged (3-hr) lag following synthesis in the nucleus (a highly stable RNA); and a fraction
which is similar to mammalian mRNA in that it appears in the cytoplasm
after a 15-30 min lag and is unstable. The latter fraction may be synthesized as a high molecular weight precursor, but there is presently
no evidence for such synthesis in insects.

   Another strange situation exists for amphibian RNA. Each lateral
loop of a lampbrush chromosome of the urodele *Triturus viridescens*
has a continuous gradient from one end to the other of short to long
RNP fibrils (Miller and Bakken, 1972). Since the lateral loops average 50 μm in length (up to 200 μm), the RNA molecules must be of similar length when transcription is completed; they thus average 50×10^6
daltons. Under the electron microscope, these RNP fibrils of lampbrush chromosomes appear to be considerably longer than comparable fibrils in HeLa cells (Miller and Bakken, 1972). It is thus likely that
hnRNA of amphibian oocytes is much larger on the average than hnRNA
of HeLa cells. The functional significance of the apparently huge
size of amphibian lampbrush RNA is not known. It appears from these
investigations that there is a tremendous range of sizes of hnRNA in
different organisms: from 5×10^5 daltons in *Dictyostelium* to 2×10^8
in *Triturus*. It is possible that in some organisms having relatively
high haploid DNA contents (e.g., *Triturus* and, to a lesser extent,
humans), most of the hnRNA molecule serves as a transcribed spacer
between sparsely packed structural genes; such a spacer may have no
further function.

   In summary, in mammalian and avian cells the average size of hnRNA
is almost certainly greater than the average size of mRNA. The real
size distribution is difficult to determine because of the possibilities that undetected internal breaks have occurred or that some RNA
molecules have aggregated to appear larger. A rigorous set of criteria is needed for the determination of size distributions under denaturing conditions; any possibility of aggregation must be eliminated
so that complete denaturation can be ensured. These investigations do

not indicate that mRNA is derived from the larger hnRNA molecules; the overlapping size distributions provide the possibility that mRNA molecules are derived from similarly sized hnRNA molecules that are not derived from larger hnRNA molecules. It is necessary to demonstrate that a specific mRNA molecule is derived from an identifiable larger nuclear precursor. The evidence for the existence of giant hnRNA molecules containing nucleotide sequences identical to smaller mRNA molecules is discussed in the next two sections.

C. Viral hnRNA and mRNA in Transformed Mammalian Cells

Viral DNA is covalently integrated into host DNA in transformed cells (Sambrook et al., 1968). Some of this integrated viral DNA is transcribed and the RNA can easily be detected by hybridization to viral DNA. Cells transformed with SV40, polyoma, or adenovirus synthesize nuclear RNA of $1-5 \times 10^6$ daltons which contains virus-specific sequences (Lindberg and Darnell, 1970; Tonegawa et al., 1970; Tsuei et al., 1972; Wall et al., 1973). The nuclear transcripts containing viral sequences are up to 4 times the length of the viral genome (Lindberg and Darnell, 1970), and some remain large even in 100% dimethyl sulfoxide (Wall et al., 1973). Polysomal RNA hybridizing to viral DNA sediments more slowly than the largest nuclear RNA (Lindberg and Darnell, 1970), sometimes in discrete size classes (Wall et al., 1973).

The following evidence indicates that adenovirus-specific hnRNA is composed of covalently linked viral- and host-specific sequences in virally transformed rat embryo cells (Wall et al., 1973). Of the giant hnRNA (sedimenting faster than 50S on a denaturing gradient) 0.22% hybridizes to viral DNA, but if its size is reduced by limited alkaline hydrolysis to an average of 16S, only 0.04% binds to DNA (hybrids were assayed without using ribonuclease, so nonhybridized sequences colinear with hybridized viral RNA were also bound and scored as hybrids). Thus the denatured giant hnRNA hybridizing to viral DNA is colinear with considerable lengths of nonviral nucleotide sequences. When these hybrids were eluted, rehybridized to viral

DNA, and treated with ribonuclease to eliminate covalently linked non-hybridizing segments, 8% of the giant RNA hybridized to viral DNA, while 43% of the alkali-degraded RNA hybridized. It was concluded that only 8-11% of the virus-specific giant hnRNA nucleotides are virus-specific sequences, and that these sequences are limited to one part of the molecule. In contrast, it was shown that at least 67% of the virus-specific polysomal RNA is composed of viral sequences (Wall et al., 1973).

If giant hnRNA selected by hybridization to viral DNA is hybridized to cellular DNA (from nontransformed cells) under conditions in which only rapidly hybridizing sequences transcribed from reiterated DNA sequences are measured, it hybridizes with kinetics very similar to cellular giant hnRNA, indicating that viral RNA sequences are covalently linked to cellular RNA sequences. Viral DNA-selected polysomal RNA does not hybridize nearly as rapidly as giant hnRNA, but more rapidly than unselected polysomal RNA. This suggests that viral mRNA may contain some sequences transcribed from the reiterated portion of cellular DNA, but not nearly so many as in hnRNA. At least some of the virus-specific hnRNA and mRNA molecules are attached to poly(A) segments (Wall et al., 1973).

It can be concluded that giant hnRNA of transformed cells contains giant molecules composed of covalently linked virus- and host-specific sequences. Viral mRNA is smaller than some of this hnRNA and has fewer host-specific sequences. It is tempting to propose that viral hnRNA is synthesized along with the host-specific sequences, most of which are eliminated by intramolecular processing to yield the smaller viral RNA. On the other hand, most of the nuclear virus-specific sequences in RNA have a size distribution comparable to that of polysomal RNA, and these molecules may serve as the sole precursors of mRNA without processing by size. The possibility of a precursor-product relationship between any size class of hnRNA and mRNA has not been established by any kind of kinetic experiment. Such kinetic analyses may be difficult because of the heterogeneity of the viral hnRNA (probably due to the heterogeneity of the host-specific RNA

components). It is, of course, possible that the processing of viral
RNA in transformed cells is not representative of RNA processing in
normal cells.

D. Evidence for mRNA Sequences in Giant hnRNA

Several recent investigations have suggested that giant hnRNA contains nucleotide sequences identical to those of much smaller mRNA.
The criteria used for these identifications (nucleic acid hybridization and translation in in vitro protein synthesis of *Xenopus* oocyte
systems) can be extremely sensitive and detect small amounts of mes-
sengerlike sequences; they are thus very susceptible to contamination
of the giant hnRNA with a slight amount of smaller mRNA sequences.
The main consideration, then, must be an elimination of the possibility of contamination.

Melli and Pemberton (1972) prepared an RNA copy of duck globin
mRNA using an RNA polymerase and found that it hybridizes to giant
duck hnRNA. The latter was not prepared under denaturing conditions
(e.g., dimethyl sulfoxide or formamide gradients), so contamination
by aggregated mRNA was possible.

Stevens and Williamson (1973a) have presented evidence that complete myeloma immunoglobulin protein specifically binds heavy-chain
mRNA, which can be precipitated by antimyeloma protein. Cytoplasmic
RNA prepared in this way after labeling with tritiated uridine sediments in two peaks; Stevens and Williamson proposed that the two peaks
are the same molecule with and without a full complement of poly(A),
but one is most likely an aggregate formed in the extraction procedure. When the preparation is injected into the *Xenopus* oocyte protein synthesizing system of Gurdon et al. (1971), both peaks direct
the synthesis of heavy-chain protein, indicating that the molecules
include mRNA sequences for heavy chain. Using the myeloma protein,
Stevens and Williamson (1973b) isolated three discrete peaks of uridine-labeled nuclear RNA. The two smaller peaks run identically to
the two cytoplasmic peaks on polyacrylamide gel electrophoresis. The
third peak, not seen in the cytoplasmic fraction, has an estimated

molecular weight of 2.4×10^6, 2-5 times larger than the cytoplasmic peaks. RNA included in this larger nuclear peak can direct the synthesis of heavy-chain protein when injected into Xenopus oocytes, but the synthesis of light-chain protein is not detected, suggesting that nuclear RNA is not contaminated with light-chain mRNA and, perhaps, any other mRNA. A specific association between nuclear RNA and heavy-chain mRNA or an mRNA-sized nuclear precursor has not been ruled out by disaggregation of RNA in denaturing conditions, but seems unlikely.

These myeloma cells were labeled for 20 min with tritiated uridine and then RNA synthesis was inhibited with actinomycin D. After 20 min of labeling no heavy-chain mRNA was detectable in the cytoplasm (suggesting that nuclear species are truly limited to the nucleus), but it accumulated over the next 30 min as the nuclear peaks, particularly the largest, disappeared. This evidence is consistent with a precursor-product relationship between nuclear and cytoplasmic molecules selected by the myeloma protein. Most of the nuclear RNA peak bound by the myeloma protein appears to enter the cytoplasm in the presence of actinomycin D rather than decaying in the nucleus. Even larger potential precursors were not detected in the nucleus. Their existence cannot be excluded because they may be processed too rapidly for detection, or they may not be recognized by the myeloma protein, perhaps because their secondary structure obscures the binding site. This is a very promising system for a detailed analysis of the kinetics of processing a probable nuclear precursor of a specific mRNA molecule. It remains to be rigorously established that the large nuclear precursor is truly larger than the mRNA (i.e., that it is not an aggregate) and that most of the radioactive RNA observed is actually specific for heavy-chain; heavy-chain-specific RNA may only be a minor fraction of RNA included in the radioactive peaks.

Williamson et al. (1973) prepared RNA extracted from whole mouse erythroblasts sedimenting more rapidly than 28S on sucrose gradients, and injected it into Xenopus oocytes. Chromatographic analysis of the proteins synthesized strongly suggests that mouse globin is synthesized. If the large RNA is prepared from brain cells mixed with reticulocyte lysate and injected into oocytes, no globin synthesis is detected. (The possibility that brain-cell material inactivates the translation of globin mRNA was not checked.) While this observation reduces the possibility of contamination of giant RNA with 9S globin mRNA, it is still possible that in erythroblasts there is a specific association between globin mRNA and an hnRNA molecule. Indeed, Macnaughton et al. (1974) present evidence that phenol extraction leads to the aggregation of duck globin mRNA molecules and their nuclear precursors. While it is not clear that the preparative procedures of Macnaughton et al. are not introducing breaks in the RNA preparation, Williamson et al. should apply their giant RNA to denaturing gradients. The possibility of a specific interaction between globin mRNA and nuclear RNA is not farfetched in view of the extensive base sequence complementarity observed between mammalian polysomal and nuclear RNA (Stampfer et al., 1972).

Macnaughton et al. (1974) used a radioactive DNA copy (cDNA) of duck 9S globin mRNA prepared by reverse transcriptase as a hybridization probe for precursors of globin mRNA in nuclear RNA. They found only a 14S peak of unlabeled nuclear RNA containing sequences complementary to globin cDNA. This RNA was estimated to be three times the length of globin mRNA and corresponds to the peak of unlabeled polyadenylated nuclear RNA on a denaturing gradient; much of the pulse-labeled hnRNA is much larger than this 14S peak, and the steady-state size distribution of (unlabeled) hnRNA was not shown. The failure to identify globin sequences in larger hnRNA at steady state may result from their failure to accumulate at a level sufficient for detection, because they have been degraded in preparation (there is also a lack of detectable polyadenylated giant hnRNA, which may be real), or because they do not exist. That there is little 9-10S globin mRNA-sized RNA in the nuclear RNA hybridizable to globin DNA indicates that the 14S hybridizable RNA does not derive from the "tailing" of 9-10S RNA into the 14S region. This report contains the only conclusive evidence that nucleotide sequences identical to those of a specific mRNA are found in larger nuclear RNA molecules; it cannot yet be concluded that these larger nuclear molecules actually serve as precursors of the globin mRNA. While the 14S peak of globin-specific nuclear

RNA is not "giant," globin mRNA is unusually small; if the largest
mRNA molecules had nuclear precursors three times their size, the
precursors would be "giant," having molecular weights of 6-8 x 10^6.

Imaizumi et al. (1973) have determined the kinetics of hybridization (C_0t curves; see Sect. IV,C) of various size fractions of nuclear
RNA from duck erythroblasts to globin cDNA. Although most newly synthesized hnRNA sediments much more slowly in 99% dimethyl sulfoxide
than in aqueous sucrose gradients, the hnRNA that remains large
(greater than 30S) hybridizes significantly with globin cDNA. From
1-3 x 10^{-4} of this RNA consists of globin mRNA sequences. While this
seems to be a very low value, it can be calculated that, based on the
fraction of DNA nucleotides transcribed that are globin specific, a
population of nuclear RNA molecules with only one mRNA precursor molecule per transcribed gene should contain 1.4 x 10^{-5} of the RNA as
globin sequence (Imaizumi et al., 1973). This estimated figure con-
firms that the small fraction of nucleotides in giant hnRNA that are
globin specific is not too small. In fact, it appears to be larger
than the prediction. This can be explained in several ways, assuming
the measurement to be accurate: There are multiple precursors of
globin mRNA in the giant hnRNA fraction; precursors of globin mRNA
are transcribed at a greater rate than other nucleotide sequences on
the average; or there is a contamination of the giant hnRNA fraction
with smaller globin sequences. The latter possibility remains, because the frequency of globin-specific sequences in giant (greater
than 50S) hnRNA on aqueous sucrose gradients is more than 100 times
that in large (greater than 30S) hnRNA on denaturing gradients.
Clearly, most globin-specific sequences in giant hnRNA are aggregates
or parts of molecules containing covert breaks indetectable on aqueous
gradients because of the secondary structure of the molecule. It is
possible that denaturation with dimethyl sulfoxide fails to remove
the fraction of a percent of globin sequences still sedimenting in
giant hnRNA; quite possibly these can be accounted for by tailing
across the gradient (see the following). It is likely that most or
all the globin-specific sequences in nuclear RNA of duck erythroblasts
at steady state are in the 14S peak found by Macnaughton et al. (1974).

In both investigations a rapidly processed nuclear precursor might
not have been detected if it had not accumulated significantly at
steady state.

Imaizumi et al. (1973) estimated that only a small fraction of
the total hnRNA might be specific for a given mRNA sequence. Davidson
and Britten (1973) have estimated that there are only about 100 copies
per cell of many nonrepetitive sequences of hnRNA in sea urchin embryos and L cells (see also Smith et al., 1974). Such levels make
detection very difficult: The DNA copies of mRNA derived by reverse
transcriptase can be used as a powerful tool for identifying and following kinetically the steps in the processing of a nuclear precursor
of a specific mRNA, but such investigations will be exceedingly difficult if the number of such precursors per nucleus is very low.

Ruiz-Carrillo et al. (1973) attempted to address themselves to the
possibility of tailing of small RNA into the fraction of giant hnRNA.
RNA was extracted from whole duck erythroblasts and fractionated on
sucrose gradients in 99% dimethyl sulfoxide. RNA sedimenting faster
than 45S was able to direct the synthesis of globin molecules in a
Krebs II ascites cell-free protein synthesizing system. In an attempt
to rule out the possibility of contamination, purified rabbit 9S globin mRNA was added to purified duck erythroblast RNA before centrifugation on sucrose gradients in dimethyl sulfoxide. No synthesis of
rabbit globin was detected for the RNA sedimenting more rapidly than
45S, while it was detected in more slowly sedimenting fractions (including the 18-42S fraction, indicating significant tailing). Unfortunately, only extensive contamination would be detected: No effort
was made to detect a low level of contamination by rabbit globin mRNA,
such as by the use of anti-rabbit-globin antibodies. A large fraction
of the protein synthesis in the presence of giant RNA is of duck globin; this is remarkable considering the small fraction of the molecules that are specific for globin in giant hnRNA according to Imaizumi et al. (1973). Perhaps only a small fraction of the hnRNA sequences can be translated in this cell-free protein synthesizing system; alternative explanations are that the giant RNA fraction of
Ruiz-Carrillo et al. (1973) is extensively contaminated with globin

mRNA, or that the giant hnRNA of Imaizumi et al. (1973) was degraded
in preparation in such a way that it no longer sedimented rapidly on
denaturing gradients.

In none of these preparations of giant RNA under denaturing conditions was the giant fraction subjected to a second fractionation on
denaturing sucrose gradients; this refractionation is necessary to reduce the significant possibility that some small RNA has tailed into
the giant fraction. The presence of 9-10S rabbit globin mRNA in the
18-42S fraction of avian erythroblast RNA (Ruiz-Carrillo et al., 1973)
casts grave doubts on the validity of the observation that globin-
specific sequences are in giant hnRNA.

In summary, there is much evidence for the presence in larger hnRNA
of nucleotide sequences of specific mRNA molecules. There is almost
certainly a 14S globin-specific sequence in duck hnRNA, about three
times longer than globin mRNA (Macnaughton et al., 1974). The identification of globin-specific sequences in giant hnRNA may be valid, but
a significant possibility remains that this giant hnRNA is contaminated by small globin-specific sequences. With the possible exception
of the investigation of Stevens and Williamson (1973b), no kinetic evidence has been presented that these giant hnRNA sequences are precursors of mRNA molecules. Spohr et al. (1974) have presented evidence,
based on crudely determined chase kinetics, that a giant nuclear RNA
molecule containing globin mRNA sequences may be a precursor of smaller putative nuclear precursors of globin mRNA in avian erythroblasts:
The larger "precursors" chase more rapidly. The authors rightly caution that their evidence does not prove that the large nuclear molecules are actual physical precursors of globin mRNA. While the use of
DNA copies of specific mRNA molecules as probes for precursors of mRNA
is attractive, the very small fraction of the hnRNA expected to correspond to these intermediates may make further experimentation extremely difficult. The only detectable nuclear precursor of ovalbumin
mRNA is indistinguishable in sedimentation from the message itself
(McKnight and Schimke, 1974); attempts were made to detect a rapidly
processed larger precursor, but the possibility of its existence cannot be ruled out. Isolated nuclei of sea urchin embryos producing

histone mRNA in a cell-free system have no detectable nuclear precursor larger than histone mRNA (Shutt and Kedes, 1974). It now appears quite possible that many mRNA molecules are not derived from larger nuclear precursors.

Since the extensive accumulation of specific mRNA in specialized cells can be accounted for by stabilization of mRNA and does not require an increased rate of synthesis (Kafatos, 1972), it is possible that in no system does a large number of precursors of specific mRNA molecules accumulate in the nucleus. Smith et al. (1974) find evidence that some mRNA molecules in sea urchins are present in considerably higher numbers than the average. One such molecule may be the histone mRNA which, in certain circumstances, such as early embryogenesis, may make up a large fraction of the total heterogeneous RNA synthesized (Kedes and Gross, 1969). As yet a DNA copy of histone mRNA has not been synthesized because of the absence of a poly(A) primer, but this problem should be technically resolvable (Kates, 1973).

E. Poly(A)

A great deal of literature on poly(A) and the polyadenylation of RNA has accumulated. It has been reviewed recently (Weinberg, 1973; Jelinek et al., 1973). A full review of poly(A) is beyond the scope of this chapter, but some of the literature must be discussed. The observation that poly(A) is found covalently linked to the 3' end of most mammalian mRNA and many (at least 20-40%) hnRNA molecules (see Sect. III,B) has been used as strong evidence for the precursor-product relationship of polyadenylated hnRNA and mRNA.

Poly(A) is probably added to RNA posttranscriptionally, although there is no conclusive evidence for this. The evidence is that polyadenylation of RNA is not immediately sensitive to inhibition of RNA chain elongation by actinomycin D (Edmonds and Caramela, 1969; Darnell et al., 1971; Nakazato et al., 1974), and there are no detectable long sequences of poly(dAT) in cellular DNA from which it could be transcribed (Philipson et al., 1971; Birnboim et al., 1973; Bishop et al., 1974). Bishop et al. (1974), however, estimate that there may be 10^4 poly (dAT) sequences in the duck genome that can be yet undetectable

in their measurements. This is somewhat less than the estimated number of different polyadenylated HeLa mRNA molecules per cell (4 x 10^4;
Bishop et al., 1974), indicating that posttranscriptional addition of
poly(A) probably occurs. Since even after very short labeling times
free poly(A) cannot be detected in nuclei (Jelinek et al., 1973) it is
probably safe to conclude that poly(A) is added stepwise to the 3'
ends of completed hnRNA molecules.

There is evidence that most poly(A) is synthesized in the nucleus.
After a brief pulse of tritiated adenosine, the majority of radioactive poly(A) is in the nucleus, while after several hours of labeling
there is several times more radioactive poly(A) in the cytoplasm than
in the nucleus (Mendecki et al., 1972; Jelinek et al., 1973; Perry et
al., 1974; Brandhorst and McConkey, 1975). This observation is consistent with a nuclear origin for most cytoplasmic poly(A), but this
precursor-product relationship has not been rigorously demonstrated
(Perry et al., 1974; Brandhorst and McConkey, 1975). Indeed, these
kinetic investigations imply that there is significant synthesis of
poly(A) in the cytoplasm. This putative cytoplasmic synthesis is probably not accounted for by the mitochondrial poly(A) synthesis which
occurs (Perlman et al., 1973; Ojala and Attardi, 1974). It is likely,
however, that nuclear poly(A) and, consequently, some polyadenylated
hnRNA molecules, serve as precursors of most cytoplasmic poly(A) and
polyadenylated mRNA.

Poly(A) appears to have an important role in the nuclear processing
or transport to the cytoplasm of some mRNA molecules. Inhibition of
poly(A) synthesis with cordycepin (3'-deoxyadenosine) does not alter
significantly the synthesis and accumulation of hnRNA, but almost completely inhibits the accumulation of mRNA in the cytoplasm (Penman et
al., 1970; Adesnik et al., 1972; Mendecki et al., 1972). Some newly
synthesized mRNA molecules do enter the cytoplasm, but these have dras-
tically shortened poly(A) tails. The effect of cordycepin is likely
to be due to its inhibition of poly(A) synthesis, since another nucleoside analog (3'-deoxycytidine) does not inhibit poly(A) synthesis and
does not inhibit the accumulation of mRNA in the cytoplasm (Abelson
and Penman, 1972). Polyadenylation cannot be absolutely required for

the exit of mRNA from the nucleus to the cytoplasm. Histone mRNA
lacks poly(A) (Adesnik and Darnell, 1972), but certainly enters the
cytoplasm. Newly synthesized histone mRNA begins to accumulate in the
cytoplasm of mammalian cells after little or no lag (Adesnik and Darnell, 1972; Schochetman and Perry, 1972b), while most mRNA accumulates
only after a lag of 15-20 min (Penman et al., 1968). The difference
in transit times may be related to the requirement for polyadenylation
of most mRNA molecules in these cells.

Observations on the accumulation of nuclear and cytoplasmic poly(A)
led Jelinek et al. (1973) to suggest that most or all nuclear poly(A)
is conserved and transported to the cytoplasm. This possible conservation was used as strong evidence that polyadenylated hnRNA is a precursor of polyadenylated mRNA (Darnell et al., 1973). Moreover, the
conservation of poly(A) suggests that polyadenylation of a presumptive mRNA molecule might obligate the entry of that molecule into the
cytoplasm: Poly(A) would serve as a marker for nuclear precursors of
mRNA. The analysis of Jelinek et al. (1973) was carried out under
conditions in which the specific radioactivity of the ATP pool, the
precursor of poly(A), changes rapidly (Brandhorst and McConkey, 1973).
Perry et al. (1974) carried out kinetic investigations on L cells in
which there was indirect and inconclusive evidence that the specific
radioactivity of the ATP pool became constant rapidly; they concluded
that much of the poly(A) synthesized in the nucleus decays in the nucleus. More recently, Brandhorst and McConkey (1975) have carried out
a similar investigation in which the specific activity of the ATP was
measured and shown to be nearly constant. While the simplest interpretation of the data is that nuclear poly(A) is not conserved, more
complex models (in which a fraction of the poly(A) in the cytoplasm
decays rapidly) fitting the data can be developed in which nuclear
poly(A) is conserved. The investigations of Perry et al. (1974) can
be similarly interpreted. LaTorre and Perry (1973) observed that in
an actinomycin plus cordycepin chase of poly(A), pulse labeled with
tritiated adenosine, more radioactivity is lost from the nucleus than
appears in the cytoplasm of L cells; their interpretation of this observation was that nuclear poly(A) is not conserved. An alternative

interpretation is that there is a class of poly(A) decaying rapidly
in the cytoplasm.

Puckett et al. (1975) have provided evidence that many polyadenylated mRNA molecules decay rapidly. This conclusion makes it possible to fit their kinetics of accumulation of nuclear and cytoplasmic
poly(A) to models in which nuclear poly(A) is or is not conserved.
For technical reasons, Puckett et al. could not estimate the accumulation curves for nuclear and cytoplasmic poly(A) before 1 hr of labeling. Brandhorst and McConkey (1975) were able to determine the
early accumulation curves and concluded that some poly(A) either is
synthesized in the cytoplasm or passes very rapidly through a small
nuclear pool, while most of the nuclear poly(A) either decays in the
nucleus with the same kinetics as hnRNA or passes more slowly into
the cytoplasm. It appears that some nuclear RNA molecules are polyadenylated sooner than others (Derman and Darnell, 1974); the latter
may pass through the nucleus to the cytoplasm very quickly after polyadenylation (Brandhorst and McConkey, 1975). Polyadenylated nuclear
precursors of mRNA may be in ribonucleoprotein particles having unique
properties (Price et al., 1974). Thus evidence is accumulating that
there may be more than one class of nuclear poly(A). Lewin (1975)
has provided a balanced review of poly(A) metabolism, which quite
clearly is complex.

It cannot be determined from present experimental evidence whether
or not nuclear poly(A) is conserved. Thus it cannot be determined
whether all polyadenylated hnRNA serves as a precursor of mRNA; there
is a distinct possibility that many polyadenylated hnRNA molecules are
completely degraded within the nucleus. It is possible that most or
all hnRNA molecules are polyadenylated at some time. Those lacking
poly(A) at a given time may be: newly synthesized or nascent molecules
that have not yet been polyadenylated, fragments cleaved from the 5'
ends of polyadenylated molecules during normal processing, or fragments arising from breakage during isolation of the hnRNA.

Since poly(A) is situated at the 3' ends of hnRNA and mRNA molecules and there is evidence that at least some mRNA molecules are derived from larger hnRNA molecules, it follows that it is the 3' end

that is conserved. The kinetics of labeling of poly(A), though, suggest that precursors of mRNA molecules are polyadenylated only shortly
before entering the cytoplasm (Perry et al., 1974; Brandhorst and
McConkey, 1975). The original 3' ends of hnRNA molecules may already
have been released. It is also possible that giant hnRNA molecules
contain sequences for several mRNA molecules, each of which becomes
polyadenylated as the giant precursor is cleaved. When hnRNA from
duck erythroblasts is denatured to eliminate intermolecular base pairing, poly(A) is detectable only in smaller hnRNA molecules, similar
to or slightly larger than mRNA (Mcnaughton, 1974). Newly synthesized poly(A) may be attached to larger molecules, but this has not
been investigated on denaturing gradients. Thus there is no conclusive evidence that presumptive mRNA lies at or only at the 3' end of
giant hnRNA. Moreover, selection of polyadenylated hnRNA molecules
may not lead to a significant enrichment for precursors of mRNA: Most
of the nuclear poly(A) sequences attached to hnRNA in L cells may decay in the nucleus (Perry et al., 1974; Brandhorst and McConkey, 1975).
The physical properties of polyadenylated hnRNA molecules are not
necessarily the physical properties of the nuclear precursors of mRNA.

A consideration of the role of poly(A) is beyond the scope of this
review. The apparent complexities of nuclear poly(A) metabolism, the
synthesis of mitochondrial poly(A), and the probable synthesis of
other cytoplasmic poly(A) (Wilt, 1973; Perry et al., 1974; Brandhorst
and McConkey, 1975) suggest that the role of poly(A) may be complex
or multiple. It is interesting that histone mRNA lacking poly(A)
appears to decay nonstochastically in L cells, while polyadenylated
mRNA most likely decays stochastically (Perry and Kelley, 1973).
Poly(A) may be related to the regulation of the degradation of heterogeneous RNA. This could explain the difference in abundance of polyadenylated and nonpolyadenylated herpes virus mRNA (Silverstein et
al., 1973).

Removal of poly(A) from a variety of mRNA molecules alters slight-
ly, if at all, the translational efficiency of the mRNA in vitro
(Williamson et al., 1974; Bard et al., 1974; Sipple et al., 1974).
When globin mRNA from which poly(A) had been removed by exonuclease

was injected into *Xenopus* oocytes, it was initially translated with
the same estimated efficiency as polyadenylated globin mRNA, but the
rate of translation of nonpolyadenylated mRNA decreased much more
rapidly (Huez et al., 1974). While it was not conclusively shown
that the differences are due only to the lack of poly(A) (critical
details of the experimental protocol were lacking), this is strong
evidence that poly(A) is related to the functional (and probably
structural) stability of mRNA. The nonpolyadenylated, nonhistone mRNA
molecules in HeLa cells and sea urchin embryos have average stabilities similar to those estimated for polyadenylated mRNA molecules (Milcarek et al., 1974; Nemer, personal communication). These estimates
of RNA stability are not accurate enough to rule out the hypothesis
that all nonpolyadenylated mRNA's, like histone mRNA, accumulate only
during the S phase of the cell cycle, being selectively degraded at
the end of S phase. This hypothesis predicts that nonpolyadenylated
mRNA should accumulate only during S phase in synchronized populations
of cells.

F. Can All hnRNA Molecules Be Precursors of mRNA?

In the preceding sections I have reviewed the evidence that mRNA
is almost certainly derived from hnRNA which probably is, in some
cases at least, larger than the mRNA. The conclusion that a large
fraction of poly(A) attached to hnRNA may decay in the nucleus (Perry
et al., 1974; Brandhorst and McConkey, 1975) suggests that some polyadenylated hnRNA molecules may not serve as precursors of mRNA. It
could be argued, though, that these hnRNA molecules were polyadenylated after the sequence for mRNA had been cleaved off and had already
served as a precursor of mRNA.

In L cells it can be estimated that about 2% of the nucleotides
incorporated into heterogeneous RNA become mRNA (Brandhorst and Mc-
Conkey, 1974). If all hnRNA molecules serve as precursors of mRNA
(i.e., if all contain a nucleotide sequence which becomes a cytoplasmic mRNA molecule), the hnRNA precursor must be reduced in size on the
average by 98% in L cells. This is not consistent with present information. The size distribution of polyadenylated mRNA in L cells
suggests that it has a mass average molecular weight of about 1.5 x

10^6, while polyadenylated hnRNA appears to have a mass average molecular weight of less than 10^7 (Greenberg and Perry, 1972); there appears to be at most a 90% difference in size, and the measurements were not done under denaturing conditions. Total hnRNA and mRNA from L cells show even less difference in size distribution (Greenberg and Perry, 1971). These comparisons are hampered by the difficulty of estimating the molecular weight of very large RNA molecules and by the unlikely possibility that giant hnRNA has been degraded in size during preparation. All available information indicates that it is extremely unlikely that all hnRNA molecules can serve as precursors of mRNA molecules in L cells. From a comparison of the size distributions, it is likely that the majority of hnRNA molecules do not contain sequences giving rise to an mRNA molecule entering cytoplasmic polysomes. This conclusion appears to have been substantiated by Galau et al. (1974) in RNA driven sequence complexity measurements on mRNA and hnRNA of sea urchin embryos (see Sect. IV,F). These observations and interpretations make the interpretation of comparisons of the physical properties of mRNA and its purported nuclear "precursor," polyadenylated hnRNA, more difficult.

IV. COMPARATIVE ANALYSES OF THE PHYSICAL PROPERTIES OF hnRNA AND mRNA

In the following sections the physical characteristics of hnRNA and mRNA are described and compared. It should be remembered that since it is almost certain that the majority of hnRNA molecules (including polyadenylated hnRNA) are not precursors of mRNA, the property of the hnRNA molecule discussed may not be a property of the nuclear precursors of mRNA. It is also useful to point out here that two important (and similar) models of the structure of hnRNA and its relationship to mRNA place mRNA sequences at the polyadenylated 3' ends of giant hnRNA molecules (Georgiev, 1972; Georgiev et al., 1972; Molloy et al., 1974). These investigators base their models on the physical similarities between mRNA and the 3' ends of hnRNA molecules and the differences between the 3' and 5' ends of giant hnRNA molecules.

A. Oligo(A)

Nakazato et al. (1973) have described a short (20-40 nucleotide) segment of oligo(A) in the hnRNA of HeLa cells. Oligo(A) was not detectable in mRNA and is not located at the 3' termini of RNA molecules. While the synthesis of large poly(A) sequences at the 3' ends of RNA molecules is very sensitive to cordycepin and not immediately sensitive to actinomycin D, the synthesis of oligo(A) is very sensitive to actinomycin D and insensitive to cordycepin (Nakazato et al., 1974). It appears, then, that oligo(A) is transcribed from DNA sequences, rather than being added posttranscriptionally. HeLa nuclear RNA appears to have sufficient oligopurine stretches to code for oligo(A) (Birnboim et al., 1973). There does not appear to be quite enough nuclear oligo(A) for there to be an oligo(A) stretch in each hnRNA molecule (Nakazato et al., 1974), although it is not clear that it has been quantitatively recovered. While there is no evidence that nuclear precursors of mRNA contain oligo(A), these observations are consistent with a model in which part of a precursor molecule is eliminated during processing.

A different situation exists in *Dictyostelium discoideum*. In mRNA there are equal numbers of poly(A) segments 100 nucleotides long and oligo(A) segments 25 nucleotides long (Jacobson et al., 1974). About one-fourth of the hnRNA molecules are attached to poly(A) segments, but there are enough oligo(A) sequences to account for an average of one per hnRNA molecule (Firtel and Lodish, 1973). The DNA contains oligo(dT) in sufficient amount, size, and distribution to code for the oligo(A) (Lodish et al., 1973). The oligo(A) sequences in mRNA appear to be located near the poly(A) sequences (Jacobson et al., 1974). It is possible then that oligo(A) is transcribed as a part of every hnRNA sequence and serves as a primer for polyadenylation of the RNA molecule. It is interesting that oligo(A) appears to be limited to nuclear RNA in mammalian cells, but is probably in all mRNA molecules in the cellular slime mold. Perhaps the oligo(A) of mammalian hnRNA is included as part of the poly(A) segment of mRNA, for which it may be a primer.

B. Oligo(U)

Molloy et al. (1972b) and Burdon and Shenkin (1972) reported the presence in hnRNA of a ribonuclease T1-resistant oligonucleotide, 30 nucleotides long on average, oligo(U). It cannot be detected in mRNA. Giant hnRNA molecules have more oligo(U) sequences per molecule than shorter molecules (Molloy et al., 1974). The spatial distribution of oligo(U) in giant (more than 20,000 nucleotides long) polyadenylated hnRNA can be determined, since it is known that poly(A) is at the 3' end. Giant polyadenylated hnRNA was broken to smaller sizes by limited alkaline hydrolysis, and the 3' ends (polyadenylated) selected by binding to poly(U)-sepharose (Molloy et al., 1974). It was found that over 90% of the oligo(U) segments are in fragments more than 12,000 nucleotides long (400,000 daltons); i.e., only 10% of the oligo(U) sequences remain associated with degraded polyadenylated hnRNA molecules shorter than 12,000 nucleotides. Thus oligo(U) segments are very far from the polyadenylated 3' ends of most giant hnRNA molecules; in fact they are probably close to the 5' end. While it cannot be said that hnRNA molecules containing oligo(U) segments are precursors of mRNA, if any of these molecules do serve as precursors of mRNA, most of the oligo(U) segments are far from the most likely site of mRNA, the 3' end.

C. RNA Sequences Transcribed from Repetitive DNA

As reviewed by Davidson and Britten (1973), nuclear DNA of animal cells contains two types of sequences: single-copy and repetitive. The single-copy or nonrepetitive DNA sequences occur in one or very few copies per haploid genome. Most repetitive DNA sequences are "middle repetitive": They occur in a few hundred to a few thousand copies per haploid genome. Some sets of middle-repetitive copies have very similar sequences, while others have very divergent sequences. They can be distinguished from highly repetitive sequences in which there is little sequence divergence. For the sake of brevity, middle-repetitive sequences will be termed repetitive, for it is doubtful that highly repetitive sequences are transcribed into RNA (David-

son and Britten, 1973). Repetitive and nonrepetitive sequences can
be distinguished because at a given concentration the former renature
more rapidly after denaturation (Britten and Kohne, 1968). As discussed below, RNA transcripts of DNA are derived from both middle repetitive and nonrepetitive sequences.

There are many investigations demonstrating the existence of numerous sequences in hnRNA transcribed from repetitive sequences of
DNA (Church and McCarthy, 1967b; Darnell and Balint, 1970; Firtel et
al., 1973; Georgiev et al., 1973; Holmes and Bonner, 1974; Melli et
al., 1971). These repetitive sequences hybridize at low C_0t, according to the definition of Britten and Kohne (1968). C_0t is defined
as the concentration of the annealing nucleic acid species multiplied
by time; it has units of moles-nucleotide-seconds per liter. The
actual fraction of repetitive sequences in hnRNA can be difficult to
determine under some experimental conditions because some of these sequences exist in double-stranded form as loops or hairpins (Kronenberg
and Humphreys, 1972; Jelinek and Darnell, 1972; Ryskov et al., 1973;
Jelinek et al., 1974). Unless special precautions are taken (Jelinek
et al., 1974) the secondary structure of repetitive sequences of hnRNA
interferes with their hybridization to DNA, though double-stranded RNA
is scored as ribonuclease resistant. Melli et al. (1971) have reported
that 15-20% of the rat myoblast nuclear RNA forms ribonuclease-resistant hybrids with DNA in vast excess at low C_0t: i.e., 15-20% of the
nucleotides are in repetitive sequences. These figures are comparable
to 8-10% observed for sea urchin embryo hnRNA (Smith et al., 1974).
If the posthybridization ribonuclease is omitted, nonrepetitive sequences covalently linked to repetitive sequences will also be scored
as hybrids. Seventy to eighty percent of the rat ascites giant hnRNA
nucleotides hybridize at low C_0t (Holmes and Bonner, 1974) and 23-28%
of the sea urchin hnRNA fragments 1100 nucleotides in length hybridize
at low C_0t (Smith et al., 1974). Darnell and Balint (1970) provided
evidence that most or all hnRNA molecules of HeLa cells contain a repetitive sequence.

Messenger RNA appears to include far fewer nucleotides transcribed
from repetitive sequences. The kinetics of hybridization of duck and

mouse globin mRNA or DNA copies thereof (Bishop et al., 1972; Bishop
and Rosbash, 1973; Harrison et al., 1972), fibroin message (Suzuki et
al., 1972), and ovalbumin message (Sullivan et al., 1973) all indicate
that there is only one or very few copies per haploid genome. These
investigations would not have revealed the presence of short repetitive sequences representing a small fraction of the total nucleotides
in the mRNA. Among the highly enriched preparations of specific mRNA,
only sea urchin histone mRNA is certainly transcribed from middle repetitive sequences present in 400-1200 copies per haploid genome in
sea urchin embryos (Kedes and Birnstiel, 1971; Weinberg et al., 1972).
The structural genes known to be present in only one copy give rise
to the specialized products of undifferentiated cells. The mRNA's for
these specialized proteins may be, indeed appear to be, highly stable,
making it possible for these cells to synthesize enough specialized
protein with a single copy of the structural gene (Suzuki et al., 1972;
Kafatos, 1972). No amplification of the DNA nucleotide sequences of
these genes is required and none is observed; a general search for
tissue-specific amplification of DNA sequences has yielded none (Kohne
and Byers, 1973). The requirements for histone synthesis are different: Histones are synthesized during the S phase of the cell cycle
when histone mRNA is detectable in mammalian cells (Borun et al., 1967;
Gallwitz and Mueller, 1969). The lifetime of histone mRNA in L cells
is nearly the same as the length of S phase (Perry and Kelley, 1973).
At times when cells are dividing very rapidly (as in cleavage stage
of sea urchin embryos), accumulation of histone mRNA (even if stabil-
ized) transcribed from a single DNA copy may not be sufficient to provide for the required rate of histone synthesis. This may account
for the repetition of histone genes. If this hypothesis is correct,
a prediction is that the extent of sequence repetition of histone DNA
in different organisms should be related to the maximum required rate
of synthesis of histone mRNA; this in turn is probably related to the
maximum rate of cell division attained during the life cycle.

Another unusual situation exists for mRNA for antibodies. Recent
hybridization evidence suggests that from 500 to 7500 genes code for
the variable regions of light chains in a myeloma (Delovitch and Bag-

lioni, 1973). This interpretation favors the germ-line hypothesis
for antibody diversity in which the variability of the primary amino
acid sequences of antibodies is carried in the DNA.

Hybridizations of total mRNA fractions indicate that, compared to
hnRNA, fewer nucleotides are in sequences transcribed from repetitive
DNA; the vast majority of nucleotides are transcribed from nonrepetitive DNA. In carefully prepared purified mRNA of sea urchin gastrulae
and HeLa cells about 3% or less and 6%, respectively, of the nucleotides hybridize to repetitive DNA (Goldberg et al., 1973; Klein et
al., 1974). In cellular slime molds about 12% of the nucleotides in
polyadenylated mRNA hybridize to repetitive DNA. The hybrids in these
investigations were assayed after ribonuclease treatment. If hybrids
are assayed without ribonuclease treatment, at least 8% of the mRNA
molecules of L cells contain repetitive sequences (Greenberg and Perry,
1971); this figure has been corrected for the contamination of polysomal RNA by RNP's of different buoyant density. Only 3% or less or
6% of the mRNA molecules from sea urchin gastrulae or HeLa cells,
respectively, contain repetitive sequences (Goldberg et al., 1973;
Klein et al., 1974). This indicates that the repetitive sequences in
these mRNA populations are in molecules transcribed entirely from repetitive DNA sequences (Klein et al., 1974). The sea urchin mRNA
molecules employed in this investigation were shorter than their normal size (Goldberg et al., 1973; Kung, 1974). Thus many nucleotide
sequences of mRNA which may be colinear with repetitive sequences
might not have been detected. A further difficulty in the interpretation of these experiments is that size reduction during the annealing and isolation of hybrids could lead to a serious underestimate of
the number of molecules containing repetitive sequences. These problems appear to have been resolved for HeLa mRNA: The size distribution is comparable to that of many other preparations of HeLa mRNA,
suggesting that it is not seriously degraded, and this size distribution is maintained even after isolation of the hybrids (Klein et al.,
1974). A further possible problem is that short double-stranded hybrids formed between mRNA and repetitive DNA sequences might not have
been bound as hybrid in the phosphate-urea-hydroxyapatite assay for

hybrids. Anything over 100 base pairs in length would be bound, but the lower limits of the assay have not been determined. This may explain why other investigators using other methods for assaying hybrids have concluded that many or most mRNA molecules contain repetitive sequences. Dina et al. (1973) conclude that most polysomal RNA molecules of *Xenopus* contain repetitive sequences; they may be too short to have been detected by Goldberg et al. (1973) or Klein et al. (1974). Dina et al. (1973) have not provided conclusive evidence that their preparation of polysomal RNA is not contaminated with hnRNA. Darnell and Balint (1970) have tentatively concluded that there may be a repetitive sequence attached to every mRNA; they used a highly purified preparation of mRNA, making significant contamination by nuclear RNA unlikely. Firtel and Lodish (1973) find that 60% of the polyadenylated mRNA nucleotides of *Dictyolstelium* anneal to DNA at low C_0t, indicating that they are contiguous to repetitive sequences. Numerous other investigations have indicated that polysomal RNA fractions contain repetitive sequences capable of competing with repetitive sequences in hnRNA in hybridization reactions. The low levels of hybridization and poorly documented purity of the polysomal RNA make it possible that these results can be accounted for by contamination by hnRNA of the polysomal fraction.

It can be concluded that hnRNA has a higher frequency than mRNA of nucleotides transcribed from repetitive DNA sequences; many mRNA molecules may contain sequences transcribed from repetitive DNA but these sequences are short, certainly shorter than the repetitive sequences interspersed with nonrepetitive sequences in hnRNA (see next paragraph). Most nucleotides in nearly all mRNA molecules are transcribed from nonrepetitive sequences.

The distribution of repetitive sequences in hnRNA has been investigated recently. The observation that only 10% of the hnRNA nucleotides of sea urchin gastrulae are transcribed from repetitive sequences, but that about a quarter of the fragments 1100 nucleotides long contain a repetitive sequence, indicates that repetitive and nonrepetitive sequences are interspersed on the same molecule (Smith et al., 1974). Holmes and Bonner (1974) have reached similar conclusions for rat ascites hnRNA. By comparing the fraction of hnRNA hybridized at low C_0t (without ribonuclease) with the fragment size of the sheared RNA, they developed a model which fits the data: 10% of the nucleotides in hnRNA are in repetitive sequences; 28% of the hnRNA is organized as repetitive sequences 200 nucleotides long interspersed with nonrepetitive sequences 1200 nucleotides long; an additional 43% of the hnRNA consists of 200 nucleotide segments of repetitive sequences interspersed with nonrepetitive sequences 4800 nucleotides long; the character of an additional 29% cannot be predicted. The model is consistent with a model for the organization of rat DNA, based on electron microscopic examination of sheared, denatured DNA reannealed at low C_0t: Interspersed sequences account for 80% of the genome; repetitive sequences averaging 100-300 base pairs in length (often in doublets) are interspersed with nonrepetitive sequences averaging 1000-2000 base pairs in length, although some are up to 16,000 base pairs in length; the remaining 20% include highly repetitive DNA sequences (such as satellite DNA), and uninterrupted lengths of single-copy or middle-repetitive sequences (Bonner et al., 1973; Holmes and Bonner, 1974).

The organization of rat DNA is remarkably similar to that of anuran (*Xenopus laevis*) and sea urchin (*Strongylocentrotus purpuratus*) DNA (Davidson et al., 1973; Graham et al., 1974). Moreover, there is evidence that this organization in which a large fraction of the DNA consists of interspersed repetitive and nonrepetitive sequences is characteristic of many, perhaps all, animals (Graham et al., 1974). The DNA of *Drosophila melanogaster* has a different pattern of sequence interspersion: Both the repetitive and nonrepetitive elements are considerably larger, on the average, than those of *Xenopus* and sea urchins (Manning et al., 1975).

Based on the annealing of DNA copies of globin mRNA to the middle-repetitive fraction of DNA, Bishop and Freeman (1973) have proposed that a repetitive sequence may exist in DNA just beyond the 3' end of the globin cistron. This interpretation, while not conclusive, is evidence that mRNA is actually transcribed from nonrepetitive DNA sequences interspersed with repetitive sequences.

Convincing evidence that many mRNA molecules are transcribed from unique DNA sequences interspersed with repetitive sequences has been provided by Davidson et al. (1975). If denatured sea urchin DNA (approximately 2000 nucleotides in length) is reannealed at low C_0t and bound to hydroxyapatite, nonrepetitive sequences contiguous to repetitive sequences (approximately 300 base pairs in length) can be isolated. If this fraction is sheared, denatured, and reannealed at low C_0t, most of the nonrepetitive sequences contiguous to repetitive sequences will not bind to hydroxyapatite and can thus be isolated. Of the sea urchin mRNA, 50% hybridizes at high C_0t to this nonrepetitive DNA contiguous to repetitive DNA, while 65% hybridizes to total DNA. It can be calculated that this level of hybridization is much higher than would be expected if mRNA were transcribed randomly from nonrepetitive DNA sequences. Since most heterogeneous RNA transcribed in sea urchin embryos decays in the nucleus (Brandhorst and Humphreys, 1972), it cannot be concluded that mRNA is derived from larger hnRNA molecules consisting of interspersed nonrepetitive and repetitive sequences, though this appears likely.

The topographical arrangement of repetitive sequences of DNA has been investigated. If double-stranded eukaryotic DNA fragments are partially digested with exonuclease III, which digests single strands of double-stranded DNA from the 3' end, the single-stranded ends will anneal intramolecularly with high probability to form circles (Thomas et al., 1970). The single-stranded ends that anneal to form "Thomas circles" are necessarily repeated sequences. While most Thomas circles undoubtedly form from the highly repetitive fraction of DNA (Schachat and Hogness, 1973; Peacock et al., 1973; Laird et al., 1973), some probably form from middle repetitive sequences (Schachat and Hogness, 1973; Laird et al., 1973; Lee and Thomas, 1973). The frequency of middle repetitive sequences in *Drosophila* polytene DNA (Laird et al., 1973), the frequency of cyclization (Lee and Thomas, 1973), the optimal DNA fragment size for cyclization (Lee and Thomas, 1973), and the periodicity of interspersed DNA (100-200 nucleotides of repetitive DNA separated by nonrepetitive sequences 750 nucleotides in length; Wu et al., 1972) have led to the speculation that groups

of three or four different repetitive sequences interspersed with
nonrepetitive sequences are themselves repeated (Laird et al., 1973).
While there is no conclusive evidence for this topography, it is the
subject of continuing investigation. It is possible that the interspersed sequences of hnRNA transcribed from repetitive DNA have an
ordered arrangement relative to one another.

The spatial location of repetitive sequences relative to the 3'
ends of polyadenylated hnRNA molecules has been investigated. Molloy
et al. (1974) fragmented and denatured radioactive hnRNA, selected
polyadenylated fragments (3' ends), and hybridized them at low C_0t
to excess DNA. The 3' end fragments 3000 nucleotides or less in
length (the size of mRNA), hybridized to the same extent as mRNA,
while fragments of increasing size hybridized to a greater extent; a
constant fraction of input radioactive RNA was hybridized for polyadenylated fragments 8000 or more nucleotides in length. Molloy et
al. (1974) interpret these results to mean that, assuming mRNA is
derived from the polyadenylated 3' ends of hnRNA, a repetitive sequence not included in the mRNA molecule lies at or near the 5' terminus of a mostly nonrepetitive mRNA sequence. Beyond 8000 nucleotides (i.e., beyond the potential 5' ends of virtually any mRNA molecule), repetitive sequences are randomly interspersed with nonrepetitive sequences. A problem with the interpretation of this investigation is that the polyadenylated fragments were not sized under denaturing conditions: The larger fragments may be aggregates of smaller fragments. A similar investigation for *Dictyostelium* indicates
that many of the repetitive sequences are in the distal 5' half of
the molecule (Firtel and Lodish, 1973).

RNA molecules are initiated by incorporation of a nucleoside triphosphate at the 5' end; these markers of true 5' ends are released
as nucleoside tetraphosphates (and triphosphates) by alkaline hydrolysis. Only giant hnRNA molecules have triphosphorylated 5' ends
(Georgiev et al., 1972), though their identification has been disputed. Smaller hnRNA molecules release only monophosphorylated nucleotides; their 5' ends presumably arise from dephosphorylation of
their 5' end, from intramolecular breakage during extraction of the

RNA, or from specific processing of giant hnRNA into smaller molecules. Triphosphorylated ends of rat liver RNA hybridize to DNA at
low C_0t 4-10 times more efficiently than monophosphorylated ends.
It can be concluded that some sequences transcribed from repetitive
DNA lie very near the 5' ends of many giant hnRNA molecules. The result is consistent with, but does not prove, a model proposed by
Georgiev et al. (1972) in which the 5' ends of giant hnRNA precursors
of mRNA are degraded, leaving the nucleotides at the 3' end; these in
turn would become mRNA. Consistent with this model are the observations that the hybridization of the repetitive sequences of rat liver
polysomal RNA to DNA is completely inhibited by giant (35-75S) hnRNA,
but polysomal RNA only slightly competes against the hybridization
between repetitive sequences of giant hnRNA and DNA. But polysomal
mRNA strongly competes with the 3' end sequences of giant hnRNA (that
were marked by reaction with tritiated sodium borohydride). While
most hybridization competition experiments aimed at evaluating the
sequence similarities of nuclear and polysomal RNA are hampered by
the low levels of hybridization and high probability of interference
by contamination of the polysomal fraction with nuclear RNA, these
data are difficult to reconcile with nonspecific leakage of hnRNA into the cytoplasm. On the other hand, the giant hnRNA is quite possibly contaminated with smaller hnRNA since giant hnRNA fractions
were not prepared under denaturing conditions. This problem makes
it impossible to accept as conclusive the interpretation of Georgiev
et al. (1972) that the repetitive sequences of mRNA molecules (if
real) are derived from the 3' ends of much larger hnRNA molecules.
The interpretation of these experiments is also difficult because of
the probable masking of many repetitive nucleotide sequences in hnRNA
by intramolecular base pairing; these sequences are not available for
hybridization to immobilized DNA if the RNA is not sheared and denatured (Jelinek et al., 1974).

     In summary, many hnRNA molecules contain nucleotide sequences
transcribed from repetitive DNA sequences interspersed with sequences
transcribed from nonrepetitive DNA sequences. There is some indication that these sequences are not randomly arranged over the whole

molecule; they appear to be sparser near the 3' end. While mRNA
clearly contains a lower frequency of nucleotides in repeptitive sequences and includes some molecules entirely transcribed from repetitive DNA, it is possible that at least some mRNA molecules of animal
cells include short repetitive sequences. There is no conclusive evidence for the location of sequences serving as precursors of mRNA at
the 3' ends of giant hnRNA molecules. Only in the cellular slime mold
Dictyostelium does it appear nearly certain that mRNA lies at the 3'
end of polyadenylated hnRNA, but the hnRNA is only 20% larger than
mRNA (Firtel and Lodish, 1973). Much of the DNA of animal cells is
organized into interspersed repetitive and nonrepetitive sequences,
consistent with the general organization of hnRNA. But a significant
fraction (up to 20-30%) of the DNA appears to consist of long stretches of uninterrupted single-copy sequences. At least a large fraction
of the mRNA in sea urchin embryos is transcribed from the nonrepetitive DNA sequences interspersed with repetitive sequences (Davidson
et al., 1974b).

D. Double-stranded RNA

hnRNA contains double-stranded regions resistant to extensive T1
and pancreatic ribonuclease treatment, but these regions are not detectable in mRNA (Jelinek and Darnell, 1972; Kronenberg and Humphreys,
1972; Ryskov et al., 1973). The base composition of double-stranded
RNA is symmetrical. The frequency of nucleotides in double-stranded
segments ($\sim$3%) is the same for 100S and 50S regions of hnRNA on an
aqueous sucrose gradient, suggesting that there are several segments
of double-stranded RNA in the longer hnRNA molecules (Jelinek and
Darnell, 1972). In Ehrlich ascites cells there are two types of
double-stranded structures: About 1% of the nucleotides of hnRNA
are in (A + U)-rich double-stranded structures, and about 2% in apparently shorter, (G + C)-rich structures. The longer segments are
found only in giant hnRNA sedimenting more rapidly than 45S, suggesting that they are not found in the immediate precursor of mRNA, which
is smaller than giant hnRNA. In sea urchin embryos, a much smaller
fraction of the nucleotides of hnRNA are in double-stranded structures

(Kronenberg and Humphreys, 1972); this may reflect the relative lack
of giant hnRNA molecules (Brandhorst and Humphreys, 1972; Kung, 1974;
Peltz, 1973).

After denaturation these double-stranded regions of RNA hybridize
rapidly and efficiently to DNA, indicating that they are transcribed
from reiterated sequences of DNA (Jelinek and Darnell, 1972; Ryskov
et al., 1973). When hnRNA molecules containing double-stranded RNA
are denatured the double-stranded regions reform extremely rapidly;
this "snapping back" suggests that the paired nucleotide sequences
are located close to one another on the same RNA molecule (Jelinek
and Darnell, 1972). After isolation and denaturation, the double-
stranded RNA regions do not rapidly reform double-stranded structures
(Jelinek and Darnell, 1972; Kronenberg and Humphreys, 1972). This
implies that the ribonuclease treatment used to eliminate single-
stranded RNA as a step in purification also degrades the intramolecular link between the two paired strands of double-stranded RNA. This
indicates that the double-stranded RNA is not a tight hairpin, but
contains at least a short, ribonuclease-sensitive loop at the end.

The existence of self-complementary sequences within the same
RNA molecule implies the existence of these sequences in DNA. They
have been detected (Wilson and Thomas, 1974) and isolated (Georgiev
et al., 1973). Denatured DNA rapidly reforms duplex regions, even
at extremely low concentrations of DNA in which only intramolecular
interactions are likely to occur. Sedimentation and electron microscopic investigation revealed that these duplexed regions are in the
form of hairpins (inverted repeats). The number of nucleotides in
the "turnaround" is very small and it is S1 nuclease resistant (Wilson and Thomas, 1974). If the double-stranded regions of hnRNA are
derived from these DNA hairpins, then they should also be in the form
of hairpins. Double-stranded RNA in long hairpins hybridizes efficiently to isolated DNA hairpins (Georgiev et al., 1973). The distribution and frequency of hairpins in DNA and hnRNA are consistent
(Wilson and Thomas, 1974; Ryskov et al., 1973). On the other hand,
the average nucleotide length of the double-stranded RNA (roughly 100
base pairs; Jelinek and Darnell, 1972; Ryskov et al., 1973) is appa-

rently less than the size distribution of hairpins observed by Wilson
and Thomas (1974) in DNA (several hundred to several thousand base
pairs). The estimation of the size distribution of double-stranded
RNA is not likely to be very accurate. The size of DNA hairpins isolated by Georgiev et al. (1973) is comparable to that of the long RNA
hairpins; no DNA hairpins comparable in size to the smaller double-
stranded hnRNA were detected. The frequency of long hairpins in DNA
and giant hnRNA is the same, about 1 per 20,000 nucleotides. Wilson
and Thomas (1974) have shown that the DNA hairpins tend to exist in
clusters of two to four, separated by intervals of 30,000 to 240,000
nucleotide pairs, depending on the species. In HeLa cells the average separation of clusters of hairpins is 38,000 nucleotide pairs.

Besides the highly ribonuclease resistant, well paired double-
stranded regions in hnRNA, there are also double-stranded regions
moderately resistant to ribonuclease (Jelinek et al., 1974). These
structures also reform rapidly after denaturation, unless the denatured RNA is broken to less than approximately 500 nucleotides in
length. Thus these structures appear to be intramolecular loops with
base-paired stems, of a total length of only a few hundred nucleotides. These regions hybridize rapidly to DNA, indicating that they
are enriched in segments transcribed from repetitive DNA. They are
only partially resistant to ribonuclease, suggesting that they are
not perfectly base paired, or the base-paired regions are short and
interspersed with unpaired regions. While a significant fraction of
the repetitive sequences in hnRNA forms these intramolecular regions
of double-strandedness, it is not clear what fraction of the repetitive sequences do not form double-stranded regions. The melting profile and reactivity with formaldehyde of bulk hnRNA of Novikoff ascites cells indicate that approximately 60% of the nucleotides of
giant hnRNA are paired (Holmes and Bonner, 1973).

It is important to determine the length of the moderately ribonuclease resistant double-stranded segments. It can be calculated
that about half of the nucleotides of random nucleotide sequences of
a hundred nucleotides or longer can be paired to form double-stranded
regions a few nucleotides in length (Gralla and DeLisi, 1974). Ribo-

somal RNA of mammals has extensive secondary structure, as evidenced
by its hyperchromicity after melting (Macnaughton et al., 1974). Prokaryotic mRNA molecules such as T4 mRNA also possess considerable
secondary structure (Ricard and Salser, 1974). It is not clear whether this secondary structure is selected for in evolution or is the
result of random base pairing, perhaps influenced by interaction with
protein. It is clear that the highly ribonuclease resistant double-
stranded regions of hnRNA are too long to be accounted for randomly.
This is not yet clear for the moderately resistant double-stranded
regions. A possible equivalence has not been evaluated between the
moderately ribonuclease-resistant RNA duplexes of Jelinek et al.
(1974) and the short double-stranded regions of Ryskov et al. (1973).

Jelinek and Darnell (1972) have pointed out that there is no reason to believe that the double-stranded regions of hnRNA exist in
vivo. Moreover, there is no evidence that they are involved in the
processing of mRNA in the nucleus, though they might serve as convenient sites of recognition for processing enzymes. Georgiev et al.
(1973) find that smaller (mRNA-sized) hnRNA molecules do not contain
long hairpins of double-stranded RNA, and that there is an average of
one long hairpin per giant hnRNA molecule. The long hairpins might
thus serve as a recognition site in the processing of a giant precursor of mRNA. In this regard, it is interesting that heterogeneous
polysomal RNA (not necessarily pure mRNA) will anneal extensively (up
to 8% of it) with nuclear RNA (Stampfer et al., 1972). Georgiev et
al. (1973) find that small hnRNA and polyadenylated RNA from polysomes
will form ribonuclease-resistant duplexes with denatured long hairpins
of giant hnRNA. These observations suggest that a relationship may
exist between the repetitive sequences that may exist in mRNA and
those of hnRNA. A potential mRNA molecule might be derived from a
larger precursor by an endonucleolytic cleavage in the single-stranded
loop of an intramolecular double-stranded region, followed by local
denaturation of the double-stranded segment to release the mRNA sequences. Other explanations for a relationship between hnRNA and
mRNA nucleotide sequences are possible (Stampfer et al., 1972; Georgiev, 1973).

E. Methylated Heterogeneous RNA

Several physical features of hnRNA [oligo(U), oligo(A), double-
stranded segments, some repetitive sequences] are apparently lacking
in mRNA. Considering the vastly greater number of nucleotides in
hnRNA decaying in the nucleus compared to mRNA, it is not surprising
that some of the hnRNA molecules and sequences include physical features missing in mRNA. Perry and Kelley (1974) have found that polyadenylated mRNA is methylated to a much greater extent than hnRNA.
Polyadenylated mRNA from L cells labeled with tritiated methionine
and [^{14}C]uridine could be purified from methylated rRNA and tRNA by
affinity chromatography on poly(dT)-cellulose. There are about 2.2
methyl groups per 1000 nucleotides, about one-sixth the frequency in
rRNA. There appear to be fewer different bases methylated in mRNA
than in rRNA. The methylation of hnRNA (larger than 45S) was only
barely detectable, appearing to be 3-5 times less per nucleotide than
in mRNA. The measurement of methylation of hnRNA is very inaccurate,
but it is probably greater than that expected (50-fold less than that
of mRNA) if the only nucleotides methylated are in nucleotide sequences actually going to the cytoplasm as mRNA. (About 2% of the
total nucleotides synthesized are estimated to become mRNA in L cells;
Brandhorst and McConkey, 1974.) The methylation of hnRNA relative
to mRNA is more consistent with the relative size distribution of
hnRNA and mRNA, suggesting that if mRNA is indeed derived from the 3'
ends of hnRNA molecules averaging 5 times larger, all these 3' ends
contain methylated bases, even though some do not serve as precursors
of mRNA which comes to function in polysomes. In any case, the methylated bases of mRNA may provide a useful handle for following mRNA
processing, just as they were useful for following the processing of
rRNA (Greenberg and Penman, 1966).

Extensive and exciting progress has been made in the characterization of methylated mRNA. A wide variety of viral and cellular polyadenylated mRNA molecules have a peculiar oligonucleotide cap at their
5' termini consisting of 7-methylguanosine linked through its 5'-
hydroxyl group via a triphosphate (or pyrophosphate) group to the
5'-hydroxyl of a 2'-O-methylnucleoside. Sometimes there is a second

2'-O-methylnucleoside adjacent to the first, and the ribose rings of all four nucleosides have been found to be methylated (Rottman et al., 1974; Griffin, 1975; Furuichi et al., 1975; Wei and Moss, 1975; Adams and Cory, 1975; Wei et al., 1975; Perry et al., 1975). The only other identified methylated base in mRNA is 6-methyladenosine, which is present in more than one residue per mRNA molecule on the average but is not part of the poly(A) segment. It is quite possible that the unusual oligonucleotide containing 7-methylguanosine at the 5' termini is a universal feature of all eukaryotic mRNA's.

The presence of a distinctive cap of methylated nucleotides introduces a variety of interesting possibilities. The presence of the 7-methylguanosine terminus appears to be necessary for translation: reovirus and VSV mRNA's synthesized in vitro lacking this group or from which it had been removed by β elimination are not translated in an in vitro protein synthesizing system (Both et al., 1975; Muthukrishnan et al., 1975). The wheat germ protein synthesizing system is capable of methylating these viral mRNA's, suggesting that methylation of the 5' terminus of mRNA may occur in the cytoplasm, perhaps upon initiation of protein synthesis.

The 5'-methylguanosine termini may serve as convenient markers for the 5' ends of mRNA molecules (curiously the unusual linkage means that the 5' end of the molecule actually has a free 3'-hydroxyl). Since there can be only one 5'-terminal oligonucleotide per chain, an upper limit on the size of mRNA molecules can be set (Adams and Cory, 1975). If all mRNA's in a cell have this peculiar 5'-terminal oligonucleotide, polyadenylated polysomal molecules lacking them would be fragments derived from intact mRNA molecules. Quite possibly these 5'-terminal oligonucleotides will serve operationally as markers of 5' ends of mRNA molecules, making possible further characterization of the topography of mRNA molecules. The 2'-O-methylribose at the 5' terminus may be an RNase-resistant marker for cleavage during processing of a larger nuclear precursor of mRNA (Adams and Cory, 1975). The methylation of nonpolyadenylated mRNA and hnRNA have not yet been fully characterized chemically.

F. Models of hnRNA and Its Relationship to mRNA

Several models, of various degrees of detail and not all mutually exclusive, have been proposed to describe the organization of hnRNA and its relationship to mRNA. Evidence consistent with one or more of the models has been presented, but no conclusive evidence exists proving or disproving the validity of any of the models. The major technical difficulty is the great heterogeneity and vast amount (relative to mRNA) of hnRNA. Evidence for or against certain aspects of these models has been presented in the preceding sections. Since no model can be singled out as most likely, none will be described or discussed in detail.

Molloy et al. (1974) have presented a model for the organization of polyadenylated hnRNA. It places the oligo(U) sequences near the 5' end of the molecule and occasional double-stranded loops and hairpins distal to the polyadenylated 3' end. The first 500-4000 nucleotides do not form these double-stranded regions because these nucleotides are part of the mRNA sequence derived from the giant hnRNA molecule by this model. Most repetitive sequences are distal to the 3' end and at least some of these are in double-stranded regions. There may be a small fraction of repetitive nucleotide sequences within the mRNA region, since some highly purified mRNA preparations bind to DNA at low $C_o t$. It is highly likely that the general features of this model of Molloy et al. (1974) describe the nucleotide arrangement of the majority of polyadenylated, and perhaps other, hnRNA molecules. It cannot be concluded, though, that mRNA is derived from the 3' end or any other part of the giant hnRNA molecule. According to this model, nuclear processing of mRNA involves the polyadenylation of the 3' end of an hnRNA molecule followed by specific enzymatic cleavage of the molecule. The physical features characteristic of the hnRNA nucleotides distal to the 3' end may serve as recognition sites for this cleavage.

Georgiev and co-workers proposed a model based more on functional considerations that is entirely consistent with the model of Molloy et al. (1974) (Georgiev, 1969; Ryskov et al., 1973; Georgiev, 1972). They proposed that mRNA sequences lie toward the 3' ends of giant

hnRNA precursors. The evidence for this proposal is mainly based on
hybridization competition experiments between repetitive sequences of
mRNA and hnRNA. Such experiments are extremely susceptible to contamination of one RNA fraction by another. According to the model, the
nucleotide sequences at the 5' end of the giant hnRNA molecule are
transcribed from regulatory sites on the DNA molecule. These must be
transcribed in order that the colinear mRNA sequences be transcribed.
The nucleotides at the 5' end are then cleaved away from the mRNA.
Neither the Georgiev model nor the Molloy et al. (1974) model rules
out the possibility that there may be several mRNA molecules per hnRNA
precursor.

Both models are consistent with the model of Firtel and Lodish
(1973) for the hnRNA of *Dictyostelium discoideum*. Firtel and Lodish
propose that mRNA is at the 3' end of polyadenylated hnRNA molecules.
About 20% of the nucleotides at the 5' end, including some transcribed
from repetitive DNA sequences, are removed in processing to form mRNA.
The evidence for the model is good, and it is consistent with the
possible location of mRNA sequences at the 3' ends of hnRNA molecules
in higher organisms. On the other hand, it does not provide support
for the conclusion that in animal cells mRNA sequences are derived
from much larger hnRNA precursor molecules. Nor does it provide an
explanation for the observation that a large fraction of nucleotides
in hnNRA decay in the nucleus in mammalian cells.

Davidson and Britten (1973) have proposed a detailed model for
the regulation of the transcription of mRNA. The most significant
feature is the proposal that structural genes consisting of nonrepetitive DNA sequences are adjacent to "receptor" sequences which respond to "activator" proteins (or RNA). Since a given external signal
may affect the transcription of several different structural genes,
these authors propose that the receptor sequences are repetitive and
adjacent to a number of structural genes forming a "battery." If
these receptor sequences are transcribed but eliminated by processing,
the abundance of repetitive sequences in hnRNA and paucity in mRNA is
explained. If there is transcription of polycistronic batteries of
structural genes interspersed with repetitive receptor sequences, the

model provides for the interspersion of repetitive and nonrepetitive sequences in large hnRNA molecules. Although Davidson and Britten feel that a receptor or a set of a small number of receptors should be immediately adjacent to a structural gene, the proposal of Georgiev that long stretches of regulatory sequences must be transcribed is not inconsistent with their arguments. To account for at least some of the RNA that decays rapidly in the nucleus, Davidson and Britten (1973) propose that it serves as a regulatory intermediate, either as activator RNA or as mRNA for activator proteins. The latter proposal requires that these activator mRNA molecules be translated. Davidson and Britten point out that hnRNA in L cells has a half-life of 23 min (Brandhorst and McConkey, 1974), stable enough to be translated one or a few times before decaying. Since a determination of the stability of hnRNA in fractionated cells is actually a measure of its average lifetime within the nucleus, it cannot be excluded that some hnRNA enters the cytoplasm, is translated, and rapidly degraded without accumulating significantly relative to mRNA for structural genes, which accumulates and decays very slowly (Greenberg, 1972; Perry and Kelley, 1973; Singer and Penman, 1973). The regulatory model of Davidson and Britten (1973) is not necessarily inconsistent with the structural model for polyadenylated hnRNA proposed by Molloy et al. (1974).

Scherrer and Marcaud (1968) have proposed that hnRNA contains transcripts of many structural genes which are potential mRNA molecules but are not selected to enter the cytoplasm to function as mRNA. They are thus proposing a posttranscriptional regulatory event. This might explain why so much hnRNA decays rapidly. Most of the available evidence for this proposal is based on old hybridization competition experiments which are difficult to interpret (see Sec. III,A, and Church and McCarthy, 1967b). Davidson and Britten (1973) have reviewed the compelling evidence for transcriptional regulation, though the existence of transcriptional regulation does not rule out the possibility of posttranscriptional regulation. The single-copy fraction of DNA can be isolated and RNA hybridized to it to determine the fraction of DNA transcribed in a given cell type. If RNA fractions from two or more types of cells are combined and a higher fraction of the

DNA is complementary to the combined RNA than to either separate preparation of RNA, it can be concluded that there are some different sequences of RNA in the two cells. These RNA-driven complexity measurements have been carried out on cellular and nuclear RNA in various
combinations and it is clear that there are distinct qualitative differences in the populations of RNA of different cells or the same cell
in a different regulatory condition (Brown and Church, 1972; Grouse
et al., 1972; Liarkos et al., 1973; Firtel, 1972; Grady and Campbell,
1973). Similar observations have been made for RNA transcribed from
repetitive sequences of DNA (see Davidson and Britten, 1973).

     Galau et al. (1974) have measured the sequence complexity of high-
ly purified mRNA from polysomes of sea urchin gastrulae. The nonrepetitive fraction of tritium-labeled DNA was purified and hybridized to
excess mRNA. The mRNA was found to include about 2.7% of the total
nonrepetitive DNA sequence complexity, or enough for 14,000 structural
genes of average length. The authors reported that preliminary measurements indicate that nuclear RNA has a sequence complexity about
10 times that of mRNA; i.e., it includes 20-30% of the sequence complexity of the genome. It is clear that most nucleotide sequences in
DNA transcribed do not enter polysomes as mRNA. While it cannot be
concluded that nonrepetitive sequences in nuclear RNA not entering
polysomes are potential mRNA sequences, these observations are consistent with the proposal of Scherrer and Marcaud (1968). Since in
sea urchin embryos, mRNA is certainly not 10 times smaller than hnRNA
(Brandhorst and Humphreys, 1972; Kung, 1974), the measurements of
Galau et al. (1974) indicate that not all hnRNA molecules give rise
to mRNA molecules, in agreement with the conclusion of Sect. III,F.
It is interesting that the fraction of nucleotide sequence complexity
transcribed which becomes mRNA is similar to the fraction of nucleotides in heterogeneous RNA synthesized which enters the cytoplasm
(Brandhorst and Humphreys, 1972); the significance of this observation is not clear, since most of the nucleotide sequence complexity
of mRNA is accounted for by a small fraction of the total nucleotides
in mRNA (Galau et al., 1974).

Bishop et al. (1974) have estimated the sequence complexity of
mRNA by determining the kinetics of hybridization of cDNA to its polyadenylated mRNA template. They conclude that about 35,000 different
polyadenylated messenger-sized RNA sequences are present in the HeLa
cytoplasm. In agreement with Davidson et al. (1975), they find that
most of the sequences are present in only a few copies each per cell,
but that a few are present in several thousand copies per cell. These
estimates of sequence complexities shed interesting light on the controversial issue of how many genes there are in a haploid genome (see,
e.g., Bishop, 1974).

The use of reverse transcriptase to make DNA copies of mRNA should
make it possible to determine if there are nuclear sequences in a giv-
en cell type A, which are specific to mRNA of cell type B but not
found in the mRNA fraction of A. This would be strong evidence for
Scherrer's proposal. A more difficult question is whether or not only
a small fraction of nuclear transcripts of a given structural gene
actually enter the cytoplasm as mRNA: The possibility of nonspecific
"wastage" has not been eliminated. Indeed, Stevens and Williamson
(1973b) have provided evidence that at different cell concentrations
different fractions of the purported nuclear precursor of immunoglobulin heavy-chain mRNA actually enter the cytoplasm. There is evidence that when 3T6 cells convert from resting to growing state, the
amount of mRNA per cell doubles without any detectable change in the
rate of synthesis of hnRNA (Johnson et al., 1974). Since mRNA is such
a small fraction of the total heterogeneous RNA synthesized (Soeiro
et al., 1968; Brandhorst and McConkey, 1974), it is unlikely that
there should have been a detectable increase in the rate of hnRNA
synthesis to account for the increased amount of mRNA. Nevertheless,
these experiments suggest the possibility of a posttranscriptional
regulation of the amount of a given mRNA which can enter the cytoplasm. Contrary to these investigations, in developing sea urchin
embryos the instantaneous rate of synthesis per nucleus of hnRNA and
mRNA change coordinately (Brandhorst and Humphreys, 1972).

I present the following brief speculations. Nuclear precursors
of mRNA are probably slightly larger than or a few times larger than

most of the mRNA molecules for which they serve as precursors; big
mRNA molecules have big nuclear precursors. The observations of Firtel and Lodish (1973) on *Dictyostelium* clearly indicate that a giant
precursor is not an absolute requirement. The nuclear precursors of
mRNA probably include at least one repetitive-sequence element when
transcribed, and possibly other sequences, which are eliminated by
intranuclear processing. Since it appears that most mRNA is transcribed from nonrepetitive DNA sequences interspersed with repetitive
DNA sequences (Davidson et al., 1975), it is likely that these repetitive sequences have some function, either in the processing of the
precursor of mRNA or in transcriptional regulation as proposed by
Davidson and Britten (1973) and Georgiev (1972). Since it can be estimated that only about 2% of the nucleotides in hnRNA of L cells
become mRNA (Brandhorst and McConkey, 1974), even if precursors of
mRNA average five times larger than the mRNA, 90% of the hnRNA nucleotides and most of the hnRNA molecules are left unaccounted for. It
is quite likely that many of these molecules contain potential mRNA
sequences that do not enter the cytoplasm, as proposed by Scherrer
and Marcaud (1968). These molecules are probably physically organized in the manner proposed by Molloy et al. (1974); it is possible
that the nucleotides of the precursors of mRNA are arranged similarly.
On the other hand, since precursors of mRNA may be polyadenylated
later in their nuclear lifetime than hnRNA molecules decaying in the
nucleus (Perry and Kelly, 1974; Brandhorst and McConkey, 1975), it
is possible that precursors of functional mRNA are a separate class
of nuclear RNA molecules having unique physical properties. It is
also possible that some hnRNA molecules serve regulatory functions,
such as those proposed for the "activator" of Davidson and Britten
(1973). The observation that most hnRNA molecules in *Dictyostelium*
appear to become mRNA molecules (Firtel and Lodish, 1973), suggests
that an extensive set of such regulatory molecules decaying rapidly
in the nucleus may not be necessary for the regulation of gene expression even in higher cells. Alternatively, perhaps in cells of
lower organisms such as *Dictyostelium,* the metabolism of RNA transcripts of structural genes and regulatory genes is the same; e.g.,

mRNA for "activator" proteins and structural proteins may be processed, transported to the cytoplasm, translated, and degraded similarly in cellular slime molds, but not in mammalian cells. The major point of this speculative discussion is that there is so much hnRNA synthesized of such heterogeneity in mammalian cells in relation to its known functions that it probably has multiple functions and falls into several classes having different physical properties. A single model for *the* structure or *the* role of hnRNA is probably an oversimplification.

V. SUMMARY

Most nucleotides in hnRNA decay rapidly without ever becoming much more stable mRNA translated in polysomes. It appears certain, though, that some nucleotide sequences in hnRNA serve as precursors of mRNA. In mammalian cells newly synthesized hnRNA is, on the average, larger than mRNA, though there is considerable overlap of size distributions and the real size distributions are subject to dispute; in lower animals the difference in size distributions is apparently less pronounced. The best evidence for a larger nuclear precursor comes from the work of Macnaughton et al. (1974) on globin mRNA and Stevens and Williamson (1973b) on heavy-chain mRNA. In these cases the identifiable probable nuclear precursors were 2-4 times larger than the mRNA molecules. The existence of even larger hnRNA molecules containing mRNA-specific sequences was not ruled out, and the precursor-product relationship between the hnRNA and mRNA sequences was not rigorously established, though Stevens and Williamson (1973b) presented evidence consistent with it. With the possible exception of work on viral RNA in transformed cells, the severe technical problem of possible slight contamination of giant hnRNA by smaller RNA molecules has not been sufficiently eliminated to permit the conclusion in several reports that mRNA sequences are found in giant hnRNA. No evidence has been presented that these sequences are included in actual precursors of mRNA. It does appear to be certain that some hnRNA molecules in mammalian cells contain no nucleotide sequences which actually become mRNA.

If giant hnRNA molecules actually contain nucleotide sequences identical to those of at least some mRNA molecules, this proposal should be conclusively demonstrable with current techniques. To demonstrate that the sequences actually serve as precursors of RNA will be more difficult.

Extensive characterization of the physical properties of hnRNA and mRNA has yielded much new information. The topographical arrangement of poly(A), oligo(A), poly(U), double-stranded RNA, and repetitive sequences in hnRNA is beginning to be described in detail. These investigations do not warrant the conclusion that mRNA is derived from the 3' end of a giant hnRNA molecule from which some physical features are eliminated by processing.

REFERENCES

Abelson, H. T., and S. Penman. 1972. Messenger RNA formation: resistance to inhibition by 3'-deoxycytidine. *Biochim. Biophys. Acta* *277:*129-133.

Adams, J. M., and S. Cory. 1975. Modified nucleosides and bizarre 5' termini in mouse myeloma mRNA. *Nature 255:*28-33.

Adesnik, M., and J. E. Darnell. 1972. Biogenesis and characterization of histone messenger RNA in HeLa cells. *J. Mol. Biol. 67:* 397-406.

Adesnik, M., M. Salditt, W. Thomas, and J. E. Darnell. 1972. Evidence that all messenger RNA (except histone messenger RNA) contains poly(A) sequences and that the poly(A) has a nuclear function. *J. Mol. Biol. 71:*21-30.

Attardi, G., H. Parnas, M.-I.H. Hwang, and B. Attardi. 1966. Giant-size rapidly labeled nuclear RNA and cytoplasmic messenger RNA in immature duck erythrocytes. *J. Mol. Biol. 20:*145-182.

Bard, E., D. Efron, A. Marcus, and R. P. Perry. 1974. Translational capacity of deadenylated mRNA. *Cell 1:*101.

Birnboim, H. C., R. E. J. Mitchel, and N. A. Straus. 1973. Analysis of long pyrimidine polynucleotides in HeLa cell nuclear DNA: absence of polydeoxythymidylate. *Proc. Nat. Acad. Sci. U.S. 70:*2189-2192.

Bishop, J. O. 1974. The gene numbers game. *Cell 2:*81-86.

Bishop, J. O., and K. B. Freeman. 1973. DNA sequences neighboring the duck hemoglobin genes. *Cold Spring Harbor Symp. Quant. Biol. 38:*707-715.

Bishop, J. O., and M. Rosbash. 1973. Reiteration frequency of duck hemoglobin genes. *Nature New Biol.* *241*:204-207.

Bishop, J. O., R. Pemberton, and C. Baglioni. 1972. Reiteration frequency of hemoglobin genes in the duck. *Nature New Biol.* *235*:231-234.

Bishop, J. O., M. Rosbash, and D. Evans. 1974. Polynucleotide sequences in eukaryotic DNA and RNA that form ribonuclease resistant complexes with polyuridylic acid. *J. Mol. Biol.* *89*:75-86.

Bishop, J. O., J. G. Morton, M. Rosbash, and M. Richardson. 1974. Three abundance classes of HeLa cell mRNA. *Nature* *250*:199-204.

Blobel, G. 1971. Release, identification, and isolation of messenger RNA from mammalian ribosomes. *Proc. Nat. Acad. Sci. U.S.* *68*: 832-835.

Bonner, J., W. T. Garrard, J. Gottesfeld, P. S. Holmes, J. S. Sevall, and M. Wilkes. 1973. Functional organization of mammalian genome. *Cold Spring Harbor Symp. Quant. Biol.* *38*:303-310.

Borun, T. W., M. D. Scharff, and E. Robbins. 1967. Rapidly labeled, polyribosome-associated RNA having the properties of histone messenger. *Proc. Nat. Acad. Sci. U.S.* *58*:1977-1983.

Both, G. W., A. K. Banerjee, and A. J. Shatkin. 1975. Methylation-dependent translation of viral mRNAs *in vitro*. *Proc. Nat. Acad. Sci. U.S.* *72*:1189-1193.

Bramwell, M. E. 1972. A comparison of gel electrophoresis and density gradient centrifugation of heterogeneous nuclear RNA. *Biochim. Biophys. Acta* *281*:329-337.

Brandhorst, B. P. 1972. Inhibition of heterogeneous RNA synthesis by low doses of Actinomycin D. *J. Cell Biol.* *55*:27a.

Brandhorst, B. P., and T. H. Humphreys. 1972. Stabilities of nuclear and messenger RNA molecules in sea urchin embryos. *J. Cell. Biol.* *53*:474-482.

Brandhorst, B. P., and E. H. McConkey. 1973. Rapid equilibration of adenosine with the ATP pool of mammalian cells. *Exp. Cell. Res.* *82*:57-62.

Brandhorst, B. P., and E. H. McConkey. 1974. Stability of nuclear RNA in mammalian cells. *J. Mol. Biol.* *85*:451-464.

Brandhorst, B. P., and E. H. McConkey. 1975. The relationship between nuclear and cytoplasmic poly(A). *Proc. Nat. Acad. Sci. U.S.*, in press.

Brawerman, G., J. Mendecki, and S. Y. Lee. 1971. A procedure for the isolation of mammalian messenger RNA. *Biochemistry* *11*:637-641.

Britten, R. J., and D. E. Kohne. 1968. Repeated sequences in DNA. *Science* *161*:529-540.

Britten, R. J., and B. J. McCarthy. 1963. The synthesis of ribosomes in E. coli I. Analysis of the kinetics of tracer incorporation in growing cells. Biophys. J. 2:49-55.

Brown, I. R., and R. B. Church. 1972. Transcription of nonrepeated DNA during mouse and rabbit development. Develop. Biol. 29:73-84.

Burdon, R. H., and A. Shenkin. 1972. Uridylate rich sequences in rapidly labelled RNA of mammalian cells. FEBS Lett. 12:141-146.

Church, R. B., and B. J. McCarthy. 1967a. RNA synthesis in regenerating and embryonic liver. II. The synthesis of RNA during embryonic liver development and its relationship to regenerating liver. J. Mol. Biol. 23:477-486.

Church, R. B., and B. J. McCarthy. 1967b. Changes in nuclear and cytoplasmic RNA in regenerating mouse liver. Proc. Nat. Acad. Sci. U.S. 58:1548-1555.

Daneholt, B. 1972. Giant RNA transcript in a Balbiani ring. Nature New Biol. 240:229-232.

Daneholt, B. 1973. The giant RNA transcript in a Balbiani ring of Chironomus tentans. In Molecular Cytogenetics (B. A. Hamkalo and J. Papaconstantinou, eds.). New York: Plenum Press, pp. 155-166.

Daneholt, B., and H. Hosick. 1973a. Evidence for transport of 75S RNA from a discrete chromosome region via nuclear sap to cytoplasm in Chironomus tentans. Proc. Nat. Acad. Sci. U.S. 70:442-446.

Daneholt, B., and H. Hosick. 1973b. The transcription unit in Balbiani ring 2 of Chironomus tentans. Cold Spring Harbor Symp. Quant. Biol. 38:629-635.

Darnell, J. E. 1968. RNA from animal cells. Bact. Rev. 32:262-290.

Darnell, J. E., and R. Balint. 1970. The distribution of rapidly hybridizing RNA sequences in heterogeneous nuclear RNA and mRNA from HeLa cells. J. Cell Physiol. 76:349-356.

Darnell, J. E., L. Philipson, R. Wall, and M. Adesnik. 1971. Polyadenylic acid sequences: role in conversion of nuclear RNA into messenger RNA. Science 174:507-510.

Darnell, J. E., W. R. Jelinek, and G. R. Molloy. 1973. Biogenesis of mRNA: genetic regulation in mammalian cells. Science 181: 1215-1221.

Davidson, E. H., and R. J. Britten. 1973. Organization, transcription, and regulation in the animal genome. Quart. Rev. Biol. 48: 565-613.

Davidson, E. H., B. R. Hough, C. S. Amenson, and R. J. Britten. 1973. General interspersion of repetitive with non-repetitive sequence elements in DNA of Xenopus. J. Mol. Biol. 77:1-23.

Davidson, E. H., B. R. Hough, W. H. Klein, and R. J. Britten. 1975. Structural genes adjacent to interspersed repetitive DNA sequences. Cell 4:217-238.

Delovitch, T., and C. Baglioni. 1973. Immunoglobulin genes: A test of somatic vs. germline hypotheses by RNA/DNA hybridization. *Cold Spring Harbor Symp. Quant. Biol. 38*:739-751.

Delovitch, T. L., B. K. Davis, G. Holme, and A. Sehon. 1972. Isolation of messenger-like RNA from immunochemically separated polyribosomes. *J. Mol. Biol. 69*:373-386.

Derman, E., and J. E. Darnell. 1974. Relationship of chain transcription to poly(A) addition and processing in hnRNA in HeLa cells. *Cell 3*:255-264.

Dina, D., M. Crippa, and E. Beccari. 1973. Hybridization properties and sequence arrangement in a population of mRNAs. *Nature New Biol. 242*:101-105.

Doyle, D., and H. Laufer. 1969. Sources of larval salivary gland secretion in the dipteran *Chironomus tentans*. *J. Cell. Biol. 40*: 61-69.

Edmonds, M., and M. G. Caramela. 1969. The isolation and characterization of adenosine monophosphate-rich polynucleotides synthesized by Ehrlich ascites cells. *J. Biol. Chem. 244*:1314-1324.

Edstrom, J. E., and R. Tanguay. 1973. Chromosome products in *Chironomus tentans* salivary gland cells. *Cold Spring Harbor Symp. Quant. Biol. 38*:693-699.

Emerson, C. P., and T. Humphreys. 1970. Regulation of DNA-like RNA and the apparent activation of ribosomal RNA synthesis in sea urchin embryos: quantitative measurements of newly synthesized RNA. *Develop. Biol. 23*:86-112.

Firtel, R. A. 1972. Changes in the expression of single-copy DNA during development of the cellular slime mold *Dictyostelium discoideum*. *J. Mol. Biol. 66*:363-377.

Firtel, R. A., and H. F. Lodish. 1973. A small nuclear precursor of messenger RNA in the cellular slime mold *Dictyostelium discoideum*. *J. Mol. Biol. 79*:295-314.

Firtel, R. A., A. Jacobson, and H. F. Lodish. 1972. Isolation and hybridization kinetics of messenger RNA from *Dictyostelium discoideum*. *Nature New Biol. 239*:225-228.

Fromson, D., and A. Duchastel. 1975. Poly(A)-containing polyribosomal RNA in sea urchin embryos: changes in proportion during development. *Biochim. Biophys. Acta 378*:294-404.

Fromson, D., and D. P. Verma. 1976. Translation of nonpolyadenylated RNA of sea urchin embryos. *Proc. Nat. Acad. Sci. U.S.,* in press.

Furuichi, Y., M. Morgan, S. Muthukrishnan, and A. J. Shatkin. 1975. Reovirus contains a methylated, blocked 5'-terminal structure. *Proc. Nat. Acad. Sci. U.S. 72*:362-366.

Galau, G. A., R. J. Britten, and E. H. Davidson. 1974. A measurement of the sequence complexity of polysomal mRNA in sea urchin embryos. *Cell 2*:9-20.

Gallwitz, D., and G. Mueller. 1969. Histone synthesis *in vitro* on HeLa cell microsomes. The nature of the coupling to DNA synthesis. *J. Biol. Chem. 244*:5947-5952.

Gamow, E., and D. M. Prescott. 1972. Characterization of the RNA synthesized by *Phycomyces blakesecamus*. *Biochim. Biophys. Acta 259*:223-227.

Georgiev, G. P. 1972. The structure of transcriptional units in eukaryotic cells. *Curr. Topics Develop. Biol. 7*:1-60.

Georgiev, G. P., A. P. Ryskov, C. Coutelle, V. L. Mantieux, and E. R. Avakeyan. 1972. On the structure of the transcriptional unit in mammalian cells. *Biochim. Biophys. Acta 259*:259-283.

Georgiev, G. P., A. J. Varshavsky, A. P. Ryskov, and R. Church. 1973. On the structural organization of the transcriptional unit in animal chromosomes. *Cold Spring Harbor Symp. Quant. Biol. 38*:869-884.

Goldberg, R. B., G. A. Galau, R. J. Britten, and E. H. Davidson. 1973. Non-repetitive DNA sequence representation in sea urchin embryo messenger RNA. *Proc. Nat. Acad. Sci. U.S. 70*:3516-3520.

Grady, L. J., and W. P. Campbell. 1973. Nonrepetitive DNA transcription in mouse cells grown in tissue culture. *Nature New Biol. 243*:195-198.

Graham, D. E., B. R. Neufeld, E. H. Davidson, and R. J. Britten. 1974. Interspersion of repetitive DNA sequences in the sea urchin genome. *Cell 1*:127-137.

Gralla, J., and C. DeLisi. 1974. Messenger RNA is expected to form stable secondary structures. *Nature 248*:330-332.

Greenberg, H., and S. Penman. 1966. Methylation and processing of ribosomal RNA in HeLa cells. *J. Mol. Biol. 21*:527-536.

Greenberg, J. 1972. High stability of messenger RNA in growing cultured cells. *Nature 240*:102-104.

Greenberg, J. R., and R. P. Perry. 1971. Hybridization properties of DNA sequences directing the synthesis of messenger RNA and hnRNA. *J. Cell Biol. 50*:774-786.

Greenberg, J. R., and R. P. Perry. 1972. Relative occurrence of polyadenylic acid sequences in messenger and heterogeneous nuclear RNA of L cells as determined by poly(U)-hydroxyapatite chromatography. *J. Mol. Biol. 72*:91-98.

Griffin, B. 1975. Enigma variations of mammalian mRNA. *Nature 255*:9.

Gross, P. R., K. Kraemer, and L. I. Malkin. 1964. Base composition of RNA synthesized during cleavage of the sea urchin embryo. *Biochem. Biophys. Res. Comm. 18*:569-575.

Grossbach, U. 1973. Chromosome puffs and gene expression in polytene cells. *Cold Spring Harbor Symp. Quant. Biol. 38*:619-627.

Grouse, L., M. P. Chilton, and B. J. McCarthy. 1972. Hybridization of RNA with unique sequences of mouse DNA. *Biochemistry 11*:798-805.

Grunstein, M., S. Levy, P. Schedl, and L. Kedes. 1973. Messenger RNAs for individual histone proteins: fingerprint analysis and *in vitro* translation. *Cold Spring Harbor Symp. Quant. Biol. 38*:717-724.

Gurdon, J. B., C. D. Lane, H. R. Woodland, and G. Marbaix. 1971. Use of frog eggs and oocytes for the study of messenger RNA and its translation in living cells. *Nature 233*:177-182.

Harris, H. 1959. Turnover of nuclear and cytoplasmic RNA in two types of animal cell, with some further observations on the nucleolus. *Biochem. J. 73*:362-368.

Harrison, P. R., H. Hell, G. D. Birnie, and J. Paul. 1972. Evidence for single copies of globin genes in the mouse genome. *Nature 239*:219-221.

Holmes, D. S., and J. Bonner. 1973. Preparation, molecular weight, base composition, and secondary structure of giant nuclear RNA. *Biochemistry 12*:2330-2338.

Holmes, D. S., and J. Bonner. 1974. Interspersion of repetitive and single copy sequences in nuclear RNA of high molecular weight. *Proc. Nat. Acad. Sci. U.S. 71*:1108-1112.

Houssais, J., and G. Attardi. 1966. High molecular weight nonribosomal-type nuclear RNA and cytoplasmic messenger RNA in HeLa cells. *Proc. Nat. Acad. Sci. U.S. 56*:616-623.

Huez, G., G. Marbaix, E. Hubert, M. Leclercq, U. Nudel, H. Soreq, R. Salomon, B. Lebleu, M. Revel, and U. Littauer. 1974. Role of the polyadenylate segment in the translation of globin mRNA in *Xenopus* oocytes. *Proc. Nat. Acad. Sci. U.S. 71*:3143-3146.

Imaizumi, T., H. Diggelmann, and K. Scherrer. 1973. Demonstration of globin messenger sequences in giant nuclear precursors of messenger RNA of avian erythroblasts. *Proc. Nat. Acad. Sci. U.S. 70*:1122-1126.

Jacobson, A., R. A. Firtel, and H. F. Lodish. 1974. Synthesis of messenger and ribosomal RNA precursors in isolated nuclei of the cellular slime mold *Dictyostelium discoideum*. *J. Mol. Biol. 82*:213-230.

Jelinek, W., and J. E. Darnell. 1972. The occurrence of double-stranded regions in HeLa cell heterogeneous nuclear RNA. *Proc. Nat. Acad. Sci. U.S. 69*:2537-2541.

Jelinek, W., M. Adesnik, M. Salditt, D. Sheiness, R. Wall, G. Molloy, L. Philipson, and J. E. Darnell. 1973. Further evidence on the nuclear origin and transfer to the cytoplasm of poly(A) sequences in mammalian cell RNA. *J. Mol. Biol. 75*:515-532.

Jelinek, W., G. Molloy, R. Fernandez-Munoz, M. Salditt, and J. E.

Darnell. 1974. Secondary structure in heterogeneous nuclear RNA: involvement of regions from repeated DNA sites. *J. Mol. Biol. 82:* 361-370.

Johnson, L. F., H. T. Abelson, H. Green, and S. Penman. 1974. Changes in RNA in relation to growth of the fibroblast. I. Amount of mRNA, rRNA, and tRNA in resting and growing cells. *Cell 1:*95-100.

Kacian, D. L., S. Spiegelman, A. Bank, M. Terada, S. Metafora, L. Dow, and P. A. Marks. 1972. *In vitro* synthesis of DNA components of human genes for globins. *Nature New Biol. 235:*167-169.

Kafatos, F. C. 1972. The cocoonase zymogen cells of silk moths: a model of terminal cell differentiation for specific protein synthesis. *Curr. Topics in Devel. Biol. 7:*125-191.

Kates, J. 1973. Detection and utilization of poly(A) sequences in messenger RNA. *Methods in Cell Biology 7:*53-65.

Kedes, L. H., and M. L. Birnstiel. 1971. Reiteration and clustering of DNA sequences complementary to histone messenger RNA. *Nature New Biol. 230:*165-169.

Kedes, L. H., and P. R. Gross. 1969. Identification in cleaving embryos of three RNA species serving as templates for the synthesis of nuclear proteins. *Nature 223:*1335-1339.

Klein, W. H., W. Murphy, G. Attardi, R. J. Britten, and E. H. Davidson. 1974. Distribution of repetitive and nonrepetitive sequence transcripts in HeLa mRNA. *Proc. Nat. Acad. Sci. U.S. 71:*1785-1789.

Kohne, D. E., and M. J. Byers. 1973. Amplification and evolution of DNA sequences expressed as RNA. *Biochemistry 12:*2373-2378.

Kronenberg, L. H., and T. Humphreys. 1972. Double-stranded RNA in sea urchin embryos. *Biochemistry 11:*2020-2025.

Kung, C. S. 1974. On the size relationship between nuclear and cytoplasmic RNA in sea urchin embryos. *Develop. Biol. 36:*343-356.

Laird, C. D., W. Y. Chooi, E. H. Cohen, E. Dickson, N. Hutchinson, and S. H. Turner. 1973. Organization and transcription of DNA in chromosomes and mitochondria of *Drosophila*. *Cold Spring Harbor Symp. Quant. Biol. 38:*311-327.

Lambert, B. 1973a. Repeated nucleotide sequences in a single puff of *Chironomus tentans* polytene chromosomes. *Cold Spring Harbor Symp. Quant. Biol. 38:*637-644.

Lambert, B. 1973b. Tracing of RNA from a puff in the polytene chromosomes to the cytoplasm of *Chironomus tentans* salivary gland cells. *Nature 242:*51-53.

Latham, H., and J. E. Darnell. 1965. Distribution of mRNA on the cytoplasmic polyribosomes of the HeLa cell. *J. Mol. Biol. 14:*1-12.

LaTorre, J., and R. P. Perry. 1973. The relationship between polyadenylated heterogeneous nuclear RNA and messenger RNA: studies with Actinomycin D and cordycepin. *Biochim. Biophys. Acta 335:*93-101.

Leder, P., J. Ross, J. Gielen, S. Parkman, Y. Ikawa, H. Aviv, and D. Swan. 1973. Regulated expression of mammalian genes: globin and immunoglobulin as model systems. *Cold Spring Harbor Symp. Quant. Biol. 38*:753-761.

Lee, C. S., and C. A. Thomas. 1973. Formation of rings from *Drosophila* DNA fragments. *J. Mol. Biol. 77*:25-42.

Lewin, B. 1975. Units of transcription and translation: the relationship between hnRNA and mRNA. *Cell 4*:1-20.

Liarkos, C. D., J. M. Rosen, and B. W. O'Malley. 1973. Effect of estrogen on gene expression in the chick oviduct. II. Transcription of chick tritiated unique DNA as measured by hybridization in RNA excess. *Biochemistry 12*:2809-2816.

Lindberg, U., and J. E. Darnell. 1970. SV40-specific RNA in the nucleus and polyribosomes of transformed cells. *Proc. Nat. Acad. Sci. 65*:1089-1096.

Lodish, H. F., R. A. Firtel, and A. Jacobson. 1973. Transcription and structure of the genome of the cellular slime mold *Dictyostelium discoideum*. *Cold Spring Harbor Symp. Quant. Biol. 38*:899-914.

McKnight, G. S., and R. T. Schimke. 1974. Ovalbumin mRNA: evidence that the initial product of transcription is the same size as polysomal ovalbumin messenger. *Proc. Nat. Acad. Sci. U.S. 71*:4327-4331.

Macnaughton, M., K. B. Freeman, and J. O. Bishop. 1974. A precursor to hemoglobin mRNA in nuclei of immature duck red blood cells. *Cell 1*:117-125.

Manning, J. E., C. W. Schmid, and N. Davidson. 1975. Interspersion of repetitive and nonrepetitive DNA sequences in the *Drosophila melanogaster* genome. *Cell 4*:141-155.

Mathews, M. B. 1973. Mammalian messenger RNA. *Essays in Biochemistry 9*:59-102.

Mayo, U. S., and S. R. de Kloet. 1971. Disaggregation of "giant" heterogeneous nuclear RNA of mouse Ehrlich ascites cells by thermal denaturation in the presence of formaldehyde. *Biochim. Biophys. Acta 247*:74-79.

Melli, M., and R. E. Pemberton. 1972. New method of studying the precursor product relationship between high molecular weight RNA and messenger RNA. *Nature 236*:172-173.

Melli, M., C. Whitfield, K. V. Rao, M. Richardson, and J. O. Bishop. 1971. DNA-RNA hybridization in vast DNA excess. *Nature New Biol. 231*:8-12.

Mendecki, J., S. Y. Lee, and G. Brawerman. 1972. Characteristics of the polyadenylic acid segment associated with messenger RNA in mouse sarcoma 180 ascites cells. *Biochemistry 11*:792-798.

Milcarek, C., R. Price, and S. Penman. 1974. The metabolism of poly(A) minus mRNA fraction in HeLa cells. *Cell 3*:1-10.

Miller, O. L., and A. H. Bakken. 1972. Morphological studies of transcription. In *Gene Transcription in Reproductive Tissue*. Stockholm: Karolinska Institute, p. 155-172.

Molloy, G. R., M. Sporn, D. E. Kelley, and R. P. Perry. 1972a. Localization of polyadenylic acid sequences in messenger RNA of mammalian cells. *Biochemistry* 11:3256-3260.

Molloy, G. R., W. L. Thomas, and J. E. Darnell. 1972b. Occurrence of uridylate-rich oligonucleotide regions in heterogeneous nuclear RNA of HeLa cells. *Proc. Nat. Acad. Sci. U.S.* 69:3684-3688.

Molloy, G. R., W. Jelinek, M. Salditt, and J. E. Darnell. 1974. Arrangement of specific oligonucleotides within poly(A) terminated hnRNA molecules. *Cell* 1:43-53.

Murphy, W., and G. Attardi. 1973. Stability of cytoplasmic messenger RNA in HeLa cells. *Proc. Nat. Acad. Sci. U.S.* 70:115-119.

Muthukrishnan, S., G. W. Both, Y. Furuichi, and A. J. Shatkin. 1975. 5'-Terminal 7-methylguanosine in eukaryotic mRNA is required for translation. *Nature* 255:33-37.

Nakazato, H., and M. Edmonds. 1972. The isolation and purification of rapidly labeled polysome-bound RNA on polythymidylate cellulose. *J. Biol. Chem.* 247:3365-3367.

Nakazato, H., D. W. Kopp, and M. Edmonds. 1973. Localization of poly(A) sequences in messenger RNA and heterogeneous nuclear RNA of HeLa cells. *J. Biol. Chem.* 248:1472-1476.

Nakazato, H., M. Edmonds, and D. W. Kopp. 1974. Differential metabolism of large and small poly(A) sequences in the heterogeneous nuclear RNA of HeLa cells. *Proc. Nat. Acad. Sci. U.S.* 71:200-204.

Nemer, M., M. Graham, and L. M. Dubroff. 1974. Co-existence of nonhistone mRNA species lacking and containing poly(A) in sea urchin embryos. *J. Mol. Biol.* 89:435-454.

Ojala, D., and G. Attardi. 1974. Identification of discrete poly(A) containing RNA components transcribed from HeLa mitochondrial DNA. *Proc. Nat. Acad. Sci. U.S.* 71:563-567.

Palacios, R., D. Sullivan, N. Summers, M. L. Kiely, and R. T. Schimke. 1973. Purification of ovalbumin messenger RNA by specific immunoadsorption of ovalbumin-synthesizing polysomes and millipore partition of RNA. *J. Biol. Chem.* 248:540-548.

Peltz, R. 1973. The integrity of "giant" nuclear RNA. *Biochim. Biophys. Acta* 308:148-153.

Penman, S. 1966. RNA metabolism in the HeLa cell nucleus. *J. Mol. Biol.* 17:117-130.

Penman, S., Y. Becker, and J. E. Darnell. 1964. A cytoplasmic structure involved in the synthesis and assembly of poliovirus components. *J. Mol. Biol.* 8:541-555.

Penman, S., I. Smith, and E. Holtzman. 1966. Ribosomal RNA synthesis and processing in a particulate site in the HeLa cell nucleus. *Science 154*:786-789.

Penman, S., C. Vesco, and M. Penman. 1968. Localization and kinetics of formation of nuclear heterodisperse RNA, cytoplasmic heterodisperse RNA and polyribosome-associated messenger RNA in HeLa cells. *J. Mol. Biol. 34*:49-69.

Penman, S., M. Rosbash, and M. Penman. 1970. Messenger and heterogeneous nuclear RNA in HeLa cells: differential inhibition by cordycepin. *Proc. Nat. Acad. Sci. U.S. 67*:1878-1885.

Perlman, S., H. T. Abelson, and S. Penman. 1973. Mitochondrial protein synthesis: RNA with the properties of eukaryotic messenger RNA. *Proc. Nat. Acad. Sci. U.S. 70*:350-353.

Perry, R. P., and D. E. Kelley. 1968. Messenger RNA-protein complexes and newly synthesized ribosomal subunits; analysis of free particles and components of polyribosomes. *J. Mol. Biol. 35*:37-59.

Perry, R. P., and D. E. Kelley. 1973. Messenger RNA turnover in mouse L cells. *J. Mol. Biol. 99*:681-696.

Perry, R. P., and D. E. Kelley. 1974. Existence of methylated messenger RNA in mouse L cells. *Cell 1*:38-42.

Perry, R. P., and P. D. Kelley. 1970. Inhibition of RNA synthesis by actinomycin D: characteristic dose-response of different RNA species. *J. Cell. Physiol. 76*:127-140.

Perry, R. P., D. E. Kelley, and J. LaTorre. 1974. On the synthesis and turnover of nuclear and cytoplasmic polyadenylic acid in L cells. *J. Mol. Biol. 82*:315-331.

Perry, R. P., D. E. Kelley, K. Friderici, and F. Rottman. 1975. The methylated constituents of L cell mRNA: evidence for an unusual cluster at the 5'-terminus. *Cell 4*:387-394.

Philipson, L., R. Wall, G. Glickman, and J. E. Darnell. 1971. Addition of polyadenylate sequences to virus-specific RNA during adenovirus replication. *Proc. Nat. Acad. Sci. U.S. 68*:2806-2809.

Price, R. P., L. Ransom, and S. Penman. 1974. Identification of a small subfraction of hnRNA with the characteristics of a precursor to mRNA. *Cell 2*:253-258.

Puckett, L., S. Chambers, and J. E. Darnell. 1975. Shortlived mRNA in HeLa cells and its impact on the kinetics of accumulation of cytoplasmic polyadenylate. *Proc. Nat. Acad. Sci. U.S. 72*:389-393.

Ricard, B., and W. Salser. 1974. Size and folding of the messenger for phage T4 lysozyme. *Nature 248*:314-317.

Ross, J., H. Aviv, E. Scolnick, and P. Leder. 1972. *In vitro* synthesis of DNA complementary to purified rabbit globin mRNA. *Proc. Nat. Acad. Sci. U.S. 69*:3620-3623.

Rottman, F., A. J. Shatkin, and R. P. Perry. 1974. Sequences containing methylated nucleotides at the 5' termini of mRNAs: possible implications for processing. *Cell 3:*197-200.

Ruiz-Carrillo, A., M. Beato, G. Schutz, P. Feigelson, and V. G. Allfrey. 1973. Cell-free translation of the globin message within polydisperse high-molecular-weight RNA of avian erythrocytes. *Proc. Nat. Acad. Sci. U.S. 70:*3641-3645.

Ryskov, A. P., G. F. Saunders, U. R. Farashyan, and G. P. Georgiev. 1973. Double-helical regions in nuclear precursor of mRNA. *Biochim. Biophys. Acta 312:*152-164.

Sambrook J., H. Westphal, P. R. Srinivasan, and R. Dalbecco. 1968. The integrated state of viral DNA in SV-40 transformed cells. *Proc. Nat. Acad. Sci. U.S. 60:*1288-1295.

Schachat, F. H., and D. S. Hogness. 1973. Repetitive sequences in isolated Thomas circles from *Drosophila melanogaster*. *Cold Spring Harbor Symp. Quant. Biol. 38:*371-381.

Scherrer, K. L., and L. Marcaud. 1968. Messenger RNA in avian erythroblasts at the transcriptional and translational levels and the problem of regulation in animal cells. *J. Cell Physiol. 72,* Suppl. *1:*181-212.

Scherrer, K., L. Marcaud, F. Zajdela, I. M. London, and F. Gross. 1966. Patterns of RNA metabolism in a differentiated cell: a rapidly labeled, unstable 60S RNA with messenger properties in duck erythroblasts. *Proc. Nat. Acad. Sci. U.S. 56:*1571-1578.

Schochetman, G., and R. P. Perry. 1972a. Characterization of messenger RNA released from L cell polyribosomes as a result of temperature shock. *J. Mol. Biol. 63:*577-590.

Schochetman, G., and R. P. Perry. 1972b. Early appearance of histone messenger RNA in polyribosomes of cultured L cells. *J. Mol. Biol. 63:*591-596.

Schutz, B., M. Beato, and P. Feigelson. 1972. Isolation of eukaryotic messenger RNA on cellulose and its translation *in vitro*. *Biochem. Biophys. Res. Comm. 49:*230-238.

Shearer, R. W., and B. J. McCarthy. 1967. Evidence for RNA molecules restricted to the cell nucleus. *Biochemistry 6:*283-289.

Sheldon, R., C. Jurale, and J. Kates. 1972a. Detection of polyadenylic acid sequences in viral and eukaryotic RNA. *Proc. Nat. Acad. Sci. U.S. 69:*417-421.

Sheldon, R., J. Kates, D. Kelley, and R. P. Perry. 1972b. Polyadenylic acid sequence on 3' termini of vaccinia messenger RNA and mammalian nuclear and messenger RNA. *Biochemistry 11:*3829-3834.

Shutt, R. H., and L. H. Kedes. 1974. Synthesis of histone mRNA sequences in isolated nuclei of cleavage stage sea urchin embryos. *Cell 3:*283-290.

Silverstein, S., S. L. Bachenheimer, N. Frenkel, and B. Roizman. 1973. Relationship between post-transcriptional adenylation of herpes virus RNA and messenger RNA abundance. *Proc. Nat. Acad. Sci. U.S. 70*:2101-2105.

Singer, R. H., and S. Penman. 1973. Messenger RNA in HeLa cells: kinetics of formation and decay. *J. Mol. Biol. 78*:321-334.

Sipple, A. E., J. G. Stavrianopoulos, G. Schutz, and P. Feigelson. 1974. Translational properties of rabbit globin mRNA after specific removal of poly(A) with ribonuclease H. *Proc. Nat. Acad. Sci. U.S. 71*:4635-4639.

Smith, M. J., B. R. Hough, M. E. Chamberlin, E. H. Davidson. 1974. Repetitive and nonrepetitive sequence in sea urchin heterogeneous nuclear RNA. *J. Mol. Biol. 85*:103-126.

Soeiro, R., M. H. Vaughan, J. R. Warner, and J. E. Darnell. 1968. The turnover of nuclear DNA-like RNA in HeLa cells. *J. Cell Biol. 39*:112-118.

Spohr, G., T. Imaizumi, and K. Scherrer. 1974. Synthesis and processing of nuclear precursor mRNA in avian erythroblasts and HeLa cells. *Proc. Nat. Acad. Sci. U.S. 71*:5009-5013.

Stampfer, M. M. Rosbash, A. S. Huang, and D. Baltimore. 1972. Complementarity between messenger RNA and nuclear RNA from HeLa cells. *Biochem. Biophys. Res. Comm. 49*:217-224.

Stevens, R. H., and A. R. Williamson. 1973a. Isolation of messenger RNA coding for mouse heavy-chain immunoglobulin. *Proc. Nat. Acad. Sci. U.S. 70*:1127-1131.

Stevens, R. H., and A. R. Williamson. 1973b. Isolation of nuclear pre-mRNA which codes for immunoglobulin heavy chain. *Nature New Biol. 245*:101-104.

Sullivan, D., R. Palacios, J. Slavneyer, J. M. Taylor, A. J. Faras, M. L. Kiely, N. M. Summers, J. M. Bishop, and R. T. Schimke. 1973. Synthesis of a DNA sequence complementary to ovalbumin messenger RNA and quantification of ovalbumin genes. *J. Biol. Chem. 248*: 7530-7539.

Suzuki, Y., and D. D. Brown. 1972. Isolation and identification of the messenger RNA for silk fibroin from *Bombyx mori*. *J. Mol. Biol. 63*:409-429.

Suzuki, Y., L. P. Gage, and D. D. Brown. 1972. The genes for silk fibroin in *Bombyx mori*. *J. Mol. Biol. 70*:637-649.

Thomas, C. A., B. A. Hamkalo, N. Misra, and C. S. Lee. 1970. Cyclization of eukaryotic DNA fragments. *J. Mol. Biol. 51*:621-632.

Tonegawa, S., G. Walter, A. Bernardini, and R. Dulbecco. 1970. Transcription of the SV40 genome in transformed cells and during lytic infection. *Cold Spring Harbor Symp. Quant. Biol. 35*:823-826.

Tsuei, D., K. Fujinaga, and M. Green. 1972. The mechanism of viral carcinogenesis by DNA mammalian viruses: RNA transcripts containing viral and highly reiterated cellular base sequences in adeno-virus-transformed cells. *Proc. Nat. Acad. Sci. U.S. 69:*427-430.

Verma, I. M., G. F. Temple, H. Fan, and D. Baltimore. 1972. *In vitro* synthesis of DNA complementary to rabbit reticulocyte IOS RNA. *Nature New Biol. 235:*163-167.

Wall, R., J. Weber, Z. Gage, and J. E. Darnell. 1973. Production of viral mRNA in adenovirus-transformed cells by the post-transcriptional processing of heterogeneous nuclear RNA containing viral and cell sequences. *J. Virol. 11:*953-960.

Wei, C.-M., and B. Moss. 1975. Methylated nucleotides block 5'-terminus of vaccinia virus mRNA. *Proc. Nat. Acad. Sci. U.S. 72:*318-322.

Wei, C.-M., A. Gershowitz, and B. Moss. 1975. Methylated nucleotides block 5'-terminus of HeLa cell mRNA. *Cell 4:*379-386.

Weinberg, E. S., M. L. Birnsteil, D. R. Purdom, and R. Williamson. 1972. Genes coding for polysomal 9S RNA of sea urchins: conservation and divergence. *Nature 240:*225-228.

Weinberg, R. A. 1973. Nuclear RNA metabolism. *An. Rev. Biochem. 42:*329-353.

Williamson, R., C. E. Drewienkiewicz, and J. Paul. 1973. Globin messenger sequences in high molecular weight RNA from embryonic mouse liver. *Nature New Biol. 241:*66-68.

Williamson, R., J. Crossley, and S. Humphries. 1974. Translation of mouse globin mRNA from which the poly(A) has been removed. *Biochemistry 13:*703-707.

Wilson, D. A., and C. A. Thomas. 1974. Palindromes in chromosomes. *J. Mol. Biol. 84:*115-144.

Wilt, F. H. 1973. Polyadenylation of maternal RNA of sea urchin eggs after fertilization. *Proc. Nat. Acad. Sci. U.S. 70:*2345-2349.

Wu, J. R., J. Hurn, and J. Bonner. 1972. Size and distribution of the repetitive segments of the *Drosophila* genome. *J. Mol. Biol. 64:*211-219.

Wu, R. S., and F. H. Wilt. 1973. Poly(A) metabolism in sea urchin embryos. *Biochem. Biophys. Res. Comm. 54:*704-714.

Chapter 2
CELLULAR ASPECTS OF HISTONE SYNTHESIS*

Thoru Pederson

The Worcester Foundation
   for Experimental Biology
Shrewsbury, Massachusetts

---

*Dedicated to the memory of my friend, Gordon M. Tomkins.

I. INTRODUCTION

We are a long way from understanding how histone proteins are mar-
shalled in chromosomes, and the roles they play there. However, the
biosynthesis of these proteins has emerged as a topic of interest in
its own right, particularly to the cell biologist. Thus, the temporal coincidence of histone synthesis and DNA replication during the
cell cycle has provided a promising model for the investigation of
intracellular coordination in general. In addition, since histones
are almost exclusively localized in the cell nucleus, it has been
logical to investigate the intracellular site of their biosynthesis.
These two "cellular" aspects of histone synthesis form the subject
of this review. Since there has been significant progress in this
field over the past several years, this is an appropriate juncture
at which to summarize the advances and some of the remaining problems.

II. INTRACELLULAR SITE OF HISTONE SYNTHESIS

In the last few years there have been several studies in which a cytoplasmic site of histone synthesis has been demonstrated beyond
reasonable doubt. In some of these, nascent histone polypeptides
were localized on a specific class of cytoplasmic polyribosomes (Robbins and Borun, 1967; Borun et al., 1967; Nemer and Lindsay, 1969;
Kedes et al., 1969; Moav and Nemer, 1971), while in other studies
histone synthesis was demonstrated in vitro by using these polysomes
in a cell-free protein synthesis system (Gallwitz and Mueller, 1969a;
Pederson and Robbins, 1970). However, these studies were not designed
to elucidate whether a low level of histone synthesis might occur
within the nucleus as well, and this remains as at least a theoretical possibility. Of course, the issue of histone synthesis in the
cell nucleus is transcended by the more fundamental question of whether nuclei can synthesize protein altogether, a topic which has been
reviewed by Goldstein (1970) and which will be discussed briefly here.

The various experiments that have been done on the intracellular
site of histone synthesis can be divided into four arbitrary but convenient groups: (a) cytochemical studies, (b) measurements of amino
acid labeling kinetics in various subcellular fractions, and the spatial redistribution of label under "chase" conditions, (c) character-

ization of nascent polypeptides on cytoplasmic polyribosomes after labeling either in vivo or in cell-free systems, and (d) experiments with isolated nuclei.

A. Cytochemical Studies

It is no surprise that the first intimations of the subcellular locus of histone biosynthesis came from cytochemistry. By taking advantage of a particularly favorable biological situation, and by employing a diverse arsenal of cytochemical methods, Bloch and Brack (1964) were able to present a fairly compelling case for a cytoplasmic site of histone synthesis. During spermatogenesis in grasshoppers, as in most organisms, the typical somatic histones of the early spermatid are gradually replaced by ones richer in arginine. That this replacement involves actual synthesis of a new histone is indicated by the concomitant appearance of [^{3}H]arginine label in the nucleus as well as a threefold increase in protein-bound arginine, measured cytophotometrically. The two key points upon which Bloch and Brack built a case for a cytoplasmic synthesis of this new histone were (a) the complete absence of RNA from the nucleus, thereby eliminating the latter as a possible site of protein synthesis and (b) the fact that following the administration of [^{3}H]arginine, grains were found first in the cytoplasm and only subsequently in the nucleus. The authors' conclusion that histones were synthesized in the cytoplasm therefore seems reasonable on the basis of the data they had at hand. But one of their conclusions, lack of nuclear RNA, is subject to certain reservations. By using relatively short pulses of [^{3}H]cytidine (5 hr) the authors were able to demonstrate that synthesis of nuclear RNA is restricted to the earlier stages of spermatogenesis and is undetectable later, when the new histone species is synthesized.* However, the question is whether any of the early

---

*There is a striking similarity between this temporal separation of RNA and protein synthesis in the maturing spermatid (see Fig. 17 in Bloch and Brack, 1964) and during erythropoiesis, in which RNA synthesis ceases prior to the maximal rate of hemoglobin production (e.g., Grasso et al., 1963; Cameron and Prescott, 1963; Grasso and Woodard, 1966; Pederson and Gelfant, 1970).

nuclear RNA persists into the later stages of spermatogenesis. Unfortunately, the authors only state that some pulse-labeled RNA does
indeed persist into late spermatids but do not specify whether the
grains in such cells were nuclear or cytoplasmic. Of course, most
cell biologists are now inclined to rule out protein synthesis in the
nucleus for entirely separate reasons (see Sect. II,D). Taken together the cytophotometric and autoradiographic data obtained by Bloch
and Brack are in fact quite suggestive of a cytoplasmic site of histone synthesis.

Shortly after this study, there appeared a similar but more abbreviated one on mouse spermatogenesis (Monesi, 1964), purporting to
have demonstrated a nuclear synthesis of the arginine-rich histone.
This conclusion was based on the observation that within 15 min after
injecting mice with [^{3}H]arginine, the nuclei of late spermatids were
heavily labeled, while there was little nuclear labeling in earlier
stages. The author argued that this labeling pattern was a manifestation of protein synthesis in the nucleus rather than transfer of
newly synthesized polypeptides from the cytoplasm, since the latter
process could not occur within 15 min. Although this might have been
a reasonable conjecture at the time (however see Dintzis, 1961), there
is growing evidence that proteins that are synthesized in the cytoplasm can reach their nuclear destinations within only a few minutes,
or less. For example, in HeLa cells, where there is compelling evidence for a cytoplasmic site of histone synthesis (see Sect. C), labeled histones are found within the nucleus after only a 2-min pulse
(Robbins and Borun, 1967). Similarly, labeled ribosomal proteins appear in nucleolar precursor particles in the HeLa cell nucleus within 5 min (Pederson and Kumar, 1971), although the present evidence
strongly suggests that these proteins are synthesized in the cytoplasm (Heady and McConkey, 1970; Craig and Perry, 1971). Finally, as
Goldstein (1970) has correctly pointed out, a radioactive amino acid
incorporated at or near the carboxyl terminus of a cytoplasmic nascent polypeptide could very well appear in the nucleus within seconds,

and Wu and Warner (1971) have indeed shown that newly synthesized polypeptides can reach the nucleus within 30 sec in HeLa cells. Another
related difficulty in Monesi's experiment is that the 15 min "pulse"
is much longer than the synthesis time of an individual histone molecule (on the order of 1 min*), so that one is visualizing sites of
histone accumulation rather than histone synthesis. However, it should
be pointed out that spermatogenesis is a slow process physiologically
and the overall tempo of protein synthesis might be slowed down significantly in relation to growing tissue culture cells. Another argument for Monesi's case was the fact that the ratio of nuclear to cytoplasmic labeling did not change with time after the pulse. This
would seem to be a crucial point, but unfortunately the author pre-
sents no data to support it. Finally, it was observed that while late
spermatids displayed heavy nuclear labeling with [^{3}H]arginine, this
was not the case with other amino acids. While it is true that histones and protamines have no tryptophan and very little tyrosine, histidine, or phenylalanine, the lack of nuclear labeling with these precursors is not by itself convincing, since there are many nuclear proteins besides histones in which these aromatic and heterocyclic amino
acids are scarce. More to the point is the absence of leucine and lysine incorporation in the nucleus. The point here is that the "arginine-rich" histone of the late spermatid is not a true protamine (i.e.,
70% arginine, with no lysine or leucine; Ando and Susuki, 1966, 1967)
but is instead a high-arginine species intermediate between somatic-
type histones and protamines.** While it is not known whether this
intermediate histone contains lysine, the analogous one in squid sperm-

---

*By extrapolation from data for globin (Hunt et al., 1969), which
has a molecular weight (16,000) roughly similar to that of histones.
**This distinction is based on alkaline (pH 8.0) fast-green stain-
ing (Alfert and Geschwind, 1953) in which true protamines are lost
from the tissue sections during the hot trichloroacetic acid hydrolysis
step, while histones are retained. As Monesi shows, the late spermatid
nuclei of mice stain intensely, indicating that their basic proteins
are not protamines but very arginine-rich histones.

atids does (Bloch and Hew, 1960; Bloch, 1962). Therefore, the absence
of lysine labeling in mouse spermatids reported by Monesi is somewhat
puzzling. But despite these technical criticisms, there is a more
pertinent problem of interpretation, which is the following. Even if
the spermatids were indeed synthesizing a true protamine (i.e., one
with 70% arginine, no leucine or lysine), the selective labeling of
nuclei with [^{3}H]arginine after a long pulse (15 min) cannot logically
be used as conclusive evidence regarding the intracellular site of
synthesis. This result is compatible not only with a nuclear synthesis, but also with a situation in which no proteins are synthesized
in the nucleus at all (thus explaining the lack of labeling with all
other amino acids tested) and where histones of very high arginine
content are synthesized in the cytoplasm and rapidly accumulate in
the nucleus.

   In addition to the two preceding studies, there have been several
cytochemical studies on the intracellular site of synthesis of nuclear
proteins in general, not specifically histones. One of the best examples is the comprehensive study of interphase growth in cultured
mouse fibroblasts by Zetterberg (1966a). He showed that the rate of
protein synthesis per cell was higher in the S phase of the cell cycle (period of DNA replication) than in the preceding G_1 phase, but
that more protein accumulated in the cytoplasm during G_1. This suggested that a relatively larger fraction of the protein synthesized
during the S phase ended up in the nucleus. This and similar experiments have been reviewed elsewhere (Goldstein, 1970), and since we
are concerned here specifically with histones, it is unnecessary to
discuss them again.

B. Pulse-chase Experiments

   We now consider experiments that assess the subcellular distribution of newly synthesized histones after pulse labels, as well as
their subsequent movements through various cellular compartments with
time.

   The first important study of this type was carried out by Birnstiel and Flamm (1964), who measured the subnuclear distribution of

newly synthesized proteins in two systems, whole tobacco cells in vivo and isolated pea nuclei in vitro, and concluded in both cases that
histones were synthesized in or near the nucleolus and subsequently
accumulated in the chromosomes. In the tobacco cells, the authors
noted amino acid incorporation into both the nucleolar and nucleo-
plasmic fractions after a 40-sec pulse. A small fraction of this protein was histone,[*] and it was argued that its presence in the nucleus
after only 40 sec of labeling manifested a nuclear site of synthesis,
since this was far too short a time for transport from the cytoplasm.
As pointed out earlier, this assumption is no longer warranted, at
least for mammalian cells (see for example Wu and Warner, 1971). But
of course this is hindsight, and the authors cannot be blamed for making an assumption that seemed reasonable at the time. What might be
criticized, however, is the pulse-chase experiment they performed to
show that transfer of new protein from cytoplasm to nucleus was not
influencing the results. This experiment is unconvincing due to (a)
the failure of [^{14}C]lysine incorporation to stop completely after resuspending the cells in fresh medium containing excess [^{12}C]lysine
and (b) the long interval (20 min) that elapsed before the first
"chase" sample was removed. Thus, between the end of the pulse (10
min) and 20 min later, total cell radioactivity increased by almost
30%, although the authors state that "...further incorporation...was
virtually abolished." Therefore, the possibility of spatial redistribution of label before the 20-min chase sample was taken was not
ruled out. On the positive side, Birnstiel and Flamm do seem to have
demonstrated, in the tobacco cells at least, that newly synthesized
histones, regardless of their origin, are found first in the nucleolar fraction and subsequently accumulate in chromatin. This is an
interesting observation that deserves further attention in additional
experimental systems. It is also to the authors' credit that they
employed a definition of histone that distinguished this class from
ribosomal proteins (i.e., solubility in $HgSO_4$-H_2SO_4); if they had not,

---

[*]Defined here as protein soluble in 0.2 N HCl and lacking sulfhydryl groups as judged by its solubility in the acid mercuric sulfate
reagent of Mirsky and Pollister (1946).

the apparent shift of protein from nucleolus to chromosomes with time
could have in fact been due to a rapid steady-state flux of ribosomal
proteins through the nucleolus superimposed upon a slower, independent accumulation of histones in the chromatin.

The experiments of Birnstiel and Flamm with isolated pea nuclei
would seem at first glance to be more cogent than those in which the
labeling was done in living cells (tobacco), since the possible contribution of the cytoplasm is presumably eliminated. However, experiments with intact cells and isolated nuclei are so fundamentally different that a separate set of criteria must be used to evaluate the
latter, and therefore the pea nuclei experiments will be discussed in
Sect. D.

The intracellular locus of histone synthesis, as well as its timing during the cell cycle, was studied in HeLa cells by Robbins and
Borun (1967). Although the major portion of this study concerns the
identification of histone-synthesizing polyribosomes in the cytoplasm
of S-phase cells, discussed in the next section, it also includes a
particularly convincing pulse-chase experiment. The authors recognized
that since histones contain abundant lysine but little or no tryptophan, it should be possible in a pulse-chase experiment to resolve
the movements of histones in the midst of nonhistone proteins by
double labeling with both amino acids; migration of histones from
one cellular compartment to another would be manifest by a flux of
lysine unaccompanied by tryptophan. Second, there is a technical advantage in that both of these amino acids are transported by HeLa
cells quickly, allowing not only rapid labeling but an effective
"chase." Thus, total cell radioactivity [trichloroacetic acid (TCA)
insoluble] changed only +6 and -3% for [^{14}C]tryptophan and [^{3}H]-
lysine, respectively, during a 10-min chase following a 2-min pulse.
There was, however, a dramatic spatial redistribution of label between
nucleus and cytoplasm. [^{3}H]Lysine counts in the HCl-soluble fraction
of the nucleus increased fourfold during the chase, while tryptophan
counts in this fraction increased only twofold, indicating that some
of the labeled proteins in this compartment were histones. (The fact
that the HCl-soluble tryptophan counts increased at all is due to the

presence of basic, nonhistone proteins in the HeLa cell nucleus.*)
It is important to note that this result, by itself, demonstrates only
a flux of new histone *into* the nuclear HCl-soluble fraction, and does
not reveal whether the source of this protein is the cytoplasm or another component of the nucleus (i.e., HCl-insoluble). This question
was resolved by the behavior of label in the HCl-insoluble fraction
during the chase, which showed a simultaneous loss of both tryptophan and lysine. Had there been a selective chase of lysine out of
this fraction it might have been concluded to be the site of histone
synthesis. The key assumption here, which is not particularly obvious in the text of the paper, is that as far as we know, any protein that contains tryptophan will also contain lysine, and therefore
the two labels leaving the HCl-insoluble fraction together must reside in the same polypeptide chains, i.e., nonhistones by definition.

C. Evidence for Histone Synthesis on Cytoplasmic Polyribosomes

The first evidence for polyribosome-directed histone synthesis
was provided by Robbins and Borun in experiments with synchronized
HeLa cells (1967). These authors recognized that the key problem was
to identify nascent histone polypeptides in the presence of many other
nonhistone proteins and that this process might be facilitated by
double labeling with [^{3}H]lysine and [^{14}C]tryptophan. Since they contain little or no tryptophan but abundant lysine, nascent histones
might be distinguished from other nonhistone, tryptophan-containing
proteins by comparison of isotope ratios. Secondly, Robbins and
Borun had found that histone synthesis is largely restricted to the
S phase of the cell cycle (DNA synthesis). Synchronization thus permitted the lysine-tryptophan labeling to be performed at a time when
all cells in the population were synthesizing histones, and at a maximal rate per cell. Like the cytochemical study of Bloch and Brack
(1964), the paper by Robbins and Borun does not provide a direct,
formal proof of cytoplasmic histone synthesis, but it offers a number

---

*For example, see Bhorjee and Pederson (1972, 1973).

of independent observations which, when taken together, make that conclusion almost inescapable.

When S-phase cells were labeled for 2 min with [^{3}H]lysine and [^{14}C]tryptophan it was found that small polyribosomes (120-200S) contained a considerable excess of lysine over tryptophan, in comparison with larger polyribosomes further down the gradient. However, the data are trickier than they appear. The increased lysine/tryptophan ratio in the light polyribosomes does not necessarily indicate the presence of nascent histone polypeptides; the key question here is the basis of the lysine-tryptophan ratio asymmetry. Close inspection of the data reveals that the [^{14}C]tryptophan specific activity (cpm/OD_{260}) decreases in the lighter polyribosomes, but it is not clear that this decrease is any greater than that which would be predicted on the basis of the well-known relationship between polyribosome size and nascent chain radioactivity per monosome (e.g., Kuff and Roberts, 1967). Second, it is also clear that the [^{3}H]lysine specific activity does not decrease in the light region as much as that of tryptophan, which means that the nascent chains of the small polyribosomes contain, on average, more lysine (per chain) than those of the larger polyribosomes. However, the distinguishing feature of histones is not lysine richness but the absence of tryptophan. Synthesis of lysine-rich, nonhistone (i.e., tryptophan-containing) proteins on small polyribosomes would be perfectly compatible with the data. What is needed is a point of reference to assess the decreased tryptophan specific activity in the light polyribosome region. Fortunately, this is provided by data from cells in the G_1 phase of the cell cycle, which do not synthesize histones. In this case, the [^{14}C]tryptophan specific activity of the light polyribosomes is clearly greater than in the S-phase cells. Further, the similarity of tryptophan/OD_{260} in heavy polyribosomes in G_1 versus S eliminates the possibility of changes in the tryptophan pool during the cell cycle. Thus, using the G_1 gradient as a reference point, it becomes clear that the lysine-tryptophan ratio asymmetry in the small polyribosomes of S cells is the combined effect of an increase in the specific activity of lysine and a decrease in that of tryptophan; the latter is the key evidence that histones are present. Moreover, the region of the S-phase gradient containing the increased lysine-tryptophan ratio is that containing the polyribosome size class expected to be synthesizing polypeptides in the molecular weight range of histones (approximately 10,000-21,000 daltons).

In addition to the foregoing, the conclusion that histones are synthesized on light polyribosomes was also based on the observation that the increased lysine-tryptophan ratio was abolished when DNA and histone synthesis was inhibited with cytosine arabinoside or FUdR. Finally, there were the convincing pulse-chase kinetics of nuclear histones discussed in the preceding section. While these combined results were quite compelling, it is to be noted that they all rested upon the histone criterion of a lack of tryptophan, rather than a definitive biochemical characterization of the nascent chains.

The heuristic impact of the Robbins and Borun study was manifest by the tentative identification of histone-synthesizing polyribosomes in early sea urchin embryos by Nemer and Lindsay (1969). It had been known that at approximately the 16-cell stage of development a prominent class of small (∿200S) polyribosomes appears (Infante and Nemer, 1967) and that these comprise the majority of polyribosomes by the 100- to 200-cell stage (morula). Recognizing that these were the same size as the polyribosomes identified by Robbins and Borun (1967) and that their presence was coincident with intense synthesis of chromosomal proteins. Nemer and Lindsay labeled sea urchin morulae with [^{14}C]arginine and [^{3}H]tryptophan, and found a distinctly increased ratio of arginine/tryptophan labeling in the light polyribosome region. While this is suggestive, it is not compelling because of the ambiguity of the decrease in tryptophan labeling in the light region; i.e., is it greater than that expected from the relationship between polyribosome size and radioactivity per monosome relationship? Again, it is not simply arginine richness but the relative absence of tryptophan which must be demonstrated.

A more comprehensive study of histone synthesis in early sea urchin embryos was published later the same year (Kedes et al., 1969). This investigation resembled that of Robbins and Borun (1967) with

HeLa cells in that the evidence for histone synthesis on light polyribosomes was again derived from a number of independent observations
which, taken together, were fairly cogent. Kedes et al. began by considering the fact that early development consists of nuclear replica-
tions rather than cell growth. Therefore, one might expect a larger
fraction of the total protein synthetic activity to be devoted to nuclear proteins in embryos than in growing cell populations. This prediction was borne out by both autoradiographic and cell fractionation
experiments, which indicated that at least 40-45% of the newly synthesized proteins of blastula embryos accumulate in the nucleus. Of
these, approximately half are basic (acid soluble), and autoradiographic inspection of metaphase cells revealed that at least some of
these nuclear proteins are truly chromosomal, which is an important
point. The magnitude of the early sea urchin's commitment to synthesis of nuclear proteins is especially striking in view of the fact
that ribosome synthesis at this time is either low or absent,* so that
ribosomal proteins are probably not a significant part of the picture.

Pulse labeling of 32-64 cell embryos with $[^{14}C]$lysine and $[^{3}H]$-
tryptophan for 10 min revealed an increased $^{14}C/^{3}H$ ratio in the light
polyribosome region; it is of interest that this increase was approximately three times the $^{14}C/^{3}H$ ratio of the heavier polyribosomes,
which is the same difference found by Nemer and Lindsay. Kedes et
al. state that the ratio asymmetry in the light polyribosomes is due
to a decrease in that of tryptophan. While the validity of their
assertion can be verified by calculations on the reader's part, the
data on this point should have been presented.

The contention that the light polyribosomes contain nascent histones was reinforced considerably by data on the effects of hydroxyurea, demonstrated by Kedes et al. to be a highly selective inhibitor
of DNA synthesis in sea urchin embryos. Exposure of morulae to this
compound for 90 min resulted in a marked reduction of polyribosome
content (OD_{260}) which was confined primarily, but not exclusively,
to the light region. That this effect was not as precisely restricted

---

*This point remains controversial.

to the light polyribosomes as it was in the case of synchronized HeLa
cells (Robbins and Borun, 1967) may be due to (a) the relatively "long"
exposure ot hydroxyurea (90 min = 2-3 cell cycles) resulting in effects
secondary to the inhibition of DNA synthesis or (b) the possibility
that the syntheses of several proteins other than histones are linked
to DNA replication in sea urchin embryos. The effects of hydroxyurea
on the lysine/tryptophan ratios of nascent polypeptides were again
primarily confined to the light polyribosomes, indicating the presence of histone proteins. (This effect is not particularly apparent
in the actual radioactivity profiles, where the reader tends to compare the light polyribosome regions of the control versus hydroxyurea
gradients; however, the meaningful comparison is the heavy versus
light region of each gradient and this can be visualized more clearly
in the normalized isotope ratios illustrated in Fig. 4b and 4d of the
Kedes et al. study.)

The effects of actinomycin on the lysine and tryptophan labeling
of nascent polypeptides resembled those of hydroxyurea. There was a
marked reduction in the lysine/tryptophan ratio asymmetry of the light
polyribosomes with relatively little effect in the heavier region of
the gradient, indicating that a major portion of histone synthesis is
directed by "new" messenger RNA in sea urchin embryos. While the authors state that "some histone-like proteins continue to be synthesized" in the presence of actinomycin, their data indicate that there
is no more histone synthesized with actinomycin than in the presence
of hydroxyurea.

Taken together, the evidence presented by Kedes et al. indicated
that histone synthesis during early sea urchin development takes place
on small cytoplasmic polyribosomes; further characterization of the
"histone-like" nascent polypeptides would have made the overall argument even more convincing.

A comprehensive study of protamine synthesis in trout testis cells
was published in 1969 by Ling et al., in which good evidence was provided for a cytoplasmic site of synthesis. When testis cell suspensions were labeled for 30 sec with [^{14}C]arginine, the specific activity of chromatographically-identifiable protamine was orders of mag-

nitude higher in the cytoplasm than in the nucleus; when the labeling
was extended to 90 sec, the specific activity of protamine increased
about tenfold in the nuclear fraction but only slightly in the cytoplasm. Since the total protamine content of these cells can be shown
to be almost exclusively nuclear, it is hard to reconcile these labeling data with any hypothesis other than a cytoplasmic site of synthesis. This interpretation was strengthened by pulse-chase kinetics
which illustrated a "chase" of labeled protamine out of the microsomal
fraction and a concomitant accumulation in the nucleus. Secondly,
when isolated subcellular fractions were incubated in vitro with $[^{14}C]$-
arginine, GTP, and ATP, the cytoplasm (mitochondrion-free) incorporated 20 times more radioactivity into protamine than did the high-speed
supernatant and 40 times more than isolated nuclei. Finally, analysis
of polyribosomes from testis cell suspensions pulsed with $[^{14}C]$argin-
ine revealed a dramatic positive deviation from the expected specific
activity of the gradient region containing dimeric polyribosomes
($\sim$120S). While this indicates only arginine richness, chromatography
revealed that labeled protamine molecules were indeed present. A
minor problem here is that while the authors provided good evidence
for the presence of labeled protamine in the 120S polyribosomes they
did not actually demonstrate that these were nascent polypeptides.
However, since dimers are the approximate size of polyribosomes expected to be coding for protamines ($\sim$5000 daltons)* it is quite reasonable to suppose that they are present as nascent polypeptides.

The most convincing evidence for a cytoplasmic site of at least
some, if not all, histone synthesis comes from studies with cell-free
systems (i.e., isolated cytoplasm or purified ribosomes), the first
of which was that of Gallwitz and Mueller (1969a). These authors prepared microsomes (110,000 x g pellet) from amethopterin-synchronized
HeLa cells in the S phase of the cell cycle. Incubation of such mi-

---

*While Ling et al. calculated 1.04 ribosomes per protamine mes-
senger RNA on the basis of the α chains of hemoglobin (141 amino acids
5 ribosomes/messenger), the observed localization of labeled protamine
on disomes could easily result from increased ribosome packing; alternatively, protamine mRNA might be dicistronic.

crosomes with 110,000 x g supernatant and an ATP-regenerating system resulted in the incorporation of labeled amino acids into histone-like polypeptides. These were characterized by their acid solubility and coelectrophoresis with markers of native histones in both pH 4.3 gels and gels containing sodium dodecylsulfate. Further, microsomes from cells in which DNA synthesis had been inhibited in vivo were incapable of carrying out histone synthesis in vitro, although other nonhistone polypeptides continued to be labeled. These data provide compelling evidence for a cytoplasmic site of histone synthesis.

Another study of histone synthesis employing a cell-free system was that of Pederson and Robbins (1970). In this case cell-free protein synthesis was programmed specifically by the class of small polyribosomes implicated previously as the in vivo site of histone synthesis (Robbins and Borun, 1967). The histone-like polypeptides synthesized in vitro were characterized by their acid solubility, relative excess of incorporated lysine over tryptophan in comparison to other labeled polypeptides, and their coelectrophoresis with nuclear histones. As was the case in the study of Gallwitz and Mueller, cell-free synthesis of histones was effectively abolished by experimental inhibition of DNA synthesis prior to cell disruption. In addition, Pederson and Robbins demonstrated that cell-free systems containing G_1-phase polyribosomes, or programmed with heavy polyribosomes of S cells, did not synthesize histone-like polypeptides in vitro. It is to be emphasized that the cell-free systems used in both papers were defective in the process of chain initiation, so that incorporation of label primarily reflects chain elongation on messengers that were being translated at the time of cell disruption. (In the case of the study by Pederson and Robbins, it was also demonstrated that release of completed polypeptides from the polyribosomes was occurring.). Thus, while these in vitro systems are not exact representations of the in vivo situations, they do suffice for the demonstration that histone synthesis initiated on polyribosomes in the living cell can culminate on those polyribosomes in vitro and thus leave little doubt that histone biosynthesis occurs in the cytoplasm.

A more complete characterization of the histone-like nascent polypeptides of sea urchin embryo "s-polysomes" was published in 1971 by
Moav and Nemer, which confirmed and considerably strengthened earlier
conclusions (Nemer and Lindsay, 1969; Kedes et al., 1969). When RNase
digests of sea urchin polyribosomes, pulse labeled in vivo, were subjected to gel filtration on Sephadex G-100, it was found that the nascent radioactivity of large polyribosomes resided mainly in proteins
of average molecular weight 57,000, while the s-polysomes contained
nascent polypeptides of approximately 9500 daltons. Based upon this
sixfold difference in nascent peptide size, and assuming identical
translation rates for large and small mRNA's, Moav and Nemer proceeded
to calculate whether the observed lysine, arginine, and tryptophan
labeling of small versus large polyribosomes deviated significantly
from that expected on the basis of the size differential. They found
that the ratio asymmetry of the s-polysomes was indeed due to a positive deviation from the expected specific activities of lysine and
arginine, and a decrease in that of tryptophan; the importance of this
consideration has been discussed earlier (pp. 78-79). Acid extraction of pulse-labeled s-polysomes followed by cation-exchange chromatography revealed that approximately 70% of the eluted protein was in
positions characteristic of native histones. Due to the tenacious
binding of about 70% of the acid-soluble proteins applied to the column, the eluted material represented only 30% of the total. Moav
and Nemer felt that the nonelutable material might represent incomplete polypeptide chains and they thus reasoned that more protein
might be eluted if the polyribosomes were first incubated in vitro to
allow completion of more chains. (The tacit assumption really is that
chain initiation will not occur in such an incubation, otherwise little
would be gained.) Chromatographic analysis of acid-soluble proteins
from s-polysomes pulsed in vivo and then incubated in vitro again revealed histone-like polypeptides, but the results were not significantly different from those obtained with unincubated polyribosomes since
the major portion of acid-soluble proteins was again bound irreversibly to the column. This, in combination with their finding that, unlike the results in the Pederson and Robbins system (1970), completed

chains were not released during incubation of the polyribosomes, suggests that relatively little was happening in this in vitro system altogether. Finally, the histone-like nature of the nascent polypeptides that cochromatographed with native histones was confirmed by selection of the appropriate column fractions followed by electrophoresis; the radioactivity coincided precisely with the stained bands of unlabeled nuclear histone included as a marker. These characterizations demonstrate beyond any reasonable doubt that histone proteins are among the nascent polypeptide chains of sea urchin embryo s-polysomes.

While the results that have been discussed in this section lead to the inescapable conclusion that histone synthesis occurs on cytoplasmic polyribosomes, none of the experiments was designed to determine whether a small fraction of the cell's total synthesis of histones occurs in the nucleus as well. This possibility is now considered.

D. Experiments with Isolated Nuclei

There have been a number of papers in which synthesis of histones has been supposedly demonstrated in isolated nuclei. The general issue of nuclear protein synthesis is controversial, to say the least, and Goldstein (1970) has reviewed the topic in a thorough and competent fashion. This author would add only two general comments before discussing specific experiments. First, there has been a great deal of controversy as to whether or not eukaryotic nuclei contain ribosomes. Since ribosomes are assembled in the nucleolus it is no surprise that particles containing ribosomal RNA are found in nuclei (Warner and Soeiro, 1967; Rogers, 1968; Liau and Perry, 1969; Pederson and Kumar, 1971). Most analyses of nuclear "ribosomes" have not been designed to distinguish between mature ribosomes and these ribosomal precursor particles. Moreover, the key question is not whether nuclei contain ribosomes, but whether they contain *polyribosomes,* which, in my opinion, has never been convincingly demonstrated. Second, most studies have been concerned with whether nuclei carry out protein synthesis as we know it, but they ignore the dis-

tinct, and intriguing, possibility of nonribosomal mechanisms of peptide bond formation. Studies on the biosynthesis of peptide antibiotics (Lipmann, 1971) have revealed enzymes that can add as many as
eight activated amino acids sequentially onto an oligopeptide; a similar mode of peptide synthesis in eukaryotic nuclei would reconcile
the paradox of amino acid incorporation (which in certain cases has
probably been demonstrated) and the absence of polyribosomes.

The major theater of studies on nuclear protein synthesis has
been the thymic lymphocyte nucleus. There is no question that "isolated" thymocyte nuclei incorporate labeled amino acids into protein,
but the problem is, of course, the question of purity. Techniques for
isolating thymocyte nuclei were summarized by Allfrey et al. (1964).
Their best results were obtained by employing hypotonicity to weaken
the plasma membrane, which permits cell disruption at reduced forces,
thereby minimizing nuclear damage. This was followed by sedimentation
through dense solutions, particularly Ficoll, in which cells and cytoplasm float away from the denser nuclei. The authors presented several lines of evidence to demonstrate that these nuclei synthesize protein in vitro. Thus, after 30 or 60 min of labeling in vitro, autoradiographs of thin sections revealed grains within the nuclei, not
just at the edges where cytoplasmic contamination might be expected.
However, after a 30- or 60-min labeling period, one may be looking
primarily at sites of protein accumulation, not synthesis.* Given
the existence of some cytoplasmic contamination, which the authors
admit and which is evident in their micrographs (e.g., Fig. 1C), there
is no way of telling whether this net, accumulated label was incorporated into protein inside or outside of the nucleus. The authors
state: "Of course, some grains can be seen in close proximity to the
nuclear envelope, and they occasionally occur in cytoplasmic tabs ad-
hering to some of the nuclei, *but the presence of a tab is not a pre-
requisite for nuclear labeling*" (italics mine). If we accept the
statement that substantial numbers of nuclei in the autoradiographs

---

*That most of such accumulation is nuclear is not surprising in a
cell with as high a nuclear:cytoplasmic volume ratio as the thymocyte.

contained internal silver grains while lacking cytoplasmic tabs, it must nevertheless be remembered that the nuclei were sedimented through Ficoll *after* the labeling. As the authors themselves demonstrate elsewhere in the study (Table IV), this step causes significant nuclear damage, which may have included stripping of some cytoplasmic tabs.

To this reviewer, the issue of protein synthesis in thymocyte nuclei boils down to (a) demonstrating that isolated nuclei incorporate more labeled amino acid than can be accounted for by the 2-5% cytoplasmic plus whole-cell contamination and (b) showing that "nuclear" protein synthesis has features that distinguish it from such synthesis in the cytoplasm. With regard to the first point, it does appear from the data of Allfrey et al. (Table IV) that the isolated nuclei incorporated considerably more label than would have been expected solely on the basis of whole-cell contamination. Thus, a preparation of nuclei containing 3.7% cells, incorporated about 13% as much label as pure (88%) cells, i.e., three times the expected amount (3.7/88 = 4.2%). But the problem here is not just the contamination of the population by whole cells (3.7%) but also of each individual nucleus by cytoplasm. This would only have had to be 10% to generate the observed results (see also calculations on p. 89). On the other hand, the dependence of "nuclear" protein synthesis on sodium ions and its unique resistance to ribonuclease (Allfrey et al., 1957, 1961) are very strong lines of evidence that a distinctive protein synthetic system exists in or about the nucleus. (The true meaning of these data again depends on nuclear purity and they may only reflect differences between cytoplasmic and perinuclear ribosomes.) It is probably wisest to adopt the view that there is a distinct possibility that living thymus lymphocytes carry out protein synthesis within their nuclei but that this has still not been demonstrated beyond all reasonable doubt. Since no other system has received the careful and thorough analysis to which the thymocyte nucleus has been subjected (Allfrey et al., 1964), it is impossible to make a firm statement concerning the generality of the phenomenon (for further discussion see Goldstein's review).

With the foregoing considerations as a backdrop, we may now pro-
cede to examine experiments dealing specifically with histone synthesis in isolated nuclei. A portion of the study by Birnstiel and Flamm
(1964), discussed in the section on pulse-chase experiments, dealt
with histone synthesis in isolated pea nuclei. The experimental details are presented in a rather abbreviated fashion, so that a critical appraisal is difficult, if not impossible. Incubation of "isolated" pea nuclei for 3 min at 37°C with 50 μCi/ml of [³H]leucine resulted in incorporation into both histone and nonhistone proteins;
subnuclear fractionation revealed most labeling in the nucleolar fraction. The problem here is lack of information by which to assess the
purity of these nuclei, and consultation of the references given for
the isolation procedure (Rho and Chipchase, 1963; Birnstiel et al.,
1962) reveals little compelling evidence that such nuclei are free of
cytoplasmic ribosomes. In addition, a special problem arises in the
isolation of plant nuclei because of the considerable nuclear damage
which accompanies the forces necessary to disrupt cell walls. Adven-
titious nuclear binding of cytoplasmic constituents, including ribosomes, could be a more frequent event with such damaged nuclei. Thus,
while there is insufficient information to state definitively that the
labeling observed by Birnstiel and Flamm was an artifact, their relatively strong conclusion regarding a nuclear site of histone synthesis seems unwarranted.

In the paper by Allfrey et al. (1964), thymocyte nuclei isolated
under either isotonic or hypotonic conditions were incubated in vitro
with one of several different labeled amino acids. *After* incubation
the nuclei were sedimented through sucrose to "...remove cells and
tabbed nuclei." Needless to say, the authors found that during the
15- to 60-min incubation radioactivity was incorporated into proteins
having the solubility and chromatographic properties of native histones. This seems a strange way to have done the experiment; why not
do the labeling with the purest possible nuclei? On the other hand,
the authors do demonstrate that the histone labeling has the property
of ribonuclease insensitivity, which characterizes general protein
synthesis by the thymocyte "nucleus." A similar result was obtained

by Reid and Cole (1964) who illustrated both RNase insensitivity and
Na$^+$ dependence of histone f1 synthesis by isolated thymocyte nuclei.
These latter authors also carried out a more thorough analysis of the
labeled histones including gel electrophoresis, ion exchange chromatography, and determination of amino acid composition. They also demonstrated internal labeling in histone f1 by peptide fingerprinting.
In a subsequent paper, Reid et al. (1968) confirmed synthesis of histone f1 using thymocyte nuclei purified by the Ficoll gradient method.
The data that are particularly pertinent to our present discussion
are those depicted in their Table V, where the question of nuclear
purity is examined. Reid et al. argue that since preparations which
have 1 and 27% whole-cell contaminations, respectively, do not show a
corresponding 27-fold difference in histone-specific activity (a
threefold difference was observed), this can only mean that some of
the synthesis is by the nuclei. Again, while they have correctly
taken into consideration whole-cell contamination, the other factor
is the average cytoplasmic contamination of each nucleus itself. For
example, assuming that each isolated nucleus contains 10% of the original cytoplasmic mass, and assuming also that in reality all protein
synthesis is cytoplasmic, we may perform the following calculation.
A preparation that was only 1% whole cells had a histone specific activity of 217 cpm/mg (Table V, Reid et al.), which according to our
assumption is due to the 10% contamination of *each* nucleus, not the
1% whole cells. An equal mass of material that is 100% whole cells
would have 10 times as much cytoplasm as the nuclei and an expected
specific activity of 2170. It follows that a preparation with 27%
whole cells will have a specific activity of 0.27 x 2170, or 586,
which is quite close to the value of 618 observed by Reid et al. In
other words, cytoplasmic contamination of the nuclei amounting to 10%
could easily give the observed results. This is not to imply that
such contamination was not below 10%; it may have been, but this is
precisely the point. *That the observed amount of labeling in isolated*
thymocyte nuclei exceeds that expected from the whole cell plus cyto-
plasmic contamination has probably never been demonstrated. We must
conclude, from the experiments of Allfrey et al. (1964), Reid and Cole

(1964), and Reid et al. (1968), that a portion of the total histone synthesis in isolated thymocyte nuclei is carried out by a unique Na^+-dependent, RNase-insensitive mechanism, but that it has not been convincingly demonstrated that this mechanism lies within the nucleus.

To summarize this section, it is clear that cytoplasmic polyribosomes are the major subcellular sites of histone biosynthesis in a variety of eukaryotic cells, although the experiments which have led to this conclusion would not have detected a low level of histone synthesis occurring simultaneously in the nucleus. Direct attempts to demonstrate a nuclear synthesis of histones, either by analysis of pulse-chase kinetics in intact cells or by incubation of isolated nuclei in vitro, have been inconclusive.

III. TIMING OF HISTONE SYNTHESIS DURING THE CELL-DIVISION CYCLE

Our understanding of the temporal relationship between histone synthesis and DNA replication is based on a number of cytochemical and biochemical studies, most of which have indicated a precise coordination between these two events. Before reviewing this literature, it may be useful to state briefly the cell cycle concept and its terminology. Although the interphase synthesis of DNA had been demonstrated in the early fifties (Swift, 1950; Howard and Pelc, 1951; Walker and Yates, 1952), it was the observation that synthesis of DNA occurred during only a portion of interphase which gave rise to the concept of a cell "cycle" (Howard and Pelc, 1953).* The discontinuous synthesis of DNA during interphase permits subdivision of the cell's life history into:

---

*While this is the usual citation given, it is of historical interest to note that in 1953, Swift published a figure (Fig. 7 in Swift, 1953) in which the original data on Feulgen-DNA content of lily meristem nuclei (*Proc. Nat. Acad. Sci. 36*:643, 1950) were replotted as a function of nuclear volume. In this system nuclear volume is directly proportional to stage of interphase (see Woodard et al., 1961). So presented, the data clearly illustrated that DNA content/nucleus did not begin to increase until after about one-third of interphase had elapsed. Thus, data showing the existence of G_1 were available as early as 1949, when Swift's original measurements were made.

G_1 ("gap" 1), which is the period after mitosis and before DNA replication; the DNA synthetic phase, called S, which is followed by another non-DNA-synthetic period, G_2; and then mitosis, M, consisting
of the classical stages of prophase, metaphase, anaphase, and telo-
phase. In addition, a special term, G_0, has been introduced to deal
with the situation of nondividing cells arrested in G_1 (Lajtha, 1963;
Quastler, 1963). Although these G_0 cells are chronologically in the
G_1 phase of the cell cycle, their biochemistry and physiology usually
differ considerably from cells that are actively traversing G_1 (for
reviews see Baserga, 1968; Epifanova and Terskikh, 1969).

A. Cytochemical Studies

Bloch and Godman (1955) measured the DNA and histone contents of
nuclei from rat liver and cultured fibroblasts by cytophotometry.
The distributions of alkaline fast-green (histone) and Feulgen (DNA)
values among fibroblast nuclei were quite similar; the greater degree
of scatter among the fast-green values did not obscure the fact that
the modal distribution coincided with the pattern of DNA contents.
Even more convincing were experiments in which DNA and histone contents were measured in the same nuclei, rather than in one population
versus another. This was accomplished by substituting 1 N trichloroacetic acid (TCA) for HCl in the Feulgen reaction. While TCA effectively removes purine bases from DNA, which is the basis of Feulgen
staining, it does not extract histones, as does the usual HCL hydrolysis. After cytophotometrically measuring the DNA contents, the nuclei
were mapped by camera lucida drawings, incubated in hot TCA to hydro-
lyze the DNA and thus liberate the bound Feulgen dye, and then re-
stained with fast green at pH 8.0 (Alfert and Geschwind, 1953). Comparison of the values revealed a fairly constant DNA:histone ratio
per nucleus; nuclei with DNA contents greater than the 2C value* also
had proportionally increased histone contents. Of course, these cytochemical data do not really demonstrate concurrent synthesis of DNA

---

*2C refers to the DNA content of a cell during the G_1 phase of
its cell cycle.

and histone, as the authors imply, but do show clearly that the two
macromolecules accumulate within the nucleus in parallel. A similar
conclusion was reached by Alfert (1955) on the basis of Feulgen and
fast-green measurements in onion-root meristem nuclei. In this system nuclear volume is a direct function of stage in interphase, so
that cells at different positions of the cell cycle can be measured
cytophotometrically as separate groups (for a detailed discussion of
the technique of "statistical seriation" see "Materials and Methods"
in Woodard et al., 1961). Alfert found that the ratio of Feulgen to
fast-green extinction was constant in all nuclear volume classes,
signifying that the two substances double at the same time. In a
similar study, Rasch and Woodard (1959) made cytophotometric measurements of fast-green-stained *Tradescantia* tissues. Data for "statis-
tically seriated" root meristem nuclei revealed that histone content
is constant for the first 40% of interphase, doubles over the next
20%, and then remains constant again until mitosis. As Rasch and
Woodard pointed out, this resembled precisely the time course of
Feulgen-DNA doubling that had been published previously (Swift, 1953,
for *Tradescantia* and Rasch et al., 1959, for *Vicia faba*). A particularly elegant demonstration of the temporal coincidence of DNA and
histone doubling was provided by the cytochemical study of macronuclear replication in the protozoan *Euplotes* by Gall (1959). In this
ciliate the macronucleus is a slender, ribbonlike structure in which
the replication of DNA begins simultaneously at each end and proceeds
toward the center. The positions of the replicating DNA zones are
demarcated by lightly staining "reorganization bands," which are found
at or just ahead of the sites of [^{3}H]thymidine labeling. Gall reasoned that if histone accumulated in the macronucleus in temporal and
spatial coordination with DNA, then the total fast-green extinction
value for a given nucleus should be inversely proportional to the
distance between the two reorganization bands. Given the fast-green
value for macronuclei with no reorganization bands, one should be able
to predict the values for replicating nuclei just by the bands' position. He found that such predicted values were in excellent agreement

with those observed, indicating that histones accumulate and double in coordination with DNA replication.

While the foregoing studies indicated simultaneous increases in DNA and histone contents during interphase, they did not actually demonstrate that the syntheses of these macromolecules were linked, since the possibility that histones are synthesized continuously throughout the cell cycle but bind to DNA only during the S phase was not eliminated. Naturally, the resolution of this question depended on the use of labeled amino acids in conjunction with selective extraction procedures for histones. Although the first studies of this kind were "biochemical" (see Littlefield and Jacobs, 1965; Prescott, 1966; Robbins and Borun, 1967), we here deviate from strict chronological sequence to discuss two pertinent autoradiographic studies (Bloch et al., 1967; Das and Alfert, 1968)[*] before proceeding on to biochemical analyses of histone synthesis in synchronized cells.

The study by Bloch et al. (1967) on onion-root meristem is a particularly clever approach to the question of the timing of histone synthesis during the cell cycle. If a population of random cells is pulse labeled with [^{3}H]thymidine, the durations of the component phases of the cell cycle can be determined from the time course of subsequent appearance and disappearance of labeled mitotic figures (Quastler and Sherman, 1959; Wimber, 1960). For example, the existence of a post-DNA-synthetic phase, G_2, is manifest in such experiments by a delay between the pulse and the first appearance of labeled mitoses. Bloch et al. reasoned that if histone synthesis were in fact temporally coordinated with the S phase, then a plot of the percentage of mitoses

---

[*]Since this review specifically concerns histone synthesis, we shall not bring into the discussion studies that dealt with autoradiographic labeling of chromosomal proteins in toto (except in the special case of Bloch et al., 1967, in the subsequent discussion). These include, among others, the analysis of nucleic acid and protein synthesis during the cell cycle of *Vicia faba* by Woodard et al. (1961), the autoradiographic study on *Euplotes* by Prescott and Kimball (1961), and the detailed interferometric and cytophotometric analyses in cultured mammalian cells by Zetterberg (1966a and b) and Seed (1966a, b, c, and d).

labeled with [^{3}H]lysine versus time after pulse should coincide with
the curve obtained in a similar experiment with [^{3}H]thymidine. Of
course, it is assumed that a pulse of [^{3}H]lysine will selectively
label histone proteins. This assumption is theoretically unwarranted,
since there are also basic, nonhistone proteins present in the nucleus (Bhorjee and Pederson, 1972). Nevertheless, Bloch et al. found
that, with the exception of a slight wave of lysine labeling during
G_2, the appearance of lysine-labeled mitoses coincided with the thy-
midine-labeling curve. Since these authors took precautions to insure
the retention of histones in their preparations (namely, used 1 N TCA
in place of HCl for Feulgen staining) and since the majority of chromosomal lysine labeling was susceptible to HCl hydrolysis *after* fixation, they concluded that in this system brief pulses with [^{3}H]lysine
do indeed label primarily histones. Concerning the wave of chromosomal lysine labeling during G_2, Bloch et al. state that this was as
evident in Carnoy-fixed (absolute ethanol:acetic acid, 3:1) squashes
as in formalin-fixed sections. They suggest that this relative acid
resistance might indicate that the G_2 protein is nonhistone. However,
the authors also state, "Preliminary experiments indicated that much
of the material staining with fast green, i.e. histone, was retained
in these preparations (Carnoy squashes)...."[*] Therefore, the Carnoy
resistance (i.e., acid resistance) of G_2 labeling cannot logically be
used as evidence that it reflects nonhistone protein. In view of the
foregoing considerations, it is not clear whether Bloch et al. demonstrated a temporal coupling of DNA and histone synthesis, or of DNA
and the majority of chromosomal protein synthesis, histone plus nonhistone.

---

[*]General agreement has not been reached on the question of whether
histones are lost by fixation in ethanol-acetic acid. De (1961) found
that Carnoy fixation resulted in lower grain densities over chromosomes of [^{3}H]arginine-labeled onion-root meristem cells and Cave (1966)
concluded that this fixative resulted in histone removal from cultured
lymphocytes on the basis of a reduced fast-green reaction. On the
other hand, Swift (1964) reported that neither Carnoy's fixative nor
the 45% acetic acid used routinely for preparing squashes of *Drosophila* salivary glands resulted in histone extraction from the polytene
chromosomes.

Das and Alfert (1968) studied the timing of histone synthesis in relation to the cell cycle in primary spermatocytes of *Urechis caupo* (phylum Annelida). Spermatocytes were incubated in vitro with [^{3}H]-arginine, lysine, phenylalanine, or tryptophan, stained with the Feulgen reaction, and autoradiographed. After grain counts were made on mapped cells, the grains were bleached with potassium ferricyanide and the DNA contents of individual nuclei determined by cytophotometry. In this way it was possible to correlate amino acid labeling with cell-cycle stage. Das and Alfert found that cells in the S phase (i.e., Feulgen-DNA values between 2C and 4C) incorporated two to five times more lysine, arginine, or phenylalanine than cells in G_1 or G_2, while the incorporation of [^{3}H]tryptophan was the same at all stages of interphase. The notion that this increased labeling in S represented histone synthesis was reinforced by the fact that approximately 50% of the lysine labeling of S cells was susceptible to 0.2 N sulfuric acid extraction, a treatment which, as the authors demonstrated, reduced fast-green staining by an equal amount and had no effect on [^{3}H]tryptophan label. Taken together, these data indicate clearly that the increased amino acid labeling of S cells is largely a reflection of histone synthesis. Since data were not presented on the effects of acid extraction in G_1 and G_2 cells, the possibility of low levels of histone synthesis at these stages was not ruled out.

B. Biochemical Studies on Synchronized Cells

We shall consider here experiments that measure the synthesis of histones in relation to the cell cycle in synchronized populations. These studies have generally indicated a precise temporal coordination between histone and DNA synthesis.

Littlefield and Jacobs (1965) studied the timing of histone synthesis in cultures of mouse L cells partially synchronized by FUdR. During exposure of cells to the antimetabolite there was a low level of histone labeling, but this did not lead to an increased cellular histone content. However, upon release of the FUdR block by addition of thymidine, a wave of histone labeling was observed that resulted in a nearly twofold increase in histone content. Unfortunately, the

degree to which this increase was coordinated with DNA replication is
not apparent since only two time points were taken after release of
the block and also because the time course of DNA synthesis was not
measured in the same experiments. Littlefield and Jacobs referred to
an earlier paper on DNA synthesis in partially synchronized L cells
(Littlefield et al., 1963) to support their contention that the time
course of DNA synthesis coincides with the synthesis of histones.
But on the basis of the data presented in the 1965 study, it is impossible to determine whether histone synthesis ends abruptly after
DNA replication or extends partially into G_2.

A more convincing demonstration of a temporal coordination between
histone and DNA synthesis was provided in a study on synchronized *Euplotes* by Prescott (1966). Cells were manually selected at cytokine-
sis, fed [^{3}H]amino acid- or [^{3}H]thymidine-labeled *Tetrahymena,* and
harvested at intervals during the subsequent cell cycle. [^{3}H]Thymi-
dine label did not appear until approximately 4 hr after division,
whie amino acid incorporation into macronuclear total proteins was
noted within 30 min. This latter observation indicated that the 4-hr
lag in the appearance of [^{3}H]thymidine incorporation was due to a
true G_1 phase, rather than to a delay in the reutilization of the
Tetrahymena macromolecular label. In contrast to total nuclear protein, labeled amino acids were incorporated into acid-soluble nuclear
proteins at a low but finite rate during G_1, and then at a greatly
increased rate during the period of DNA synthesis. The key evidence
that amino acid incorporation represented histone synthesis was that
the incorporation of [^{3}H]tryptophan into the acid-soluble nuclear
proteins did not change significantly at the G_1/S transition. More
convincing evidence for the lack of histone synthesis during G_1 was
obtained by culturing cells for a full cycle on amino acid-labeled
Tetrahymena and then transferring them to nonradioactive food at the
time of division. It so happened that the time required for complete
utilization of the *Euplotes'* intracellular free amino acid label was
equal to the duration of G_1, so that at the time replication bands
first appeared (S), incorporation of label into total cell protein
had ceased. Since in this case there was no incorporation of label

into histone during S, Prescott correctly concluded that there had
been no synthesis of histone during G_1, to be subsequently transferred
to the nucleus during S. Taken together, these data cannot easily
be reconciled with any hypothesis other than a strict temporal coupling of DNA and histone synthesis.

Additional studies were published subsequently employing synchronized mammalian cells in culture (Spalding et al., 1966; Robbins and
Borun, 1967; Gurley and Hardin, 1968, 1969; Sadgopal and Bonner, 1969;
Gurley et al., 1972). While these have generally confirmed the S-
phase-dependent synthesis of histones reported by Littlefield and
Jacobs (1965) and Prescott (1966), in some instances they have suggested the possibility of low levels of histone synthesis during other
phases as well. Naturally, the accurate detection of G_1 or G_2 histone
synthesis requires excellent cell synchrony and this may have been a
kay fault in some studies, as we shall see. However, in at least one
case (Gurley et al., 1972) histone synthesis was observed to occur
during G_1 at a rate greater than that expected from the number of contaminating S-phase cells. Another general finding has been that there
is some synthesis of histones when DNA synthesis is inhibited in randomly growing cells by excess thymidine. I do not believe that this
exceeds the amount of residual DNA synthesis that occurs in the presence of the inhibitor, as will be shown.

Spalding et al. (1966) studied the metabolism of three electrophoretic groups of basic (acid-soluble) nuclear proteins in synchronized HeLa cells. They observed that although the rate of histone
labeling was accelerated during S, there was also a significant turnover of histone when DNA synthesis was inhibited in random populations
with amethopterin or thymidine. Due to their experimental design it
is difficult to relate the magnitude of this turnover to the active
synthesis of histones during the S phase. For example, when random
cultures were treated for 16 hr with either thymidine or amethopterin
and labeled simultaneously with [^{14}C]leucine, almost half as much
radioactivity was incorporated into each of the three bands of basic
nuclear proteins as was incorporated in untreated cells. On the other
hand, when synchronized S-phase cells were treated with the inhibitors

and then pulsed for 1 hr, the labeling of the histones was only about 20% of the control cells, in contrast to the 50% value obtained with the 16-hr continuous label in random cultures. Another uncertainty is the nature of the three electrophoretic groups of acid-soluble proteins; there are no data to define the extent to which each is contaminated by basic, nonhistone proteins. If we assume that such nonhistones are synthesized continuously throughout the cell cycle, their relative abundance in the three electrophoretic regions analyzed by Spalding et al. would constitute a baseline of radioactivity that would be resistant to experimental inhibition of DNA synthesis. However, while nonhistone, basic proteins comprise 15-20% of the total basic nuclear proteins in HeLa cells (Bhorjee and Pederson, 1972), it does not seem that their continued synthesis during thymidine or amethopterin treatment could entirely explain the relatively high turnover rates of the three electrophoretic bands observed by Spalding et al. In a subsequent study, Robbins and Borun (1967) used HeLa cells synchronized by the selective detachment of mitotic cells from monolayers and detected a small amount of histone synthesis in cells pulse labeled in very late G_1. In this case the most likely explanation would seem to have been slight contamination of the late G_1 sample by S-phase cells rather than an actual G_1-phase synthesis, but a definitive answer to this point is lacking since no data were presented for cells pulsed in early G_1. In a subsequent paper (Borun et al., 1967) histone synthesis was studied over a full cell cycle using 20-min pulse labels, followed by extraction from isolated nuclei and gel electrophoresis. The rate of synthesis of all histone fractions in G_1 and G_2 was at most 3-4% of the peak rate during mid-S, an amount that did not exceed the low level of DNA synthesis observed in the same G_1 population. These latter data indicate that the high rate of synthesis of histones observed by Spalding et al. during an imposed inhibition of DNA synthesis is much more than can be expected from simply arresting cells at some point as they traverse G_1 or at the G_1/S boundary. The notion that histone synthesis during the inhibition of DNA synthesis is greater than in unperturbed cells traversing G_1 is also borne out by experiments of Sadgopal and Bonner (1968) and

Gurley and Hardin (1968, 1969). These studies made use of cells synchronized by a double thymidine blockade (Puck, 1964). In this method
cells released from the second block display a highly synchronized S
phase followed by a fairly well resolved G_2 period (although see Pederson and Robbins, 1971). By waiting until the entire population
has completed division, usually 11-12 hr after release, it is possible
to obtain a moderately synchronized G_1 sample, especially for the
first 3 hr following mitosis. Although the mitotic selection technique is always preferable for studies in G_1, both because of the
higher degree of synchrony and the relatively unperturbed state of
the cells, many biochemical analyses requiring large G_1 samples have
been profitably carried out employing the double thymidine procedure.
In the case of Sadgopal and Bonner, HeLa cells were labeled for 4 1/2
hr with [^{14}C]amino acids at different intervals during "G_1" and "S."
The results indicated that the lysine-rich histones were synthesized
during G_1 at 18-25% of the rate in S, while the arginine-rich species
were made at almost half the S-phase rate. In reading the text, one
has the distinct impression that this experiment was done by labeling
one sample soon after release from the second thymidine block ("S")
and another ("G_1") at some time after mitosis, e.g., 14 hr after release. However, the experimental details reveal that the "G_1" sample
was labeled for 4 1/2 hr *during* the second thymidine blockade. The
situation here is thus analogous to the experiments of Spalding et
al. in which histone synthesis was studied in cells blocked for 16 hr
with amethopterin or thymidine. It is therefore not surprising that
Sadgopal and Bonner report histone synthesis at 25-50% of the S-phase
rate. In the experiments of Gurley and Hardin (1968), who used Chinese hamster cells, synthesis of all histone fractions was observed
during thymidine blockade. This again amounted to about 50% of the
rate of histone synthesis in exponential cultures. Since Gurley and
Hardin prepared "chromatin" by homogenization of whole cells, their
histone preparations may have had more contamination by cytoplasmic
basic proteins than those obtained by Spalding et al., who used nuclei,
and Sadgopal and Bonner, who used chromatin purified from isolated
nuclei. Nevertheless, since all three groups reported approximately

the same amount of histone synthesis during thymidine blockade, it is
unlikely that such contamination was an important factor. *It does*
seem clear therefore that histone synthesis in thymidine-treated cells
is paradoxically high. One possible explanation is that an experimental inhibition of DNA synthesis differs in some way from that which
the cell imposes upon itself during G_1 and G_2 and that the former
causes a derangement of the intracellular control mechanism coupling
DNA and histone synthesis. Alternatively, compounds like thymidine
and amethopterin may not be as effective in curtailing DNA synthesis
as is commonly supposed. In this connection it may be noted that none
of the three groups discussed above took the seemingly obvious pre-
caution of making parallel measurements of DNA synthesis. In an experiment done by this author several years ago, the rate of DNA synthesis in thymidine-blocked HeLa cells was found to be 15% of that
during the peak of the subsequent S phase. This was done by using
[^{3}H]deoxycytidine (Müller, 1966) and making corrections for incorporation into RNA, which is 20% of the total. The 15% value is almost
certainly an underestimate since the dCTP pool is increased substan-
tially in thymidine-blocked cells (e.g., Bell and Wolff, 1966). Recently, the existence of significant DNA synthesis during thymidine
blockade has been demonstrated unequivocally in Chinese hamster cells
by Meyn et al. (1970) and in mouse L cells and Chinese hamster cells
by Bostock et al. (1971).[*] In both studies evidence was obtained indicating a slow traversal of the G_1/S boundary during exposure to 2
mM thymidine. The amount of DNA synthesis in blocked cells was esti-

---

[*]It would be a mistake to conclude from these studies that thymidine is an unacceptable tool for cell synchronization. While allowing significant traversal of the G_1/S boundary, the agent does *dif-*
ferentially affect the component phases of the cell cycle and thus
can induce a useful degree of synchrony, particularly in repeated
schedules. In fact, it was undoubtedly the existence of significant
DNA synthesis during blockage that led Puck (1964) to develop the
"double thymidine" procedure. The compound may also be profitably
employed to "refocus" synchrony in populations synchronized initially
by another method (Pederson and Robbins, 1971).

mated by Meyn et al. to be 25% of untreated, exponential cells by BUdR density labeling, and by Bostock et al. at 33% by measurements of ^{32}P incorporation. Of course, both values are remarkably close to the levels of histone synthesis observed in the three studies discussed above. While no one has yet measured both DNA and histone synthesis in the same thymidine-blocked population, it does seem quite clear that most, if not all, of the histone synthesis in such cells is accompanied by a parallel synthesis of DNA. Until the necessary comparative measurements are made, previous data on the existence of histone synthesis in thymidine-blocked cells cannot be used as evidence for its temporal dissociation from DNA synthesis.

Related to the foregoing considerations is the question of histone turnover. Gurley and Hardin (1969, 1970) have shown that with the exception of the lysine-rich species, fl, histones do not turn over in exponentially growing Chinese hamster cells but may be considered as stable as DNA. On the other hand, in thymidine-blocked cells they observed significant turnover of the normally stable histones, as well as faster turnover of fl. This introduces yet another factor into the observed labeling of histones during thymidine blockade, and leads to the question of whether DNA turnover is also potentiated under this circumstance. The data of Bostock et al. indicate that at least some of the DNA synthesis occurring in 2-mM thymidine is chromosome replication, since it results in a foreshortened S phase and a premature arrival of some cells at mitosis. But their data do not rule out some DNA turnover as well. We may thus begin to grasp how complex this issue is. (a) DNA replication occurs in thymidine-blocked cells and is accompanied by histone synthesis, via the normal cellular regulatory signals. (b) There is, in addition, an increased turnover of histones (Gurley and Hardin, 1968), but whether this increase is accompanied by DNA turnover has not been examined. The point which is relevant to the present discussion is simply that, on the basis of the uncertainties outlined in the preceding paragraphs, it is not possible to state that histone synthesis is dissociated from DNA synthesis during thymidine blockade.

We may now proceed to consider the evidence for a low level of histone synthesis during the G_1 phase of the cell cycle (Littlefield and Jacobs, 1965; Spalding et al., 1966; Robbins and Borun, 1967; Borun et al., 1967). As pointed out earlier, such a conclusion would ideally require the ultimate degree of cell synchrony. Alternatively, it would be necessary to demonstrate that there is more histone synthesis in G_1 cells than could be expected on the basis of S-cell contamination. This point was not directly examined in the first three papers, but in the case of Borun et al. (1967) the rate of histone synthesis observed during G_1 was sufficiently accounted for by the rate of DNA synthesis. Gurley et al. (1972) have recently measured both DNA and histone synthesis in G_1 Chinese hamster cells synchronized in mitosis by selective detachment. While histone synthesis was found to be 14-21% of an exponential culture, DNA synthesis was only 0.8%. When these data were expressed in relation to values calculated for S-phase cells, histone synthesis in G_1 was 3.8-5.3% and DNA synthesis only about 0.2%. The only apparent complication in these experiments is that DNA synthesis was measured with 15-min pulses while a 2-hr labeling period was used to measure histone synthesis. However, a rather contrived argument is required to show that this procedure could have led to either an overestimate of G_1 histone synthesis or an underestimate of G_1 DNA synthesis. The data of Gurley et al. thus suggest a low level of histone synthesis in G_1 Chinese hamster cells, which is not accompanied by DNA synthesis. Further work is obviously necessary to establish the generality of this conclusion.

We have seen that the evidence for a dissociation of DNA and histone synthesis during experimental inhibition of the former is not compelling. As a separate issue (biologically), there may be a low level of DNA-independent histone synthesis during the G_1 phase of the cell cycle (Gurley et al., 1972). It is now relevant to discuss other experimental systems in which the temporal relationship of histone and DNA biosynthesis has been evaluated.

C. Experiments with Regenerating Liver

There have been a number of studies on the time course of DNA and histone synthesis in regenerating rat liver. Following partial hepatectomy, there is a parasynchronous response in which approximately 30% of the parenchymal cell population undergoes a single round of replication (Grisham, 1962; Fabrikant, 1968).* The events leading up to mitosis may be subdivided into a "prereplicative period" of 12-18 hr, a phase of DNA synthesis lasting about 14 hr, followed by cell division that usually reaches a peak value some 30 hr after hepatectomy. These parameters obviously pertain to the organ and may be contrasted with cell-cycle kinetics of individual hepatocytes, which at 20-36 hr posthepatectomy are: $S = 8.0$-8.5 hr, $G_2 = 3.0$ hr, and mitosis about 60 min (Fabrikant, 1967, 1968).** It is therefore clear that this system is poorly synchronized in relation to mammalian tissue-culture cells, in which the durations of G_1, S, and G_2 for synchronized populations are only slightly longer than values measured for individual cells (e.g., Terasima and Tolmach, 1963; Pederson and Robbins, 1971). Nevertheless, one may still inquire as to whether the broad wave of DNA synthesis in regenerating liver is coincident with a parallel, and equally heterodisperse, wave of histone synthesis. Unfortunately, despite many studies, general agreement has not been reached on this point. One of the earliest papers (Holbrook et al., 1962) reported that the peak rate of histone synthesis preceded that of DNA synthesis by some 6 hr, the maxima occurring at approximately 19 and 26 hr posthepatectomy, respectively. However, in an accompanying paper (Evans et al., 1962) these authors showed that an increase in histone content per nucleus was coincident with the rise

---

*For a review of both the biological and biochemical aspects of liver regeneration the reader is referred to Bucher and Malt (1971). A valuable article on the biochemistry of normal and stimulated cell division has been written by Baserga (1968).

**The timing of G_1 by the labeled-mitoses technique requires that daughters of one division proceed into a second cell cycle. Since this occurs infrequently in regenerating rat liver, the duration of G_1 cannot be precisely determined.

in nuclear DNA content. It is certainly possible that histone synthesis could precede its association with nuclear DNA by some hours (although see Weintraub, 1973). But the unusual histone extraction procedures employed by these authors, as well as the gross heterogeneity
of the histones upon polyacrylamide gel electrophoresis (Irvin et al.,
1963), suggest that both histones and nonhistones contributed to the
pattern of nuclear protein labeling observed after hepatectomy. Another uncertainty lies in the imprecision with which the time course
of DNA synthesis was determined. Although several points were taken
on the ascending face of the protein synthesis curve, which encom-
passed 15-22hr posthepatectomy, DNA synthesis was measured at 19 hr
and then not again until 27 hr. It is thus not convincingly demonstrated that the ascent of DNA synthesis lags behind the rise in protein synthesis. A more complex view of the situation was offered by
Umana et al. (1964) who found two waves of histone synthesis following hepatectomy, one preceding DNA synthesis and the other coincident
with it. These authors also measured the histone:DNA ratio and found
a sharp decrease shortly after hepatectomy. They concluded that immediately after hepatectomy there is a catabolism of nuclear histones,
then a replacement of these by new histones (thus explaining why the
first histone labeling wave precedes DNA synthesis), and then finally
a coordinate increase of DNA and histone in that portion of the population which is destined for mitosis (about 30%). While this is an
intriguing suggestion, other workers have not confirmed the biphasic
histone synthesis curves reported by Umana et al. Two other claims
of histone synthesis prior to DNA replication in regenerating liver
are those of Butler and Cohn (1963) and Orlova and Rodionov (1970).
Butler and Cohn did not make parallel measurements of DNA synthesis,
so their case is relatively unconvincing. They found a broad maximal
rate of histone synthesis encompassing 12-22 hr posthepatectomy, and
then relied upon the literature to state that DNA synthesis occurs
at 18-20 hr. Obviously DNA could have been synthesized slightly earlier in their particular experiments. On the other hand, a similar
study by Takai et al. (1968) demonstrated an extremely precise temporal coupling between the two processes, as precise in fact as that

in synchronized tissue-culture cells. This reviewer sees no obvious explanation for the conflicting results, and perhaps it is best to simply state that the resolution of this issue will require further study. Seed (1966e) attempted to resolve the discrepancies by pointing out that since only 30% of the hepatocytes are stimulated to synthesize DNA, the "early" wave of histone synthesis might represent the activity of the nonreplicating portion of the population. But if nonreplicating cells synthesize histones after hepatectomy, *this is still histone synthesis without DNA synthesis*. The key point would be to determine whether the "early" synthesis reported by Holbrook et al. (1962), Butler and Cohn (1963), Umana et al. (1964), and Orlova and Rodniov (1970) reflects proteins that are actually histones or only histonelike. As a final note, it would be well to keep in mind the extremely provocative finding of Gurley et al. (1970) that histone fl may be a normal component of ribosomes. If this is generally true, since the synthesis of new ribosomes is an early biochemical event in regenerating liver (Bucher and Malt, 1971), it is possible that newly synthesized, nonchromatin histone fl would be present in nucleoli, possibly in an acid-extractable form, prior to the beginning of DNA synthesis. This could be checked by measuring the relative amounts of labeled histone extracted from nucleoli versus chromatin after hepatectomy. Perhaps only the latter would correlate with DNA replication.

D. Studies with X-irradiated or Virus-infected Cells

A study has appeared indicating that DNA and histone synthesis are uncoupled in HeLa cells following x-irradiation (Bases and Mendez, 1971). These authors noted that while [^{3}H]thymidine incorporation was inhibited by approximately 30% during the first hour after 1000 rad, histone synthesis continued unabated. However, in studies on histone turnover and phosphorylation in x-irradiated Chinese hamster cells (800 rad), Gurley and Walters (1971, 1972) noted that even though [^{3}H]thymidine incorporation into DNA declined following irradiation, DNA content as measured by OD_{260} continued to increase, sug-

gesting that the decrease in [^{3}H]thymidine labeling was due to either
an expanded intracellular pool of TTP or a reduced permeability of
irradiated cells to the nucleoside. This explanation was confirmed
by the fact that no decrease in DNA synthesis was observed when ^{32}P
was used as a precursor. The significance of the results of Bases
and Mendez (1971) are therefore in doubt.

Conflicting results concerning the temporal association of DNA
and histone synthesis have also been obtained in experiments with
viruses that influence, either positively or negatively, the normal
events of the cell-division cycle. Shimono and Kaplan (1969) reported
that DNA and histone synthesis increased in parallel in mouse embryo
cells infected with polyoma virus, and Winocour and Robbins (1970)
found a similar temporal coupling in mouse kidney and BHK cells infected with polyoma, and in mouse 3T3 cells infected with SV40. I
found that the rates of DNA and histone synthesis decreased in a
strictly parallel fashion during the shutoff of host-cell metabolism
that occurs in poliovirus-infected HeLa cells (Pederson and Contreras,
1969). Holoubek and Rueckert (1964) used a different picornavirus,
Maus-Eberfield, and found an apparent dissociation of DNA and histone
synthesis in infected Ehrlich ascites cells. Histone synthesis increased soon after infection, while DNA synthesis remained constant
until the onset of cytopathogenicity. The labeling of DNA and histones in this case was carried out in vitro, and a subsequent report
(Holoubek, 1970) revealed that uninfected Ehrlich ascites cells under-
went a similar apparent uncoupling of DNA and histone synthesis upon
removal from the peritoneal cavity and incubation in vitro. It thus
seems that the effect observed initially may have been totally unrelated to the infectious cycle of ME virus. Finally, infection of mitotic HeLa cells with adenovirus type 2 results in an inhibition of
cellular DNA synthesis in the subsequent interphase, and a parallel
decrease in histone synthesis (Hodge and Scharff, 1969). It thus
appears that, with the exception of the report by Holoubek and Rueckert, the concept of a tight coupling of DNA and histone biosynthesis
is supported by studies with animal viruses.

To summarize this section on the timing of histone synthesis during the cell cycle, there is abundant evidence that the biosynthesis
of histone proteins is coupled to the replication of DNA in a wide
variety of experimental systems. Many of the claims for a temporal
dissociation of the two processes do not hold up under close scrutiny;
others remain tantalizingly unconfirmed. Until there is a convincing
demonstration to the contrary, it seems best to proceed with the hypothesis that the intracellular regulatory mechanism that coordinates
DNA and histone synthesis remains intact under a variety of experimental and biological conditions.

IV. HISTONE MESSENGER RNA AND THE CONTROL OF HISTONE SYNTHESIS

A. Histone Messenger RNA's

There can be no doubt that the most exciting development in the
field of histone biosynthesis has been the description of messenger
RNA's for histones (Borun et al., 1967; Kedes and Gross, 1969; Kedes
and Birnstiel, 1971). It is not always appreciated that these studies
have a bearing not only on the nature of the controls that coordinate
histone synthesis and DNA replication, but also on the prospects for
studying the intracellular life history of a specific messenger RNA.
While efforts along these lines are just now intensifying, it seems
appropriate to conclude this review with a brief discussion of the
progress that has been made so far.

Borun et al. (1967) studied the size distribution of pulse-labeled
RNA extracted from the light polyribosomes of synchronized HeLa cells.
In contrast to the heterodisperse pattern of RNA species observed in
the heavy polyribosomes or monosomes of S-phase cells, as well as in
the light polyribosomes of G_1 cells, RNA obtained from S-phase light
polyribosomes displayed a prominent radioactivity peak at approximately 8-9S when electrophoresed in dilute polyacrylamide gels containing
sodium dodecyl sulfate. The facts that the relative abundance of
this RNA class coincided with the rise and fall of histone synthesis

during S, that its labeling was effectively abolished when DNA synthesis was inhibited by cytosine arabinoside, and that in the presence
of actinomycin D, both 8-9S polyribosomal RNA and nascent histone polypeptides displayed half-lives of 1 hr, all suggest that this RNA is
indeed histone messenger RNA. One might ask why this approach and
this particular system led to such spectacular success when so many
other early (1960-1968) attempts to resolve individual messenger RNA's
for specialized cell products had failed. At the peak of DNA synthesis during S, histone proteins constitute at best 10% of the HeLa cell
translational activity. While the use of cells synchronized in S certainly helped, this procedure would have increased the relative abundance of histone mRNA in the population only about threefold (since
S is about one-third of the HeLa cell cycle). Therefore, it does not
seem that the degree to which the cells were committed to histone synthesis is an entirely satisfactory explanation for the results. Similarly, the possibilities that 8-9S is a unique size class of mRNA, or
that the light polyribosomes from which such mRNA was extracted were
exclusively engaged in histone synthesis are contradicted by the
available evidence. Rapidly labeled HeLa cell, polyribosome-associated RNA is reproducibly 8-30S (Penman et al., 1968). The 8-9S "histone messenger RNA" is thus within the size range of HeLa mRNA. With
regard to the light polyribosomes, it can be estimated from the data
of Robbins and Borun (1967) that, at best, 10% are engaged in the
synthesis of S-phase-dependent, cytosine-arabinoside-sensitive polypeptides (i.e., histones). The successful detection of 8-9S histone
mRNA appears in retrospect to have been the result of its more rapid
exit from the nucleus and association with polyribosomes than the bulk
of the messenger RNA (see Schochetman and Perry, 1972; Adesnik and
Darnell, 1973). Thus, while the appearance of newly synthesized RNA
in HeLa polyribosomes usually displays a distinct lag of about 15 min,
labeled 8-9S histone mRNA can be found from almost the outset of the
labeling period. An interesting, and perhaps related, finding is that
histone mRNA of both L and HeLa cells lacks the 3'OH terminal polyadenylate sequences that characterize many other mRNA's (Adesnik and
Darnell, 1972; Greenberg and Perry, 1972).

The findings of Borun et al. were subsequently confirmed by Kedes and Gross (1969) in early sea urchin embryos. The latter authors found 8-9S RNA in the light polyribosomes that had been previously implicated as sites of histone synthesis in this system (Kedes et al., 1969) and demonstrated that its presence in these polyribosomes depended on DNA synthesis. By electrophoresis in relatively long, 8% polyacrylamide gels the 8-9S RNA was resolved into three species, perhaps representing the messengers for individual histone fractions (see also Weinberg et al., 1972).

B. Control of Histone Synthesis in Relation to DNA Replication

It is important to realize that the results of Borun et al. demonstrated not that the transcription of histone messenger RNA occurs only during the S phase of the cell cycle, but rather that its association with polyribosomes is restricted to S. While it was perhaps less likely a priori, their data did not exclude the possibility of a continuous transcription of histone messenger throughout the cell cycle, combined with a translational control permitting its utilization only during S. Indeed, the existence of a translational-level control mechanism coordinating histone mRNA utilization with DNA replication has been intimated by the results of studies with inhibitors of DNA and RNA synthesis (Borun et al., 1967; Gallwitz and Mueller, 1969b). Borun et al. found that when S-phase cells were treated with actinomycin D the amount of prelabeled 8-9S RNA in light polyribosomes decayed with a half-life of approximately 1 hr, although the majority of HeLa mRNA has an apparent half-life of 3-4 hr under similar experimental conditions (Penman et al., 1963).[*] However, since DNA synthesis itself has a 1-hr half-life in actinomycin-treated cells, it is

---

[*]The actual half-life of HeLa mRNA is 24-72 hr (Singer and Penman, 1973; Murphy and Attardi, 1973). The original estimates of 3-4 hr reflected an effect of actinomycin on protein synthesis, probably specifically initiation (Singer and Penman, 1972). There is evidence suggesting that RNA species yet to be defined play a role in polypeptide chain initiation (McCormick and Penman, 1969; Fan and Penman, 1970; Leibowitz and Penman, 1971; Reichman and Penman, 1973).

not certain whether the observed decay of histone mRNA is the result
of its degradation by an extrinsic factor that is produced in increasing amounts as DNA synthesis wanes, or simply the dissociation of histone mRNA from polyribosomes in undegraded form, again through the
action of a factor produced in response to the cessation of DNA synthesis. The latter two possibilities would of course be translational
control mechanisms operating in temporal coordination with DNA synthesis. A related observation made by Borun et al. was that in contrast
to a half-life of 1 hr, in cells treated with cytosine arabinoside,
histone mRNA had a half-life of only about 20 min, and DNA synthesis
a half-life of about 10 min. Similar results were obtained by Gallwitz and Mueller (1969b) using a slightly different approach. These
authors treated S-phase HeLa cells with inhibitors of DNA or RNA synthesis and then assayed the effects on histone synthesis by incubat-
ing microsomes in a cell-free system (Gallwitz and Mueller, 1969b).
They confirmed the finding of Borun et al. that inhibition of DNA
synthesis led to a more rapid decay of histone synthesis than did
blocking RNA synthesis. In addition, they performed an experiment to
determine the metabolic fate of histone mRNA in cells treated with
DNA inhibitors. When amethopterin-synchronized S cells were treated
with hydroxyurea for 1 hr and then resuspended in fresh medium, histone synthesis returned to almost normal levels within 1 hr, in parallel with the reversal of the drug's inhibitory effect on DNA synthesis. However, this restoration of histone synthesis did not occur
when an RNA inhibitor was added at the time of hydroxyurea removal.
This result indicates either that histone mRNA is actually degraded
when DNA synthesis is interrupted or, alternatively, that another RNA
factor must be continually produced for existing histone messenger to
be reutilized (see footnote p. 109). The fact that Kedes and Gross
(1969) did not find prelabeled 8-9S RNA moving into single ribosomes
or postribosomal RNP after hydroxyurea inhibition of RNA synthesis
suggests, but does not prove, degradation. It should be evident from
the foregoing discussion that the possibility of a translational control mechanism is real and should be explored further.

One way of investigating this possibility is to use cell-free protein synthesis systems. Pederson and Robbins (1970) studied the effects of cytoplasmic factors from G_1, G_2, or metaphase HeLa cells on
the cell-free synthesis of histone polypeptides on S-phase light polyribosomes. They found that the amount of radioactivity incorporated
into histones was not influenced by the presence of soluble cytoplasm
from cells not engaged in histone synthesis. As pointed out earlier
(Sect. II,B) it was demonstrated that this cell-free system carried
out chain elongation and release only, not initiation. Therefore,
the conclusion that a translational control of histone synthesis cannot be demonstrated in vitro applies only to the first two parameters.
This problem was rectified by a later study (Jacobs-Lorena et al.,
1973), in which purified histone mRNA was assayed in a cell-free system demonstrably capable of at least limited polypeptide chain initiation. There was no inhibitory effect on histone mRNA translation
by postribosomal supernatant from S-phase cells treated with cytosine
arabinoside. Thus, the apparent translational effect of this inhibitor on histone synthesis observed in the intact cell cannot be dupli-
cated in vitro, at least not with the experimental conditions employed.
The possibility of a short-lived inhibitor of histone mRNA translation, or its production in only stoichiometric amounts with respect
to histone mRNA, is not ruled out by these cell-free experiments.

The most suggestive, although still indirect evidence for translational control of histone synthesis comes from the metabolic behavior of histone mRNA. In both HeLa cells and mouse L cells, most
mRNA decays with first-order kinetics (Greenberg, 1972; Singer and
Penman, 1972, 1973; Perry and Kelley, 1973). In other words, the
probability of a given mRNA molecule decaying is independent of its
"metabolic age." In contrast, histone mRNA in mouse L cells displays
turnover kinetics that can best be described by an equation for zero-
order decay (Perry and Kelley, 1973). The estimated functional lifetime of 11 hr,[*] taken with the cell doubling time and the durations

---

[*]In zero-order decay, "half-life" is not as meaningful a term as
"lifetime."

of the component phases of the cell cycle, can be used to show that
the most likely model is one in which histone mRNA persists for a
fixed portion of the cell cycle (e.g., the S phase) and then undergoes
rapid, catastrophic degradation (see Appendix, Perry and Kelley, 1973)
This situation predicts a complete resynthesis of the cell's histone
mRNA each generation. Thus, the metabolic behavior of histone mRNA
points to a life history in which transcription of histone genes is
switched on at the onset of S but in which the cessation of histone
synthesis at the end of the S phase is due to a translational control,
which culminates in the destruction of histone mRNA.

Another example of translational control of histone synthesis
comes from studies of histone mRNA in sea urchin oocytes and early
embryos. Histone synthesis in early embryos is only partially blocked
by actinomycin (Kedes et al., 1969; Ruderman and Gross, 1974), suggesting that it is programmed, at least in part, by a postfertiliza-
tion recruitment of preexisting (maternal) histone mRNA. This interpretation receives strong support from the observation that nonpoly-
somal RNP from unfertilized sea urchin oocytes can program histone
synthesis in vitro (Gross et al., 1973).

C. Prospects for Purifying Histone mRNA

All the experimental information reviewed in the foregoing discussion has been obtained without an actual purification of histone
mRNA as mass. Eventually such purification will be required for a
full understanding of histone mRNA's function. Clearly, the most immediate thought is to exploit histone mRNA's lack of poly(A), and
thereby isolate it from other 8-9S mRNA's, all of which would presumably contain poly(A) sequences. The problem is that it is not
really clear that most mRNA contains poly(A), since the measurements
of poly(A) content are always made on pulse-labeled, not steady-state-
labeled, mRNA. A second approach might be to inquire as to whether
histone mRNA has unique features as RNP. For example, since at least
one of the major proteins in HeLa mRNP particles is bound to poly(A)
(Blobel, 1973; Pederson and Kumar, 1975), histone mRNP might contain

less protein, or proteins more easily removed, than bulk mRNP. A third approach might make use of the infrequency of repeated sequence transcripts in mRNA. In sea urchins at least, histone mRNA is transcribed from repeated DNA sequences; the estimates range from 400 to 1200 gene copies for each of the five histone species present in the cell (Kedes and Birnstiel, 1971; Weinberg et al., 1972). Since bulk mRNA in sea urchins is almost entirely transcripts of single-copy DNA (Goldberg et al., 1973), it might be possible to choose C_0t's (product of RNA concentration x time) at which histone mRNA is preferentially hybridized. For preparative utility, RNA-excess hybridization would be warranted, followed by isolation of the hybrids by hydroxylapatite chromatography. After thermal elution, the RNA would require separation from the DNA, for example by cesium sulfate centrifugation. The drawback of this method is that there would be very low recoveries, since at RNA excess only a small amount of the input RNA forms hybrids. DNA-excess reactions are theoretically of greater preparative value (for RNA), but are technically not feasible since at a preparative (RNA) scale they would require gram quantities of DNA.

In summary, it appears likely that, in growing mammalian cells at least, histone synthesis is restricted to the S phase of the cell cycle by a combination of transcriptional and translational control. Further developments in this area will depend on direct purification of histone mRNA (as mass), as well as general progress in the area of transcriptional and translational control. Whether histone mRNA itself can be a vehicle for such progress remains to be seen.

V. PERSPECTIVES

This review has dealt with "cellular" aspects of histone synthesis. We have considered the intracellular site of synthesis of these nuclear proteins and the interrelationship of histone synthesis with DNA replication and the cell-division cycle. It seems clear that questions of mechanism and function will dominate the landscape in the future. How is the transcription of histone mRNA switched on at a precise time in the cell cycle? How do newly synthesized histones reach the cell nucleus, and how do they combine in spatial precision with replicating DNA units at the chromosomal level? Are histones involved in "chromosome dynamics," that is, the coiling cycle of visible mitosis and the more subtle changes in chromosome architecture that transpire during interphase (Pederson and Robbins, 1971; Pederson, 1972)? With regard to the concepts discussed in this review, perhaps the most provocative question for the future is still the nature of the cellular control mechanism that couples DNA replication and histone synthesis. Its elucidation would be a significant stride forward in our understanding of the cell cycle and, in a broader sense, would offer new insight into the problem of intracellular coordination.

ACKNOWLEDGMENTS

Research cited from this laboratory was supported by grants to the author from the National Institutes of Health (GM 21595, CA 13400) and the American Cancer Society (VC-119) and by a Scholar Award from the Leukemia Society of America. I am grateful to Dr. James E. Darnell, Jr., and Dr. Lawrence Gurley for providing manuscripts in advance of publication and to Dr. Hewson Swift and Dr. Jay R. Greenberg for helpful comments. Ms. Phyllis Carey provided conscientious assistance in preparing the manuscript.

This review was completed in 1973 and was updated in 1974. The author regrets any inaccuracies created by the appearance of new literature after mid-1974, during production delays.

REFERENCES

Adesnik, M., and J. E. Darnell. 1972. Biogenesis and characterization of histone messenger RNA in HeLa cells. *J. Mol. Biol. 67:* 297-406.

Alfert, M., and I. I. Geschwind. 1953. A selective staining method

for the basic proteins of cell nuclei. *Proc. Nat. Acad. Sci. U.S.*
39:991-999.

Allfrey, V. G., A. E. Mirsky, and S. Osawa. 1957. Protein synthesis
in isolated cell nuclei. *J. Gen. Physiol. 40*:451-490.

Allfrey, V. G., R. Meudt, J. W. Hopkins, and A. E. Mirsky. 1961.
Sodium-dependent "Transport" reactions in the cell nucleus and
their role in protein and nucleic acid synthesis. *Proc. Nat. Acad.
Sci. U.S. 47*:907-932.

Allfrey, V. G., V. C. Littau, and A. E. Mirsky. 1964. Methods for
the purification of thymus nuclei and their application to studies
of nuclear protein synthesis. *J. Cell Biol. 21*:213-231.

Ando, T., and K. Suzuki. 1966. The amino acid sequence of the second component of clupeine. *Biochim. Biophys. Acta 121*:427-429.

Ando, T., and K. Suzuki. 1967. The amino acid sequence of the third
component of clupeine. *Biochim. Biophys. Acta 140*:375-377.

Baserga, R. 1968. Biochemistry of the cell cycle: a review. *Cell
and Tissue Kinetics 1*:167-191.

Bases, R., and F. Mendez. 1971. Dissociation of histone and DNA
synthesis in x-irradiated HeLa cells. *Exp. Cell Res. 69*:289-294.

Bell, S., and S. Wolff. 1966. Effects of FUdR and thymidine on incorporation of deoxycytidine into DNA of *Vicia faba*. *Exp. Cell
Res. 42*:408-412.

Bhorjee, J. S., and T. Pederson. 1972. Non-histone chromosomal proteins in synchronized HeLa cells. *Proc. Nat. Acad. Sci. U.S. 69*:
3345-3349.

Bhorjee, J. S., and T. Pederson. 1973. Chromatin: its isolation
from cultured mammalian cells with particular reference to contamination by nuclear ribonucleoprotein particles. *Biochemistry 12*:
2766-2773.

Birnstiel, M. L., and W. G. Flamm. 1964. Intranuclear site of histone synthesis. *Science 145*:1435-1437.

Birnstiel, M. L., M. I. Chipchase, and R. J. Hayes. 1962. Incorporation of L-(^{14}C) leucine by isolated nuclei. *Biochim. Biophys.
Acta 55*:728-733.

Blobel, G. 1973. A protein of molecular weight 78,000 bound to the
polyadenylate region of eukaryotic messenger RNA's. *Proc. Nat.
Acad. Sci. U.S. 70*:924-928.

Bloch, D. P. 1962. Histone synthesis in non-replicating chromosomes.
J. Histochem. Cytochem. 10:137-144.

Bloch, D. P., and S. D. Brack. 1964. Evidence for the cytoplasmic
synthesis of nuclear histone during spermiogenesis in the grass-
hopper *Chortophaga viridifasciata* (de Geer). *J. Cell Biol. 22*:
327-340.

Bloch, D. P., and G. C. Godman. 1955. A microphotometric study of the syntheses of desoxyribonucleic acid and nuclear histone. *J. Biophys. Biochem. Cytol. 1:*17-28.

Bloch, D. P., and H. Y. C. Hew. 1960. Changes in nuclear histones during fertilization and early embryonic development in the pulmonate snail, *Helix aspera. J. Biophys. Biochem. Cytol. 8:*69-81.

Bloch, D. P., R. A. MacQuigg, S. D. Brack, and J. R. Wu. 1967. The synthesis of deoxyribonucleic acid and histone in the onion root meristem. *J. Cell Biol. 33:*451-467.

Borun, T. W., M. D. Scharff, and E. Robbins. 1967. Rapidly labeled, polyribosome associated RNA having the properties of histone messenger. *Proc. Nat. Acad. Sci. U.S. 58:*1977-1983.

Bostock, C. J., D. M. Prescott, and J. B. Kirkpatrick. 1971. An evaluation of the double thymidine block for synchronizing mammalian cells at the G_1-S border. *Exp. Cell Res. 68:*163-168.

Bucher, N. L. R., and R. A. Malt. 1971. *Regeneration of Liver and Kidney.* Boston: Little, Brown.

Butler, J. A. V., and P. Cohn. 1963. Studies on histone. 6. Observations on the biosynthesis of histones and other proteins in regenerating rat liver. *Biochem. J. 87:*330-334.

Cameron, I. L., and D. M. Prescott. 1963. RNA and protein metabolism in the maturation of the nucleated chicken erythrocyte. *Exp. Cell Res. 30:*609-612.

Cave, M. 1966. Incorporation of tritium-labeled thymidine and lysine into chromosomes of cultured human leukocytes. *J. Cell Biol. 29:* 209-222.

Craig, N., and R. P. Perry. 1971. Persistent cytoplasmic synthesis of ribosomal proteins during the selective inhibition of ribosomal RNA synthesis. *Nature 229:*75-80.

Das, N. K., and M. Alfert. 1968. Cytochemical studies on the concurrent synthesis of DNA and histone in primary spermatocytes of *Urechis caupo. Exp. Cell Res. 49:*51-58.

De, D. N. 1961. Autoradiographic studies of nucleoprotein metabolism during the division cycle. *The Nucleus* (India) *4:*1-24.

Dintzis, H. M. 1961. Assembly of peptide chains of hemoglobin. *Proc. Nat. Acad. Sci. U.S. 47:*247-261.

Epifanova, O. I., and V. V. Terskikh. 1969. On the resting periods in the cell life cycle. *Cell and Tissue Kinetics 2:*75-93.

Evans, J. H., D. J. Holbrook, Jr., and J. L. Irvin. 1962. Changes in content of nuclear proteins and nucleic acids in regenerating liver. *Exp. Cell Res. 28:*126-132.

Fabrikant, J. I. 1967. The effect of prior continuous irradiation on the G_2, M, and S phases of proliferating parenchymal cells in the regenerating liver. *Rad. Res. 31:*304-314.

Fabrikant, J. I. 1968. The kinetics of cellular proliferation in regenerating liver. *J. Cell Biol. 36:*551-565.

Fan, H., and S. Penman. 1970. Regulation of protein synthesis in mammalian cells. II. Inhibition of protein synthesis at the level of initiation during mitosis. *J. Mol. Biol. 50:*655-670.

Gall, J. G. 1959. Macronuclear duplication in the ciliated protozoan *Euplotes*. *J. Biophys. Biochem. Cytol. 5:*295-308.

Gallwitz, D., and G. C. Mueller. 1969a. Histone synthesis *in vitro* by cytoplasmic microsomes from HeLa cells. *Science 163:*1351-1353.

Gallwitz, D., and G. C. Mueller. 1969b. Histone synthesis *in vitro* on HeLa cell microsomes. *J. Biol. Chem. 244:*5947-5952.

Goldberg, R. B., G. A. Galau, R. J. Britten, and E. H. Davidson. 1973. Nonrepetitive DNA sequence representation in sea urchin embryo messenger RNA. *Proc. Nat. Acad. Sci. U.S. 70:*3516-3520.

Goldstein, L. 1970. On the question of protein synthesis by cell nuclei. *Adv. Cell Biol. 1:*187-210.

Grasso, J. A., and J. W. Woodard. 1966. The relationship between RNA synthesis and hemoglobin synthesis in amphibian erythropoiesis. Cytochemical evidence. *J. Cell Biol. 31:*279-294.

Grasso, J. A., J. W. Woodard, and H. Swift. 1963. Cytochemical studies of nucleic acids and proteins in erythrocytic development. *Proc. Nat. Acad. Sci. U.S. 50:*134-140.

Greenberg, J. R. 1972. High stability of messenger RNA in growing cultured cells. *Nature 240:*102-104.

Greenberg, J. R., and R. P. Perry. 1972. Relative occurrence of polyadenylic acid sequences in messenger and heterogeneous nuclear RNA of L cells as determined by poly(U)-hydroxylapatite chromatography. *J. Mol. Biol. 72:*91-98.

Grisham, J. W. 1962. Morphologic study of deoxyribonucleic acid synthesis and cell proliferation in regenerating rat liver: Autoradiography with thymidine-H^3. *Cancer Res. 22:*842-849.

Gross, K. W., M. Jacobs-Lorena, C. Baglioni, and P. R. Gross. 1973. Cell-free translation of maternal messenger RNA from sea urchin eggs. *Proc. Nat. Acad. Sci. U.S. 70:*2614-2618.

Gurley, L. R., and J. M. Hardin. 1968. The metabolism of histone fractions. I. Synthesis of histone fractions during the life cycle of mammalian cells. *Arch. Biochem. Biophys. 128:*285-292.

Gurley, L. R., and J. M. Hardin. 1969. The metabolism of histone fractions. II. Conservation and turnover of histone fractions in mammalian cells. *Arch. Biochem. Biophys. 130:*1-6.

Gurley, L. R., and J. M. Hardin. 1970. The metabolism of histone fractions. III. Synthesis and turnover of histone f1. *Arch. Biochem. Biophys. 136:*392-401.

Gurley, L. R., and R. A. Walters. 1971. Response of histone turnover and phosphorylation to x-irradiation. *Biochem.* *10*:1588-1593.

Gurley, L. R., and R. A. Walters. 1972. The metabolism of histone fractions. V. The relationship between histone and DNA synthesis after x-irradiation. *Arch. Biochem. Biophys. 153*:304-311.

Gurley, L. R., R. A. Walters, and M. D. Enger. 1970. Isolation and characterization of histone f1 in ribosomes. *Biochem. Biophys. Res. Commun. 40*:428-436.

Gurley, L. R., R. A. Walters, and R. A. Tobey. 1972. The metabolism of histone fractions. IV. Synthesis of histones during the G_1-phase of the mammalian life cycle. *Arch. Biochem. Biophys. 148*: 633-641.

Heady, J. E., and E. H. McConkey. 1970. Completion of nascent HeLa ribosomal proteins in a cell-free system. *Biochem. Biophys. Res. Commun. 40*:30-36.

Hodge, L. D., and M. D. Scharff. 1969. Effect of adenovirus on host cell DNA synthesis in synchronized cells. *Virology 37*:554-565.

Holbrook, D. J., Jr., J. H. Evans, and J. L. Irvin. 1962. Incorporation of labeled precursors into proteins and nucleic acids of nuclei of regenerating liver. *Exp. Cell Res. 28*:120-125.

Holoubek, V. 1970. Dissociation of histones and DNA synthesis after a transfer of Ehrlich ascites cells from mouse to *in vitro* culture. *J. Cell Biol. 47*:90a (abstract).

Holoubek, V., and R. R. Rueckert. 1964. Studies on nuclear protein metabolism after infection of Ehrlich ascites cells with Maus-Eberfield (ME) virus. *Biochem. Biophys. Res. Commun. 15*:166-171.

Howard, A., and S. R. Pelc. 1951. Nuclear incorporation of P^{32} as demonstrated by autoradiographs. *Exp. Cell Res. 2*:178-187.

Howard, A., and S. R. Pelc. 1953. Synthesis of desoxyribonucleic acid in normal and irradiated cells and its relation to chromosome breakage. *Heredity Suppl. 6*:261-273.

Hunt, T., T. Hunter, and A. Munro. 1969. Control of hemoglobin synthesis: rate of translation of the messenger RNA for the α and b chains. *J. Mol. Biol. 43*:123-133.

Infante, A. A., and M. Nemer. 1967. Accumulation of newly synthesized RNA templates in a unique class of polyribosomes during embryogenesis. *Proc. Nat. Acad. Sci. U.S. 58*:681-688.

Irvin, J. L., D. L. Holbrook, Jr., J. H. Evans, H. C. McAllister, and E. P. Stiles. 1963. Possible role of histones in regulation of nucleic acid synthesis. *Exp. Cell Res.* Suppl. 9:359-366.

Jacobs-Lorena, M., F. Gabrielli, T. W. Borun, and C. Baglioni. 1973. Studies on the translational control of histone synthesis. I. Translation of histone messenger RNA by heterologous cell-free systems prepared from cells inactive in DNA synthesis. *Biochim. Biophys. Acta 324*:275-281.

Kedes, L. H., and M. L. Birnstiel. 1971. Reiteration and clustering of DNA sequences complementary to histone messenger RNA's. *Nature New Biol. 230*:165-169.

Kedes, L. H., and P. R. Gross. 1969. Identification in cleaving embryos of three RNA species serving as templates for the synthesis of nuclear proteins. *Nature 223*:1335-1339.

Kedes, L. H., P. R. Gross, G. Cognetti, and A. L. Hunter. 1969. Synthesis of nuclear and chromosomal proteins on light polyribosomes during cleavage in the sea urchin embryo. *J. Mol. Biol. 45*:337-351.

Kuff, E. L., and N. E. Roberts. 1967. *In vivo* labeling patterns of free polyribosomes: relationship to tape theory of messenger ribonucleic acid function. *J. Mol. Biol. 26*:211-225.

Lajtha, L. G. 1963. On the concept of the cell cycle. *J. Cell. Comp. Physiol. 62* (Suppl. 1):143-145.

Leibowitz, R., and S. Penman. 1970. Regulation of protein synthesis in HeLa cells. III. Inhibition during poliovirus infection. *J. Virol. 8*:661-668.

Liau, M. C., and R. P. Perry. 1969. Ribosome precursor particles in nucleoli. *J. Cell Biol. 42*:272-283.

Ling, V., J. R. Trevithick, and G. H. Dixon. 1969. The biosynthesis of protamines in trout testis. I. Intracellular site of synthesis. *Can. J. Biochem. 47*:51-60.

Lipmann, F. 1971. Attempts to map a process evolution of peptide biosynthesis. *Science 173*:875-884.

Littlefield, J. W., and P. S. Jacobs. 1965. The relation between DNA and protein synthesis in mouse fibroblasts. *Biochim. Biophys. Acta 108*:652-658.

Littlefield, J. W., A. P. McGovern, and K. B. Margeson. 1963. Changes in the distribution of polymerase activity during DNA synthesis in mouse fibroblasts. *Proc. Nat. Acad. Sci. U.S. 49*:102-107.

McCormick, W., and S. Penman. 1969. Regulation of protein synthesis in HeLa cells: translation at elevated temperatures. *J. Mol. Biol. 39*:315-333.

Meyn, R. E., R. M. Humphrey, and R. R. Hewitt. 1970. Evaluation of the cell cycle position obtained by synchronization with excess thymidine. *J. Cell Biol. 47*:139a (abstract).

Mirsky, A. E., and A. V. Pollister. 1946. Chromosin. A desoxyribose nucleoprotein complex of the cell nucleus. *J. Gen. Physiol. 30*:117-148.

Moav, B., and M. Nemer. 1971. Histone synthesis. Assignment to a special class of polyribosomes in sea urchin embryos. *Biochemistry. 10:*881-888.

Monesi, V. 1964. Autoradiographic evidence of a nuclear histone synthesis during mouse spermiogenesis in the absence of detectable quantities of nuclear ribonucleic acid. *Exp. Cell Res. 36:*683-688.

Müller, D. 1966. Autoradiographische Untersuchungen an HeLa-Zellen über die Brauchbarkeit des H³-deoxycytidins als Spezifischer Vorläufer der Desoxyribonukleinsäure. *Exp. Cell Res. 44:*627-631.

Murphy, W., and G. Attardi. 1973. Stability of cytoplasmic messenger RNA in HeLa cells. *Proc. Nat. Acad. Sci. U.S. 70:*115-119.

Nemer, M., and D. T. Lindsay. 1969. Evidence that the s-polysomes of early sea urchin embryos may be responsible for the synthesis of chromosomal histones. *Biochem. Biophys. Res. Commun. 35:*156-160.

Orlova, L. V., and V. M. Rodionov. 1970. The time of histone synthesis in regenerating rat liver. *Exp. Cell Res. 59:*329-333.

Pederson, T. 1972. Chromatin structure and the cell cycle. *Proc. Nat. Acad. Sci. U.S. 69:*2224-2228.

Pederson, T., and G. Contreras, unpublished results, 1969.

Pederson, T., and S. Gelfant. 1970. Macromolecular synthesis in dogfish peripheral blood cells. *J. Cell Biol. 45:*183-187.

Pederson, T., and A. Kumar. 1971. Relationship between protein synthesis and ribosome assembly in HeLa cells. *J. Mol. Biol. 61:*655-668.

Pederson, T., and A. Kumar. 1975. Fractionation of hnRNP particles from HeLa cell nuclei on oligo(dT)-cellulose. *Proc. 9th Meeting of Federation of European Biochemical Societies,* North Holland Publ. Co., Amsterdam, 69-74.

Pederson, T., and E. Robbins. 1970. Absence of translational control of histone synthesis during the HeLa cell life cycle. *J. Cell Biol. 45:*509-513.

Pederson, T., and E. Robbins. 1971. A method for improving synchrony in the G_2 phase of the cell cycle. *J. Cell Biol. 49:*942-945.

Pederson, T., and E. Robbins. 1972. Chromatin structure and the cell division cycle. Actinomycin binding in synchronized HeLa cells. *J. Cell Biol. 55:*322-327.

Penman, S., K. Scherrer, Y. Becker, and J. E. Darnell. 1963. Polyribosomes in normal and poliovirus-infected HeLa cells and their relationship to messenger-RNA. *Proc. Nat. Acad. Sci. U.S. 49:*652-662.

Penman, S., C. Vesco, and M. Penman. 1968. Localization and kinetics of formation of nuclear heterodisperse RNA, cytoplasmic heterodisperse RNA, and polyribosome-associated messenger RNA in HeLa cells. *J. Mol. Biol. 34:*49-69.

Perry, R. P., and D. E. Kelley. 1973. Messenger RNA turnover in mouse L cells. *J. Mol. Biol.* 79:681-696.

Prescott, D. M. 1966. The syntheses of total macronuclear protein, histone, and DNA during the cell cycle in *Euplotes eurystomus.* *J. Cell Biol.* 31:1-9.

Prescott, D. M., and R. F. Kimball. 1961. Relation between RNA, DNA, and protein synthesis in the replicating nucleus of *Euplotes.* *Proc. Nat. Acad. Sci. U.S.* 47:686-693.

Puck, T. T. 1964. Phasing, mitotic delay, and chromosome aberrations in mammalian cells. *Science 144:*565-566.

Quastler, H. 1963. The analysis of cell population kinetics. In *Cell Proliferation.* Oxford: Blackwell.

Quastler, H., and F. G. Sherman. 1959. Cell population kinetics in the intestinal epithelium of the mouse. *Exp. Cell Res.* 17:420-438.

Rasch, E., and J. W. Woodard. 1959. Basic proteins of plant nuclei during normal and pathological cell growth. *J. Biophys. Biochem. Cytol.* 6:263-276.

Rasch, E., H. Swift, and R. M. Klein. 1959. Nucleoprotein changes in plant tumor growth. *J. Biophys. Biochem. Cytol.* 6:11-34.

Reichman, M., and S. Penman. 1973. Stimulation of polypeptide initiation *in vitro* after protein synthesis inhibition *in vivo* in HeLa cells. *Proc. Nat. Acad. Sci. U.S.* 70:2678-2682.

Reid, B. R., and R. D. Cole. 1964. Biosynthesis of a lysine-rich histone in isolated calf thymus nuclei. *Proc. Nat. Acad. Sci. U.S.* 51:1044-1050.

Reid, B. R., R. H. Stellwagen, and R. D. Cole. 1968. Further studies on the biosynthesis of very lysine-rich histones in isolated nuclei. *Biochim. Biophys. Acta 155:*593-602.

Rho, J. H., and M. I. Chipchase. 1963. Incorporation of tritiated cytidine into ribonucleic acid by isolated pea nuclei. *J. Cell Biol.* 14:183-192.

Robbins, E., and T. W. Borun. 1967. The cytoplasmic synthesis of histones in HeLa cells and its temporal relationship to DNA replication. *Proc. Nat. Acad. Sci. U.S.* 57:409-416.

Rogers, M. E. 1968. Ribonucleoprotein particles in the amphibian oocyte nucleus. Possible intermediates in ribosome synthesis. *J. Cell Biol.* 36:421-432.

Ruderman, J. V., and P. R. Gross. 1974. Histones and histone synthesis in sea urchin development. *Dev. Biol.* 36:286-298.

Sadgopal, A., and J. Bonner. 1969. The relationship between histone and DNA synthesis in HeLa cells. *Biochim. Biophys. Acta 186:* 349-357.

Schochetman, G., and R. P. Perry. 1972. Early appearance of histone messenger RNA in polyribosomes of cultured L cells. *J. Mol. Biol.* *63*:591-596.

Seed, J. 1966a. The synthesis of DNA, RNA, and nuclear protein in normal and tumor strain cells. I. Fresh embryo human cells. *J. Cell Biol. 28*:233-248.

Seed, J. 1966b. The synthesis of DNA, RNA, and nuclear protein in normal and tumor strain cells. II. Fresh embryo mouse cells. *J. Cell Biol. 28*:249-256.

Seed, J. 1966c. The synthesis of DNA, RNA, and nuclear protein in normal and tumor strain cells. III. Mouse ascites tumor cells. *J. Cell Biol. 28*:257-261.

Seed, J. 1966d. The synthesis of DNA, RNA, and nuclear protein in normal and tumor strain cells. IV. HeLa tumor strain cells. *J. Cell Biol. 28*:263-275.

Seed, J. 1966e. Synthesis of nucleic acids and nuclear protein in replicating animal cells. *Nature 210*:993-995.

Shimono, H., and A. S. Kaplan. 1969. Correlation between the synthesis of DNA and histones in polyoma virus-infected mouse embryo cells. *Virology 37*:690-694.

Singer, R. H., and S. Penman. 1972. Stability of HeLa cell mRNA in actinomycin. *Nature 240*:100-102.

Singer, R. H., and S. Penman. 1973. Messenger RNA in HeLa cells: kinetics of formation and decay. *J. Mol. Biol. 78*:321-334.

Spalding, J., K. Kajiwara, and G. C. Mueller. 1966. The metabolism of basic proteins in HeLa cell nuclei. *Proc. Nat. Acad. Sci. U.S. 56*:1535-1542.

Swift, H. 1950. The desoxyribose nucleic acid content of animal nuclei. *Physiol. Zool. 23*:169-200.

Swift, H. 1953. Quantitative aspects of nuclear nucleoproteins. *Intern. Rev. Cytol. 2*:1-76.

Swift, H. 1964. The histones of polytene chromosomes. in *The Nucleohistones*. San Francisco: Holden-Day.

Takai, S., T. W. Borun, J. Muchmore, and I. Lieberman. 1968. Concurrent synthesis of histone and deoxyribonucleic acid in liver after partial hepatectomy. *Nature 219*:860-861.

Terasima, T., and L. J. Tolmach. 1963. Growth and nucleic acid synthesis in synchronously dividing populations of HeLa cells. *Exp. Cell Res. 30*:344-362.

Umana, R., S. Updike, J. Randall, and A. L. Dounce. 1964. Histone metabolism. in *The Nucleohistones*. San Francisco: Holden Day.

Walker, P. M. B., and H. B. Yates. 1952. Nuclear components of dividing cells. *Proc. Roy. Soc. London, Ser. B 140*:274-299.

Warner, J. R., and R. Soeiro. 1967. Nascent ribosomes from HeLa cells. *Proc. Nat. Acad. Sci. U.S. 58:*1984-1990.

Weinberg, E. S., M. L. Birnstiel, I. F. Purdom, and R. Williamson. 1972. Genes coding for polysomal 95 RNA of sea urchins: conservation and divergence. *Nature 240:*225-228.

Weintraub, H. 1973. The assembly of newly replicated DNA into chromatin. *Cold Spring Harbor Symp. Quant. Biol. 38:*247-256.

Wimber, D. E. 1960. Duration of the nuclear cycle in *Tradescantia paludosa* root tips as measured with H^3-thymidine. *Am. J. Bot. 47:* 828-834.

Winocour, E., and E. Robbins. 1970. Histone synthesis in polyoma- and SV_{40}-infected cells. *Virology 40:*307-315.

Woodard, J., E. Rasch, and H. Swift. 1961. Nucleic acid and protein metabolism during the mitotic cycle in *Vicia faba*. *J. Biophys. Biochem. Cytol. 9:*445-462.

Wu, R. S., and J. R. Warner. 1971. Cytoplasmic synthesis of nuclear proteins. Kinetics of accumulation of radioactive proteins in various cell fractions after brief pulses. *J. Cell Biol. 51:*643-652.

Zetterberg, A. 1966a. Synthesis and accumulation of nuclear and cytoplasmic proteins during interphase in mouse fibroblasts *in vitro*. *Exp. Cell Res. 42:*500-511.

Zetterberg, A. 1966b. Nuclear and cytoplasmic nucleic acid content and cytoplasmic protein synthesis during interphase in mouse fibroblasts *in vitro*. *Exp. Cell Res. 43:*526-533.

velop. In this case the
lso be explained by an

interpreted as an
accompanied by a

YOTES

-inhibition of

gical Sciences
New York at Stony Brook

M.-T. Huang*

Department of Pharmacology
Albert Einstein College of Medicine
Bronx, New York

*Dr. Huang's present address: Department of Biochemistry and Drug Metabolism, Hoffmann-La Roche Inc., Nutley, New Jersey.

I. INTRODUCTION

A decade ago, the only well-established inhibitors of protein synthesis in eukaryotic cells were cycloheximide, puromycin, and sodium
fluoride. Subsequently, a number of cytotoxic drugs (see Grollman,
1974), toxins (Olsnes and Pihl, 1973), and various other agents (see
Pestka, 1971; Grollman and Huang, 1973) have been shown to act in
this manner. Not all these inhibitors have been studied in detail;
the present review emphasizes those agents whose biochemical effects
are relatively well documented.

Protein synthesis in eukaryotes involves a linked series of enzymatic and nonenzymatic reactions. In the complex and incompletely
understood biochemistry of protein biosynthesis, it is frequently
difficult to determine if a specific enzymatic process or ribosomal
function is affected by a given inhibitor. Inhibitors are classified
according to the primary reaction they appear to affect but it should
be realized that the assignment proposed may require revision as further experimental data become available.

An understanding of present concepts of protein synthesis and ribosomal function is essential to any discussion of inhibitors. Since
several excellent reviews (Lucas-Lenard and Lipmann, 1971; Haselkorn
and Rothman-Denes, 1973) and pertinent sections of this volume are
readily available, these subjects will not be discussed in detail
here. A comprehensive review of inhibitors of ribosome function has

been published (Pestka, 1971) and several recent symposia have been
devoted to this topic (Bücher and Sies, 1969; Munoz et al., 1972).

Inhibitors of eukaryotic protein synthesis are conveniently
grouped as inhibitors of initiation, peptide chain elongation, or
chain termination, according to the functional events that they per-
turb. Initiation of protein synthesis includes those events that occur prior to formation of the complex between ribosome, mRNA, and
methionyl-tRNA (Fig. 1, stage III). Chain elongation involves binding
of tRNA to the ribosome, peptide bond formation, and translocation, a
series of reactions (Fig. 1, stages IV-VI) that repeats, sequentially,
until the peptide chain is complete. Chain termination occurs when
peptides are released from the ribosome and the ribosome-mRNA complex
dissociates into its component parts.

Several proteins and cofactors involved in this series of reactions have been distinguished. An initiation factor, M1, forms a complex with GTP and methionyl-tRNA^{fMet} which functions as an intermediate in transferring this initiator tRNA to the smaller ribosomal subunit (Dettman and Stanley, Jr., 1973). Globin messenger RNA binds
to the ribosome in the presence of a second initiation factor, termed
M3; the larger ribosomal subunit is then added to complete the initiation complex (Kaempfer and Kaufman, 1972; Crystal and Anderson,
1972).

Binding of aminoacyl-tRNA, the first step in chain elongation, requires a protein termed EF-I. Peptide bond formation is catalyzed by
a ribosomal protein located on the larger subunit, and translocation
requires the presence of a soluble protein, EF-II. Release factors
have also been detected in extracts prepared from eukaryotic cells.
Hydrolysis of two molecules of GTP is needed for peptide bond formation; however, the precise role of GTP in protein synthesis remains
obscure (see Haselkorn and Rothman-Denes, 1973).

Alternative schemes for classification of inhibitors that emphasize binding to ribosomes (Vazquez and Battaner, 1972) or interference with specific steps and reactions in protein synthesis (Pestka,
1971) have been proposed. Regardless of the classification used,
the factors noted below should be considered in interpreting experimental data and specifying modes of action.

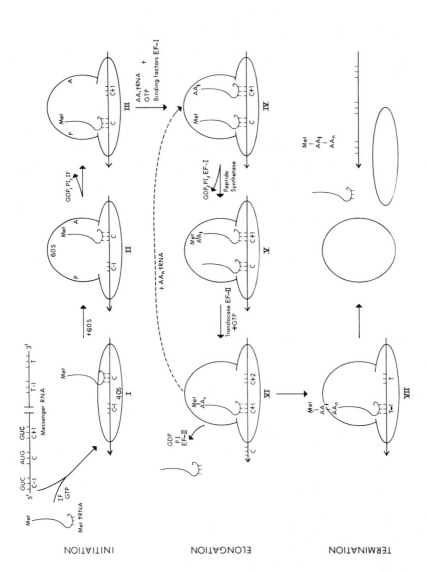

FIG. 1. Schematic representation of the reactions involved in protein synthesis. Key: IF, initiation factors; EF, elongation factors; fMet, formylmethionyl; tRNA, transfer RNA; C, initiation codon; T, termination codon; PI, inorganic phosphate; AA, amino acid residue; P, peptidyl (or donor) site; A, aminoacyl (or acceptor) site.

1. Some inhibitors affect several steps, frequently at different concentrations of the drug. Such multiple effects may reflect overlapping ribosomal sites or conformational changes induced by the inhibitor. It is difficult to distinguish between primary and secondary effects; assay of isolated reactions may result in assigning the mode of action to a secondary step, overlooking the primary effect.

2. The various phases of protein synthesis include analogous steps; for example, both initiation and elongation involve binding of aminoacyl-tRNA to the aminoacyl (A) site on the ribosome. An inhibitor that affects this site could inhibit both processes. Similarly, chain termination involves the peptidyl transferase center on the larger ribosomal subunit. Hydrolysis of peptidyl-tRNA has requirements similar to those of peptide bond formation; thus, inhibitors of peptide bond formation may also affect chain termination.

3. Our present knowledge of protein synthesis and ribosomal function in eukaryotes is incomplete. Various models have been proposed, based partly on analogy to *E. coli;* such models have not been universally accepted. It is evident that precise mechanisms of inhibitor action cannot be determined until the molecular biology of protein synthesis is firmly established.

4. Experimental data can be misleading when the effect of inhibitors on model reactions are studied. Unwarranted conclusions have been drawn from studies of initiation in partially reconstituted systems, utilizing the "fragment reaction" to measure peptide bond formation and polynucleotides to replace natural messenger RNA or tRNA's other than methionyl-tRNA^{fMet}.

5. Although structurally related compounds generally act in a similar manner, small modifications in structure occasionally alter the mode of action. Results with a given inhibitor should not be assumed to apply to its analogs.

6. Displacement of inhibitors from the ribosome by other compounds does not necessarily imply identical or even similar modes of action. Ribosomes assume various conformational states and the apparent competition may reflect an allosteric interaction.

II. INHIBITORS OF INITIATION

A. Aurintricarboxylic Acid and Related Triphenylmethane Dyes

Aurintricarboxylic acid (ATA) (1) is an anionic triphenylmethane dye which, in the form of its ammonium salt, has been used as a standard analytical reagent for the detection of aluminum. The sodium

salt is marketed under the name of chrome violet (C.I. 43810). The
effects of ATA on protein synthesis in prokaryotes (Grollman and Stewart, 1968) and eukaryotes (Grollman, 1968a; Stewart et al., 1971;
Huang and Grollman, 1972a) are exhibited by certain other triphenylmethane dyes (Grollman, 1968a; Liao et al., 1974c; Huang and Grollman,
1973).

(<u>1</u>)

Aurintricarboxylic acid and related triphenylmethane dyes are the
only small organic molecules presently known that affect formation of
the initiation complex (Grollman and Stewart, 1968). It has been vari
ously proposed that ATA inhibits binding of messenger RNA (Grollman
and Stewart, 1968; Stewart et al., 1971; Marcus et al., 1970a; Hoerz
and McCarty, 1971) or tRNA (Crystal and Anderson, 1972; Zasloff and
Ochoa, 1973; Marcus et al., 1970b) to ribosomes; the dye also prevents
formation of the M1-GTP-methionyl-tRNA^{fMet} complex (Dettman and Stanley, Jr., 1973). Aurintricarboxylic acid may affect several reactions
involved in protein synthesis; indeed, triphenylmethane dyes prevent
interactions between many polynucleotides and proteins (Huang and
Grollman, 1972a; Liao et al., 1974a; Liao et al., 1975; Blumenthal
and Landers, 1973; Roberts and Colemen, 1971).

A functional effect of ATA on initiation of protein synthesis was
suggested by experimental observations with reticulocyte lysates (Stewart et al., 1971; Huang and Grollman, 1972a; Grollman and Huang, 1973
Agents that selectively affect chain initiation can be distinguished
by their delayed onset of inhibition. As illustrated in Fig. 2, a

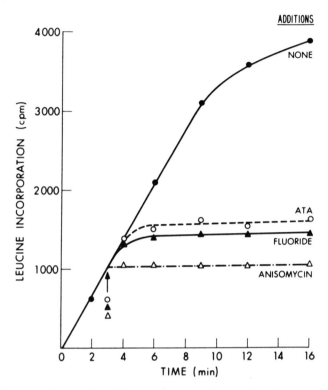

FIG. 2. Effect of inhibitors on globin synthesis in reticulocyte lysates. Four reaction mixtures, containing 10 mM tris HCl, pH 7.4; 75 mM KCl; 1 mM ATP; 0.2 mM GTP; 15 mM creatine phosphate; 2 mM MgCl$_2$; 1.7 μM [^{14}C]leucine, 240 Ci/mM; 6 mM 2-mercaptoethanol; a mixture of 19 amino acids (Lingrel and Borsook, 1963); 0.9 mg/ml creatine phosphokinase; and 0.1 ml lysate were incubated at 33° for the times indicated in a final volume of 0.5 ml. ATA (0.1 mM), NaF (25 mM), or anisomycin (0.25 mM), was present where indicated. Aliquots of 50 μl were removed from the reactions at the times indicated for determination of radioactivity. Data reprinted from Stewart et al. (1971) with permission of *Proceedings of the National Academy of Sciences.*

lag of several minutes occurs before inhibition by ATA is observed, which is comparable to the effect observed with sodium fluoride, another inhibitor of initiation. By contrast, anisomycin, an inhibitor of chain elongation, inhibits protein synthesis immediately after the drug is added to the reaction. Further evidence that ATA inhibits initiation is its ability to inhibit incorporation of formylmethionine from yeast formylmethionyl-tRNA^{fMet} into globin (Lodish et al., 1971).

Single ribosomes and ribosomal subunits formed in the presence of
ATA sediment more slowly than those formed in its absence. It was
suggested that this phenomenon indicates a conformational change induced by ATA or one of its components when bound to 40S and 80S particles (Huang and Grollman, 1972a). Alternatively, ATA may prevent
initiation factors or other proteins from binding to eukaryotic ribosomes or, if already bound, it may displace them (Ayuso-Parilla et
al., 1973). If this is the case, the observed decrease in sedimentation constant may reflect, in part, the lower molecular weight of
such particles.

Binding of ATA to reticulocyte polyribosomes, ribosomes, and ribosomal subunits has been studied by Huang and Grollman (1972a). When
a crude lysate is incubated with [^{3}H]ATA at 0° and the components separated on sucrose density gradients, ATA binds to polysomes, monosomes, and the smaller (40S) ribosomal subunit but not to the larger
(60S) subunit (Fig. 3A and B). Polyribosomes dissociate when incubated with components required for protein synthesis in the presence
of the dye; under these conditions, [^{3}H]ATA remains bound to single
ribosomes and the smaller ribosomal subunit (Fig. 3C and D). Binding
of ATA to ribosomes has also been determined by measuring retention
of the complex on Millipore membrane filters; saturation occurs when
the molar ratio of dye to ribosomes approaches 700. Binding of ATA
to ribosomes is independent of temperature (between 0 and 35°) and of
pH (over a range of 6-9) and is slightly inhibited by high concentrations of polyuridylic acid (Huang and Grollman, 1972a).

Divalent cations (magnesium, calcium, and manganese) stimulate
maximal binding of ATA to ribosomes at a concentration of 6 mM (Huang
and Grollman, 1972a). Spermine and spermidine are equally effective
in promoting binding of the dye as are higher concentrations (0.1 M)
of monovalent cations (potassium, ammonium, and sodium). Ethylenediaminetetraacetate also increases binding of ATA. These observations
may reflect alterations in net ribosomal charge or unfolding of ribosomes to expose additional binding sites.

ATA may bind to ribosomal protein, ribosomal RNA, or both; available evidence is inconclusive. Polyuridylic acid forms a complex

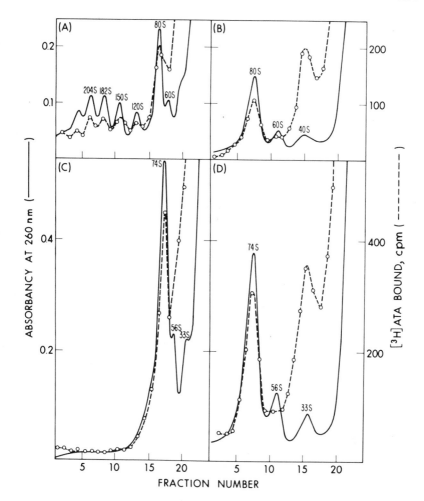

FIG. 3. Binding of [³H]ATA to polyribosomes and ribosomal subunits. ³H-Labeled ATA (25,300 cpm), at a final concentration of 0.1
mM, was added to the standard reaction mixture (0.125 μl) described
by Huang and Grollman (1972). Mixtures used in the reactions, shown
in (A) and (B), were chilled immediately, while those for (C) and (D)
were incubated at 35° for 4 min. All four samples were diluted to
1.0 ml with reticulocyte standard buffer, then subjected to sucrose
density gradient centrifugation. Gradients shown in (A) and (C) were
centrifuged for 2.5 hr and those in (B) and (D) for 7.8 hr. Fractions (1.4 ml) were collected directly into scintillation vials, and
the radioactivity determined. Data reprinted from Huang and Grollman
(1972a) with the permission of *Molecular Pharmacology*.

with proteins washed from ribosomes or 40S ribosomal subunit; this
interaction is effectively blocked by ATA (Roberts and Coleman, 1971).
Strong binding of ATA to bovine serum albumin has been reported (Lindenbaum and Schubert, 1956). ATA significantly alters the T_M of E.
$coli$ rRNA (Stewart-Blair and Grollman, unpublished).

Observations have been made on the purity of ATA and other dyes
that could affect interpretation of certain published results. Com-
mercial preparations of the dye are grossly impure and vary widely in
their ability to inhibit protein synthesis (Huang and Grollman, 1972a).
Thus, the observed effects of ATA may represent different activities
of one or more of the multiple components. Aurintricarboxylic acid
has been prepared in a chemically pure form (Huang and Grollman,
1972a); unfortunately, the purified dye proved to be unstable.

Other triphenylmethane dyes, related in structure to ATA, have
also been shown to inhibit globin synthesis in reticulocyte lysates
(Liao et al., 1975). One of these, pyrocatechol violet (2), has been
shown to specifically inhibit initiation (Huang and Grollman, 1973).
Ambiguities might be avoided in future experimental studies if pyrocatechol violet, which is available in purified form and is stable,
is used instead of ATA.

(2)

B. Pactamycin

Pactamycin (3) is a cytotoxic antibiotic isolated from $Streptomy$-
ces $pactum$ (Bhuyan et al., 1961), the structure of which has been
elucidated (Wiley et al., 1970).

(**3**)

Low concentrations of pactamycin inhibit amino acid incorporation in cell-free extracts prepared from reticulocytes (Colombo et al., 1966). Inhibition follows a brief lag period characteristic of inhibitors of initiation; polysomes break down to smaller units and eventually dissociate to form single ribosomes. Concurrently, completed α- and β-globin chains are released from the ribosome. Thus, pactamycin inhibits initiation and prevents formation of new polyribosomes while allowing translation of globin mRNA.

Polyribosome decay and release of nascent globin peptides are completed within 2 min after addition of pactamycin to reticulocyte lysates; these processes are not coordinated, suggesting premature release of incomplete polypeptide chains (Goldberg et al., 1973). At higher concentrations of drug, polyribosome structure is frozen, indicating that chain elongation is also affected.

Pactamycin binds to the smaller ribosomal subunit and the 80S ribosome but not to the larger subunit or polyribosomes (Goldberg et al., 1973). When single 80S ribosomes are derived from polyribosomes by treatment with ribonuclease, they do not bind pactamycin; when produced by dissociation with sodium fluoride, pactamycin is readily bound. The ability of a "dissociation factor" to convert 80S ribosomes into subunits is not affected by the antibiotic. Pactamycin does not prevent binding of initiation factors to 40S subunits, attachment of synthetic mRNA to the ribosome, or formation of the Ml-GTP-methionyl-tRNA^{fMet} complex (see Goldberg et al., 1973; Dettman and Stanley, 1973).

In the presence of pactamycin, the smaller ribosomal subunit is
either unable to join with the 60S subunit to form an active 80S ribosome or forms an inactive and unstable monosome structure. Methio-
nyl-tRNA^{fMet} is bound to free 40S particles that accumulate at the
expense of the 80S monosome (Goldberg et al., 1973).

Pactamycin does not prevent formation of the initiation complex,
but conversion of initiator methionyl-tRNA into a puromycin-reactive
form on ribosomes is blocked by the drug (Goldberg et al., 1973). In
reticulocyte extracts, methionyl-puromycin formation is inhibited if
pactamycin is present when methionyl-tRNA is bound. If the antibiotic
is added after the binding reaction is complete, the reaction with
puromycin is not significantly affected.

The accumulated data suggest that pactamycin binds preferentially
to free smaller ribosomal subunits and causes destabilization and dissociation of the initiation complex. At higher concentrations of
drug, chain elongation is also inhibited.

C. Harringtonine and Bruceantin

Although structurally unrelated, these two alkaloids will be considered together since their reported effects on protein synthesis
are similar. Harringtonine (4) is a cytotoxic alkaloid isolated from
seeds of *Cephalotaxus harringtonia* (Powell et al., 1969), which inhibits protein synthesis; several isomers, including isoharringtonine
and homoharringtonine, are obtained from the same source. Hydrolysis
of harringtonine yields cephalotaxine and an unsaturated fatty acid,

(4)

neither of which is inhibitory. Isoharringtonine and homoharringtonine are as active as the parent compound (Huang, 1975). Small
modifications in the unsaturated side chain abolish inhibitory activity. The structure of bruceantin (5) has been established by Kupchan et al. (1973).

(**5**)

Harringtonine and bruceantin display general characteristics of
inhibitors of chain initiation when tested in HeLa cells or rabbit
reticulocyte lysates. Thus, after the drugs are added, there is a
short delay before amino acid incorporation is affected; breakdown of
polyribosomes and release of completed globin chains occur during this
interval (Huang, 1975; Liao et al., 1974b). Harringtonine and bruceantin do not inhibit binding of synthetic polynucleotides or tRNA
to reticulocyte ribosomes, and chain elongation is unaffected by
these alkaloids even at high concentrations of drug. Unlike ATA,
both compounds inhibit protein synthesis in intact HeLa cells and
rabbit reticulocytes.

D. Sodium Fluoride

The effects of sodium fluoride on protein synthesis have been most
extensively studied in rabbit reticulocytes. Fluoride induces breakdown of polyribosomes to monosomes with concomitant release of nascent
globin peptide chains (Marks et al., 1963; Marks et al., 1965; Lin
et al., 1966). Removal of this inhibitor allows polyribosomes to reform and protein synthesis to be fully restored (Marks et al., 1965).

Lin et al. (1966) reported that sodium fluoride did not affect the
rate of amino acid incorporation into previously initiated peptide
chains, indicating a fluoride-sensitive reaction related to initiation
of new peptide chains. Kinetics of inhibition in reticulocyte lysates
(Fig. 2) support this conclusion (Stewart et al., 1971; Hoerz and
McCarty, 1971). Sodium fluoride may prevent formation of the initiation complex by blocking addition of 60S ribosomal subunits to messenger RNA (Hoerz and McCarty, 1969).

E. Edeine

Edeine, a polypeptide antibiotic (Hettinger and Craig, 1970) isolated from *Bacilus brevis,* contains two major components, edeine A
and edeine B. Edeine inhibits initiation of protein synthesis in
bacteria (Szer and Kurylo-Borowska, 1970); one detailed study of its
action on reticulocyte lysates has been published (Obrig et al., 1971).

Edeine inhibits binding of N-acylated phenylalanyl-tRNA to the P
site and of phenylalanyl-tRNA to the A site on the ribosome. Complexes formed by mixing deacylated tRNA, poly(U), and ribosomes or
by treating reticulocytes with sodium fluoride are resistant to the
action of the drug. Edeine prevents binding of phenylalanyl-tRNA,
methionyl-tRNA, or A_pU_pG to 40S ribosomal subunits and stimulates
binding of methionyl-tRNA to recombined ribosomal subunits.

It is not clear whether the primary effect of edeine is to prevent binding of mRNA or tRNA. The antibiotic blocks interactions of
tRNA with ribosomes in the absence of added mRNA. The available data
suggest that edeine interferes with formation of the initiation complex by acting at a site prior to the reaction inhibited by sodium
fluoride (see Sec. II,D).

F. Trichothecene Toxins and Related Compounds

This family of sesquiterpenoid fungal antibiotics (Bamburg and
Strong, 1971) includes 12,13-epoxytrichothecene compounds that are
toxic to eukaryotic cells. Ueno et al. (1968) first reported that
one of these toxins, nivalenol, inhibited protein synthesis. Fusare-

none X, a related toxin, was shown to disaggregate polyribosomes in
mouse fibroblasts (Medappa et al., 1971); nivalenol, verrucarrin A,
T-2 toxin, diacetoxyscirpenol (anguidine) (6) and HT-2 toxin have similar effects on HeLa cells (Cundliffe et al., 1974). Although structurally related, the individual toxins affect protein synthesis in
different ways. Trichodermin, trichodermol, and trichothecin block
chain termination and/or peptide bond formation (see Sec. VI), while
other trichothecene toxins inhibit chain initiation.

(**6**)

Breakdown of polyribosomes induced by nivalenol, T-2 toxin, and
verrucarrin is prevented by anisomycin (Cundliffe et al., 1974). This
observation is consistent with a primary effect on chain initiation;
supporting evidence for this conclusion would be desirable since the
same antibiotics were claimed by other investigators to show a primary
effect on peptide bond formation (Carrasco et al., 1973).

Diacetoxyscirpenol (anguidine), a cytotoxic antitumor agent, has
been studied in some detail (Liao et al., unpublished). Anguidine
blocks protein synthesis in HeLa cells, rabbit reticulocytes, and reticulocyte lysates. At concentrations of 1 µM, the drug induces dissociation of polyribosomes with concomitant release of completed globin
chains; at 1 mM, it prevents polyribosome breakdown. Anguidine does
not affect binding of polyuridylic acid or tRNA to ribosomes. Thus,
diacetoxyscirpenol may inhibit initiation at low concentrations of
drug and chain elongation at higher concentrations.

III. INHIBITORS OF BINDING OF AMINOACYL-tRNA

A. Tetracycline

The chemistry, biology, and earlier work on the effects of tetracycline (7) on protein synthesis is reviewed elsewhere (Laskin, 1967).
Although not as extensively studied as in prokaryotes, the action of
tetracycline and its analogs on eukaryotic protein synthesis has been
established. Franklin (1963, 1964) showed that this antibiotic inhibited protein synthesis in cell-free extracts prepared from rat liver
but did not affect charging of tRNA. Clark and Chang (1965) reported
that tetracycline inhibited polyuridylic-acid-directed EF-I-dependent
binding of phenylalanyl-tRNA to reticulocyte ribosomes. Puromycin-
induced release of peptides was not affected.

Tetracycline binds both to the 40S and 60S subunits of rat liver
ribosomes, predominantly to the smaller subunit (Day, 1966; Connamacher and Mandel, 1968). Part of this binding is nonspecific and reversible; the irreversible binding involves one molecule of tetracycline per ribosome.

In prokaryotes, the primary action of tetracycline on binding of
aminoacyl-tRNA seems clear although there is some evidence for effects
on peptide bond formation and termination; in eukaryotes, it has not
been definitely shown that processes other than tRNA binding are affected by the drug.

B. Chartreusin

The action of chartreusin on polypeptide synthesis has been studied in reticulocyte extracts using purified EF-I and EF-II and polyuridylic acid as mRNA (Gregg and Heintz, 1972). Chartreusin (8) was
the only one of a number of antibiotics screened that inhibited enzymatic and nonenzymatic binding of phenylalanyl-tRNA to ribosomes.
Relatively high levels of the antibiotic (1 mM) are required, but the
reaction appears to be specific. Chartreusin binds to ribosomes, inactivating them, but has no effect on formation of phenylalanylpuro-
mycin or diphenylalanine if added after phenylalanyl-tRNA; it was concluded by Gregg and Heintz (1972) that peptide bond formation and

translocation were unaffected by this drug. Chartreusin inhibits aminoacyl-tRNA: mRNA-dependent GTPase activity of EF-I, presumably, by preventing binding of tRNA or a tRNA-GTP-factor complex to ribosomes.

(**7**)

(**8**)

IV. INHIBITORS OF PEPTIDE BOND FORMATION

A. Anisomycin

Anisomycin, a pyrrolidine antibiotic isolated from cultures of various species of *Streptomyces,* is toxic to plants, certain protozoa, yeast, and animal cells (Lynch et al., 1954). The chemistry of anisomycin (**9**) and related compounds has been reported (Beereboom et al., 1965).

(**9**)

Anisomycin effectively inhibits protein synthesis in HeLa cells and intact rabbit reticulocytes at concentrations of 0.1 μM (Grollman 1967). In cell-free extracts prepared from rabbit reticulocytes, somewhat higher concentrations of anisomycin are required to achieve the same degree of inhibition.

Inhibition of protein synthesis by anisomycin is reversible (Fig. 4), provided that the antibiotic is rapidly removed from the cells

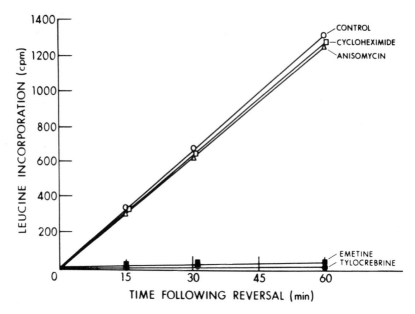

FIG. 4.   Reversal of inhibition of protein synthesis induced by
various inhibitors. HeLa cells (5 x 10^5 ml), suspended in complete
Eagle's medium containing 5% horse serum, were incubated for 5 min
in the presence of the following inhibitors: 1 μM tylocrebrine, 10
μM cycloheximide, 10 μM emetine, and 10 μM anisomycin. The control
contained no inhibitors. All cultures, including the control, were
washed three times with Earle's buffer at 37° and resuspended in fresh
leucine-depleted medium. One nmole of [^{14}C]leucine (200 mCi/mmole)
was added, and the rate of amino acid incorporation into acid-insol-
uble material was determined. Data reprinted from Grollman and Huang
(1973) with the permission of Federation Proceedings.

(Grollman, 1967). HeLa cells, exposed to anisomycin for 2, 30, 60
and 120 min, were washed and resuspended in fresh medium. Following
the reversal procedure, the rate of protein synthesis was 95, 83, 78,
and 69%, respectively, of that observed in uninhibited cultures subjected to similar manipulations.

Addition of anisomycin to growing cultures of HeLa cells or reticulocytes results in a decrease in the number of single ribosomes and
a concomitant increase in polyribosomes. When cells are exposed to
radioactive amino acids for several minutes, nascent peptides can be
detected in the polyribosome region; if anisomycin is present when
the amino acids are added, polyribosome-bound radioactivity is not

found. If the antibiotic is added after the amino acids, incorporation of radioactivity ceases but nascent peptides remain bound to the
polyribosomes (Fig. 5).

In the presence of anisomycin, puromycin fails to release nascent
peptide chains, suggesting that formation of peptidyl puromycin is
prevented (Fig. 5). This observation has been used to distinguish
anisomycin from other inhibitors of chain elongation (Grollman,
1968b; Grollman and Huang, 1973).

Using the two-step assay described by McKeehan and Hardesty (1969)
and the puromycin reaction of Monro and Marcker (1967), a primary effect of anisomycin on peptide bond formation has been demonstrated
(Grollman and Huang, 1973). Battaner and Vazquez (1971) reported that
anisomycin inhibits the "fragment reaction," potentially another measure of this effect. While providing useful confirmatory evidence,
studies using inhibitors with this model system have not always been
consistent with observations using polyribosomes or intact cells. It
should be noted that the fragment reaction (Monro et al., 1968) requires the presence of 33% methanol, which is known to significantly
affect ribosomal activity (Ballesta and Vazquez, 1972) and involves
two artificial substrates, CACCA-leu-Ac and puromycin.

B. Sparsomycin

Sparsomycin (10), a sulfur-containing antibiotic obtained from
Streptomyces sparsogenes, is toxic to eukaryotic cells (Owen et al.,
1962). Colombo et al. (1966) reported that sparsomycin inhibited
protein synthesis and prevented breakdown of polyribosomes in L cells;
subsequently, sparsomycin was shown to inhibit peptide bond formation
(Goldberg and Mitsugi, 1967a, 1967b; Goldberg et al., 1973).

(10)

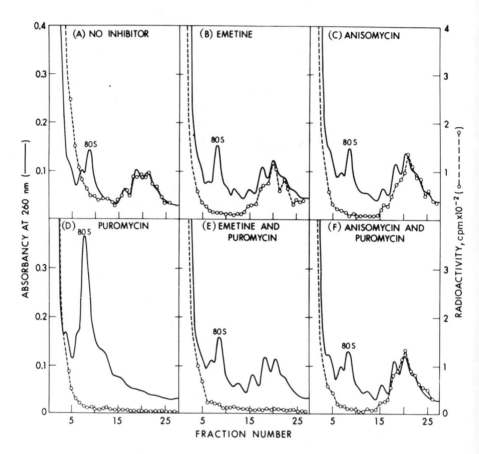

FIG. 5. Effect of puromycin on nascent peptide bound to polyribosomes in the presence of emetine and anisomycin. Six standard reaction mixtures (Huang and Grollman, 1972a), each containing 5 nmoles of [^{14}C]leucine (300 mCi/mmole), were incubated at 33° for 1 min, and inhibitors added at the following concentrations: emetine, 0.1 mM; anisomycin, 0.1 mM; and puromycin, 0.2 mM. Reaction A was chilled immediately; reactions B, C, D, E, and F were incubated 4 min longer. All reactions were terminated by addition of 0.8 ml reticulocyte standard buffer and then layered on 10-25% sucrose gradients prepared in the same buffer. Gradients were centrifuged at 27,000 rpm for 2.5 hr and the absorbancy at 260 nm was determined. Fractions (1.2 ml) were collected from the bottom of the tube for determination of radioactivity and sufficient trichloroacetic acid was added to each fraction to bring the final concentration to 5%. After heating at 95° for 15 min, samples were chilled in an ice water bath, precipitates were collected on Millipore filters, and the radioactivity was determined. Data reprinted from Grollman and Huang (1972a) by permission of *Federation Proceedings*.

The so-called "fragment reaction" in which a short N-substituted fragment of tRNA is transferred to puromycin or other acceptor, involves peptidyl transferase and occurs in the presence of the larger ribosomal subunit. This reaction, which requires certain artificial conditions (see Sect. IV,A), is sensitive to the effects of sparsomycin.

By analogy to more extensive studies with bacterial systems (Goldberg and Mitsugi, 1967b) it has been suggested that sparsomycin interferes with the "acceptor" recognition site in peptidyl transferase located on the larger ribosomal subunit. Interaction with the aminoacyl end of an incoming aminoacyl-tRNA or with the amino group of puromycin is blocked, fixing peptidyl-tRNA in the donor site.

The structural basis for the action of sparsomycin is not known but Goldberg et al. (1973) speculate that the pseudouridinelike moiety of sparsomycin, which is attached to a peptidelike grouping, may play a role in its biological activity. It is of interest that the loop of tRNA that contains pseudouridine has a complementary sequence on 5S RNA, which has been implicated in the function of peptidyl transferase (Erdmann et al., 1971).

C. Gougerotin, Amicetin, Blasticidin, and Other Aminosugar Nucleoside Antibiotics

A number of aminosugar nucleoside antibiotics are known (Suhadolnik, 1970), several of which have been found to inhibit protein synthesis. Puromycin, the best known member of this group, acts by virtue of its structural similarity to the aminoacyl-adenyl terminus of aminoacyl-tRNA. Puromycin acts as an acceptor for the growing peptide chain and is incorporated into peptides that are prematurely terminated and released from the ribosome; as a result, the drug produces breakdown of polyribosomes. The action of puromycin has been reviewed in detail (Pestka, 1971; Pestka et al., 1973) and need not be repeated here.

Several aminohexose pyrimidine nucleoside antibiotics, including blasticidin (11), amicetin (12), and gougerotin (13), inhibit protein synthesis in an essentially similar manner. Gougerotin inhibits protein synthesis in cell-free extracts obtained from mouse liver (Sino-hara and Sky-Peck, 1965) and rabbit reticulocytes (Clark and Chang, 1965; Casjens and Morris, 1965). Clark and Chang (1965) reported that gougerotin inhibited puromycin-induced release of nascent peptide chains from reticulocyte ribosomes and that the antibiotic prevented the release of peptides. In contrast to puromycin, polyribosome structure is maintained in the presence of gougerotin (Casjens and Morris, 1965).

(11)

(12)

(13)

Neth et al. (1970) have examined the effect of gougerotin on the "puromycin reaction" using CAACCA-fMet and puromycin as substrates with human tonsil ribosomes. High concentrations of this antibiotic partially inhibit the reaction. CAACCA-Phe accepts polypeptides from peptidyl-tRNA bound to human placental ribosomes; this reaction is inhibited by gougerotin (Pestka et al., 1972).

Blasticidin S and amicetin are structurally similar to gougerotin and appear to have analogous effects on protein synthesis in bacteria. These inhibitors have not been studied in sufficient detail in eukaryotes for firm conclusions to be drawn regarding their mode of action. Their effects on the puromycin reaction, as studied with human tonsil ribosomes (Neth et al., 1970), and peptidyl-tRNA transfer on placental ribosomes (Pestka et al., 1972) are similar to those described for gougerotin.

A possible ribosomal site for aminonucleoside antibiotics has been studied by determining their ability to displace bound anisomycin. Gougerotin, blasticidin, and amicetin displace labeled anisomycin from tonsil ribosomes; however, as noted in Sect. I, such experiments do not necessarily prove similarities in mode of action or even that the drugs bind to the same site on the ribosome.

D. Others

L-Tenuazonic acid, an antibiotic obtained from culture filtrates of *Aspergillus* and other fungi, is cytotoxic to animal cells and inhibits protein synthesis in cell-free preparations from eukaryotes (Shigeura and Gordon, 1963). More recently, tenuazonic acid was reported to affect the "fragment reaction" using ribosomes obtained from human tonsils or pig liver (Carrasco and Vazquez, 1973a). The drug inhibits binding of CACCA-Leu to the acceptor site of human ribosomes. Tenuazonic acid has lesser inhibitory effects on plant ribosomes and practically no effect on *Saccharomyces cerevesiae* ribosomes (Vazquez, 1973).

Several trichothecene mycotoxins were included in Sect. II,F as probable inhibitors of chain initiation. At similar concentrations, other members of this group, trichodermin, trichothecin, and trichodermol, appear to block chain elongation. This mode of action is suggested by their stabilizing action on polyribosome structure in HeLa cells (Cundliffe et al., 1974) and inhibition of N-acetylphenylalanylpuromycin formation in the "fragment reaction" (Carrasco et al., 1973).

As noted previously, conclusions drawn from such studies must be quali-
fied, particularly since the "fragment reaction" is inhibited by ver-
rucurrin A and other compounds that appear to primarily inhibit initiation in intact cells. Trichodermin has also been shown to inhibit
chain termination; however, such an effect is not inconsistent with a
primary or secondary effect on peptide bond formation (see Sect. VI).

V. INHIBITORS OF TRANSLOCATION

A. Emetine

     Emetine ($\underline{14}$), a major member of the family of ipecac alkaloids
(Openshaw, 1970), is used therapeutically to treat amebiasis. Its
biological properties and toxic effects on various eukaryotes have
been reviewed (Grollman and Jarkovsky, 1974).

($\underline{14}$)

     When emetine is added to a suspension of HeLa cells or reticulo-
cytes, polyribosomes increase and single ribosomes decrease (Grollman,
1968b). Protein synthesis is inhibited irreversibly (Fig. 4). In the
presence of emetine, nascent peptides remain attached to polyribosomes;

if puromycin is then added, nascent peptides are released but polyribosome structure remains intact (Fig. 5).

When reticulocytes are incubated in the presence of 0.01 M sodium
fluoride, polyribosomes dissociate. When such fluoride-treated cells
are washed and resuspended in a medium that supports synthesis of hemoglobin, polyribosomes reform. If this experiment is conducted in
the presence of emetine, dimeric, trimeric and tetrameric ribosomes
form, but complete polyribosomes do not reassemble (Grollman and
Huang, 1973).

Specific protein factors EF-I, EF-II, and peptidyl transferase
are required for binding of aminoacyl-tRNA, translocation of peptidyl-tRNA, and formation of peptide bonds, respectively. Puromycin
reacts with peptidyl-tRNA located at the P site on the ribosome, causing premature termination and release of peptide chains. This reaction, which does not require supernatant factors or GTP, has been used
to study the action of inhibitors of chain elongation. Established
inhibitors of peptide bond formation, such as anisomycin (Grollman,
1967; Grollman and Huang, 1973), inhibit puromycin-induced release of
nascent peptides. Emetine does not prevent this reaction nor does it
affect the "fragment reaction," indicating that the drug does not affect peptide bond formation. Activity of the binding enzyme EF-I is
also unaffected by emetine (Grollman and Huang, 1973).

The effect of emetine on translocation of peptidyl-tRNA has been
determined by a two-step assay, originally described by McKeehan and
Hardesty (1969). This assay is based on the assumption that peptidyl-
tRNA will be located at the peptidyl site following incubation of ribosomes bearing nascent chains with EF-II and GTP. Radioactive aminoacyl-tRNA is then bound to the aminoacyl site on the ribosomes in
the presence of sodium fluoride, an agent that prevents further initiation of globin peptides. Under these experimental conditions, incorporation of labeled amino acids into peptide is blocked by inhibitors of peptide bond formation, but not by inhibitors of translocation.
Emetine has no effect if added after translocation; however, if the
drug is introduced prior to addition of EF-II and GTP, incorporation
of amino acids is inhibited by approximately 50% (Grollman and Huang,

1973). Assuming that nascent peptide is equally distributed between the aminoacyl and peptidyl sites on the ribosome, such partial inhibition would be anticipated by an agent that blocks some phase of translocation.

Puromycin releases 90% of nascent peptide chains formed on polyribosomes in intact cells, even in the presence of emetine. If translocation of peptide is inhibited by the drug, some peptidyl-tRNA should accumulate in the aminoacyl site, and thus be able to react with puromycin. Under these conditions, it would appear that emetine chiefly affects an aspect of translocation that involves movement of mRNA along the ribosome.

B. Cycloheximide and Related Glutarimide Antibiotics

The biological properties of cycloheximide (15) are shared by a number of other glutarimide antibiotics, including acetoxycycloheximide, streptimidone, and streptovitacin. The chemistry of this family of compounds is discussed in a comprehensive article by Johnson (1971); the biochemical pharmacology and effects on eukaryotes have been reviewed by Sisler and Siegel (1967).

(15)

Cycloheximide is toxic to a wide variety of eukaryotes, including fungi, higher plants, and mammals, but is inactive against prokaryotes. Its mechanism of action has been extensively studied in yeast. The inhibitory effects of the drug (in tissue culture) are readily reversed by washing the cells and resuspending them in fresh media (Fig. 4).

Yeast grown in the presence of cycloheximide usually develops resistance to the antibiotic. A variety of genes confer resistance; in addition, recessive modifier genes have been found (Wilkie and Lee, 1965; Caskey and Beaudet, 1972). Resistance associated with the 60S ribosomal subunit (Rao and Grollman, 1967) may reflect alterations in ribosomal proteins, as in streptomycin-resistant strains of *E. coli,* or changes in ribosomal conformation that decrease the binding affinity for cycloheximide.

Lin et al. (1966) reported that cycloheximide inhibited both peptide chain initiation and elongation. Later, Obrig et al. (1971) showed that the drug blocks at or beyond the step in initiation that is sensitive to sodium fluoride. Inhibition of chain elongation, which is observed at high concentrations of cycloheximide (Obrig et al., 1971), is manifested by effects similar to those described for emetine (Sect. V,A). Selective effects of cycloheximide on translocation were originally described by McKeehan and Hardesty (1969). Baliga et al. (1969) suggested that cyloheximide acted on EF-II, based, in part, on the protective effect of sulfhydryl compounds.

C. Tylocrebrine

The phenanthroidolizidine alkaloids tylophorine and tylocrebrine and the phenanthroquinolizidine alkaloid cryptopleurine are vesicants isolated from *Tylophora crebiflora* and *Cryptocarya pleurosperma.* The structure of tylocrebrine (16) was established by Gellert et al. (1962).

Donaldson et al. (1968) reported that the phenanthrene alkaloids affected protein synthesis in animal cells; detailed studies of the mode of action of tylocrebrine were subsequently reported by Huang and Grollman (1972b). The effects of tylocrebrine on protein synthe-

sis generally resemble those of emetine and cycloheximide; however,
certain differences have been observed (Huang and Grollman, 1972b).
Cycloheximide reversibly inhibits protein synthesis while the action
of tylocrebrine is irreversible (Fig. 2). Under certain conditions,
cycloheximide affects chain initiation; tylocrebrine does not appear
to have this effect. Cycloheximide inhibits binding of deacylated
tRNA to ribosomes, while tylocrebrine markedly stimulates this reaction. Cycloheximide inhibits polyphenylalanine synthesis on reticulocyte ribosomes, while tylocrebrine does not affect this process.

(<u>16</u>)

Ribosomes and EF-II require sulfhydryl compounds for maximal activity. The presence of mercaptoethanol and other sulfhydryl compounds diminishes the inhibitory effects of tylocrebrine (Huang and
Grollman, 1972b). Sulfhydryl compounds have also been reported to
protect against inhibitory effects of cycloheximide in cell-free systems prepared from rat liver (Baliga et al., 1969). The exact mechanism of this effect remains obscure.

D. Fusidic Acid

Fusidic acid (<u>17</u>) is the best-known member of a group of steroidal
antibiotics, some of which are utilized as antifungal agents. These
compounds have been shown to inhibit protein synthesis in prokaryotes
by interfering with the ribosome-dependent activity of G factor (hydrolysis of GTP to GDP and inorganic phosphate) and translocation of
peptidyl-tRNA from the A site to the P site on the ribosome (Tanaka
et al., 1969a; Bodley et al., 1970a, 1970b). However, evidence suggests that the primary effect of fusidic acid on bacterial protein

synthesis results, not from blocking translocation, but by inhibition
of aminoacyl-tRNA binding to the A site (Miller, 1972; Richman and
Bodley, 1972; Cabrer et al., 1972).

(17)

In eukaryotes, fusidic acid interferes with the GTPase activity
of EF-II and ribosomes (Malkin and Lipmann, 1969; Tanaka et al.,
1969b, 1970). Bodley and Lin (1970) demonstrated formation of an
EF-II ribosome-GDP-fusidic acid complex. Fusidic acid does not significantly affect the reaction of puromycin with polyphenylalanyl-
tRNA or acetyl phenylalanyl-tRNA, which are presumably attached to
the P site on the ribosome. However, the puromycin reaction, enhanced
by GTP and EF-II, was inhibited by fusidic acid (Tanaka et al., 1970).
It was suggested that fusidic acid selectively inhibits translocation
of peptidyl-tRNA from the A site to the P site on reticulocyte ribosomes (Tanaka et al., 1970).

Fusidic acid has been reported to inhibit binding of aminoacyl-
tRNA to ribosomes as well as translocation. Complexing of ribosomes
with EF-II, GTP, and fusidic acid prevented or inhibited EF-I-dependent and -independent binding of aminoacyl-tRNA (Richter, 1973; Modolell and Vazquez, 1973; Carrasco and Vazquez, 1973b).

VI. INHIBITORS OF CHAIN TERMINATION

It is difficult to distinguish inhibitors of chain termination from agents that primarily affect chain elongation. As noted in Sect. IV, compounds that block peptide bond formation also affect termination. The only inhibitor that seems to affect termination with minimal effects on elongation is the trichothecene toxin, trichodermin.

Stafford and McLaughlin (1973) used a temperature-sensitive mutant of *Saccharomyces cerevesiae* that was defective in initiation at the restrictive temperature (36°) to distinguish between inhibitors of termination and elongation. In their experimental system, a culture of spheroplasts prepared from this mutant is shifted to the permissive temperature of 23° after being held at 36° for 5 min. Under these conditions, initiation is a rate-limiting step. Inhibitors of termination should prevent normal recycling of ribosomes, increase the rate of polyribosomal reformation, and (in the presence of [^{14}C]amino acids) increase the specific activity of polyribosomes. Opposite effects would be expected in the presence of an inhibitor of chain elongation. Using this assay, Stafford and McLaughlin (1973) showed that trichodermin acts as an inhibitor of termination; it is the only compound thus far reported to selectively effect this process. Hansen and Vaughn (1973) reached similar conclusions.

VII. TOXINS

A. Diphtheria Toxin

Diphtheria toxin is a protein of 63,000 molecular weight, elaborated by a lysogenic strain of *Corynebacterium diphtheriae*. Along with puromycin, diphtheria toxin is one of the few inhibitors of protein synthesis whose biochemical action on eukaryotes can be described with some precision (Pappenheimer, Jr. and Gill, 1973). As initially shown by Honjo et al. (1968) this toxin acts enzymatically to catalyze the following enzymatic reaction:

$$\text{EF-II} + \text{NAD}^+ \rightleftarrows \text{ADPR-EF-II} + \text{nicotinamide} + \text{H}^+$$

Intact cells contain free and ribosomal-bound EF-II; the latter is not susceptible to inactivation by toxin. Since free EF-II is in excess, the lag observed before protein synthesis is inhibited by diphtheria toxin is thought to represent the time required for depletion of this factor.

Enzymatic activity of diphtheria toxin is observed only after treatment with proteolytic enzymes. "Nicking" of toxin by treatment with low concentrations of trypsin followed by treatment with sulfhydryl agents produces the enzymatically active "fragment A" (molecular weight 24,000). Fragment A is presumably liberated inside the cells; the remaining "fragment B" is required for binding of toxin to the cell membrane.

The current view of the action of diphtheria toxin may be summarized as follows: Toxin binds to the outside of the cell membrane, ultimately penetrating to the interior of the cell. Fragment A is liberated in the cytoplasm, then reacts with free EF-II to form ADPR-EF-II and nicotinamide. This covalent complex is unable to bind to ribosomes or to support protein synthesis and does not catalyze hydrolysis of GTP in the presence of ribosomes (Raeburn et al., 1968). As ribosomes dissociate during protein synthesis, bound EF-II is released and, subsequently, inactivated by toxin. When the supply of EF-II becomes rate limiting, protein synthesis decreases.

The functional effect of diphtheria toxin is on translocation. Guanosine triphosphate binds to ADP-ribosylated EF-II; this complex binds to ribosomes and GDP is formed (Bermek and Matthaei, 1971). It is, therefore, not GTP hydrolysis, but coupling to translocation, that is inhibited by the toxin.

B. Ricin

The biological effects of ricin,[*] a toxic protein obtained from the castor bean *Ricinus communis,* have been investigated for more than 80 years (see Funatsu, 1972; Grunfeld, 1974). This toxin has

---

[*]Abrin, a closely related toxic protein isolated from *Abrus precatorius* appears to have similar effects on protein synthesis.

been shown to act as an inhibitor of protein synthesis in eukaryotic
cells (Lin et al., 1971; Grunfeld, 1974; Grollman et al., 1974). Ricin (molecular weight 60,000) is composed of two peptide chains; one
chain binds specifically to cell surfaces, the other is responsible
for inhibiting ribosomal protein synthesis (Olsnes and Pihl, 1973).

In HeLa cells, protein synthesis is inhibited by ricin only after
an extended lag period (Fig. 6). This delay may reflect the time required for uptake of the toxin or for some intracellular modification
of the toxin. The inhibitory effects of ricin are associated with a
breakdown of polyribosomes, similar to those observed with inhibitors
of chain initiation. Inhibition of protein synthesis in HeLa cells
is blocked if D-galactose or galactose-containing saccharides are

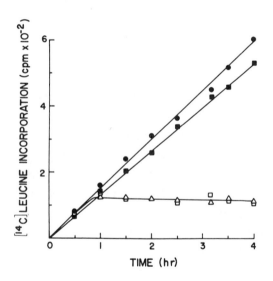

FIG. 6. Effect of saccharides on inhibition of protein synthesis by ricin. Purified ricin toxin was incubated for 5 min at room
temperature in the presence or absence of saccharides. The mixture
was added to a suspension of HeLa cells (4 x 10^5/ml) and incubated at
37°. Protein synthesis was determined by measuring the incorporation
of [^{14}C]leucine into hot trichloroacetic-acid-insoluble material.
( ) 1 µg/ml ricin, no saccharide; (△) 1 µg/ml ricin, 10 mM cello-
biose; (●) no ricin, no saccharide; (■) 1 µg/ml ricin, 10 mM lactose. Data reprinted from Grunfeld (1974).

added to the medium.   If the cells are exposed to ricin for several
minutes, the effect cannot be reversed by adding saccharides.

In reticulocyte lysates, a short lag period is observed, consistent with an effect on initiation.   However, polyribosome structure
is preserved, an effect that more closely resembles that produced by
inhibitors of chain elongation.   The inhibitory effects of ricin on
lysates are not blocked by D-galactose.

Lin et al. (1972) proposed that ribonuclease activity, which is
increased in the livers of animals treated with ricin, is responsible
for breakdown of polyribosomes; however, no such increase was detected in ricin-treated HeLa cells.   Inhibitors of peptide chain elongation (emetine and anisomycin) prevented ricin-induced breakdown of
polyribosomes, suggesting that dissociation involves "run-off" and
not destruction of messenger RNA (Grunfeld, 1974).

As noted above, nascent peptide remains attached to polyribosomes
in ricin-treated lysates; such peptide is released by addition of puromycin, indicating that peptide bonds can be formed in the presence
of the toxin (Grunfeld, 1974).   This observation is consistent with
the results of Montanaro et al. (1973).   Ricin-treated polyribosomes
dissociate in the presence of puromycin; an effect that distinguishes
ricin from other inhibitors that affect some aspect of translocation
(see Sect. V).

Treatment with mercaptoethanol increases the inhibitory activity
of ricin more than tenfold (Olsnes and Pihl, 1972).   Ricin does not
prevent charging of tRNA nor does it catalyze incorporation of ADP
ribose into EF-II, as does diphtheria toxin.   Montanaro et al. (1973),
using partially purified ricin, found that the toxin inhibits polyuridylic-acid-directed synthesis of polyphenylalanine.   Increasing
the concentration of EF-I did not influence this reaction, while an
increase in EF-II resulted in a slight decrease of inhibition.   Ricin
did not inhibit binding of labeled phenylalanine to previously trans-
located ribosomes, nor did it inhibit formation of peptidyl puromycin.

Ricin blocks ribosome-dependent hydrolysis of GTP by EF-II.   This
inhibitory effect on GTPase can be demonstrated by adding ricin to
the reaction mixture or by preincubating ribosomes in the presence of

the toxin. Ricin does not inhibit the GTPase activity of isolated
EF-II or ribosomes and has no influence on binding of GTP to EF-II
(Montanaro et al., 1973).

When treated with ricin or derived from ricin-treated ribosomes,
60S ribosomal subunits do not support polyuridylic-acid-directed polyphenylalanine synthesis, while 40S subunits are unaffected by similar
treatment. Cofactors are not required for this inactivation (Montanaro et al., 1973).

Although the effects of ricin on protein synthesis have been confirmed by several laboratories (Lin et al., 1971; Olsnes and Pihl,
1973; Montanaro et al., 1973; Grunfeld, 1974; Grollman et al., 1974),
its precise mode of action cannot be specified at present. It is
claimed that 10 toxin molecules per cell (each containing approximately 1 million ribosomes) are lethal to a culture of HeLa cells
(Olsnes et al., 1974), suggesting that a catalytic mechanism is involved. Thus, ricin, like diphtheria toxin, may act as an enzyme by
inactivating some ribosomal component involved in protein synthesis.

C. Pederine

Pederine is a poisonous substance extracted from the insect *Pae-
deris fuscies*. The chemical structure of the toxin (18) was determined by Cardani et al. (1965). Brega et al. (1968) and Perani et
al. (1968) described the inhibitory action of pederine on protein synthesis in various eukaryotes; Jacobs-Lorena et al. (1971) and Carrasco and Vazquez (1972) compared the effects of pederine to those of
other inhibitors of protein synthesis.

(18)

Pederine binds to the ribosome, irreversibly inhibiting initiation of protein synthesis at low concentrations of toxin, and inhibiting chain elongation at high concentrations (Jacobs-Lorena et al., 1971). In reticulocyte lysates, pederine inhibits the reaction of yeast-N-formylmethionyl-tRNA^{fMet} with puromycin, a model for chain initiation (Housman et al., 1970). Inhibition of elongation is evidenced by failure of polyribosomes to disaggregate in the presence of 20 ng/ml pederine. Nascent peptide remains bound to polyribosome in the presence of pederine; only 50% of the peptide is released by puromycin. This result would be anticipated if the toxin blocked translocation of peptidyl-tRNA from the A to the P site on the ribosome. Such an action may account for the effect with the model N-formylmethionylpuromycin reaction since translocation (or entry) of initiator tRNA may be inhibited by the toxin. Using another model system, Carrasco and Vazquez (1972) also concluded that pederine inhibits translocation.

REFERENCES

Ayuso-Parilla, M., C. A. Hirsch, and E. C. Henshaw. 1973. Release of the nonribosomal proteins from the mammalian native 40 S ribosomal subunit by aurintricarboxylic acid. *J. Biol. Chem. 248:* 4394-4399.

Baliga, B. S., A. W. Pronczuk, and H. N. Munro. 1969. Mechanism of cycloheximide inhibition of protein synthesis in a cell-free system prepared from rat liver. *J. Biol. Chem. 244:*4480-4489.

Ballesta, J. P., and D. Vazquez. 1972. Elongation factor T-dependent hydrolysis of guanosine triphosphate resistant to thiostrepton. *Proc. Nat. Acad. Sci. US 69:*3058-3062.

Bamburg, J. R., and F. M. Strong. 1971. 12,13-Epoxytrichothecenes. in: *Microbial Toxins* (S. Kadis, A. Geigler, and S. J. Ajl, eds.). New York: Academic Press, pp. 207-292.

Battaner, E., and D. Vazquez. 1971. Inhibitors of protein synthesis by ribosomes of the 80S type. *Biochim. Biophys. Acta 254:*316-330.

Beereboom, J. J., K. Butler, F. C. Pennington, and I. A. Solomons. 1965. Anisomycin, I. Determination of the structure and stereochemistry of anisomycin. *J. Org. Chem. 30:*2334-2342.

Bermek, E., and H. Matthaei. 1971. Interactions between human translocation factor, guanosine triphosphate and ribosomes. *Biochemistry 10:*4906-4912.

Bhuyan, B. K., A. Dietz, and C. G. Smith. 1961. Pactamycin, a new antitumor antibiotic I. Discovery and biological properties. in: *Antimicrobial Agents and Chemotherapy* (M. Finland and G. M. Savage, eds.). Ann Arbor: Am. Soc. Microbiology. pp. 184-190.

Blumenthal, T., and T. A. Landers. 1973. The inhibition of nucleic acid-binding proteins by aurintricarboxylic acid. *Biochem. Biophys. Res. Commun. 55:*680-688.

Bodley, J. W., F. J. Zieve, and L. Lin. 1970a. Studies on translocation IV. The hydrolysis of a single round of guanosine triphosphate in the presence of fusidic acid. *J. Biol. Chem. 245:*5662-5667.

Bodley, J. W., F. J. Zieve, L. Lin, and S. T. Zieve. 1970b. Studies on translocation, III. Conditions necessary for the formation and detection of a stable-ribosome-G factor-guanosine diphosphate complex on the complex of fusidic acid. *J. Biol. Chem. 245:*5656-5661.

Bodley, J. W., and W. Lin. 1970. Studies on translocation, V: Fusidic acid stabilization of a eukaryotic ribosome-translocation factor-GDP complex. *FEBS Lett. 11:*153-156.

Brega, A., A. Falaschi, L. De Carli, and M. Pavan. 1968. Studies on the mechanism of action of pederine. *J. Cell Biol. 36:*485-496.

Bücher, T., and H. Sies. 1969. *Inhibitors: Tools in Cell Research.* New York: Springer-Verlag.

Cabrer, B., D. Vazquez, and J. Modollel. 1972. Inhibition by elongation factor EF-G of aminoacyl-tRNA binding to ribosomes. *Proc. Nat. Acad. Sci. US 69:*733-736.

Cardani, C., D. Ghiringhelli, R. Mondelli, and A. Quilico. 1965. The structure of pederin. *Tetrahedron Lett. 29:*2537-2545.

Carrasco, L., and D. Vazquez. 1972. Survey of inhibitors in different steps of protein synthesis by mammalian ribosomes. *J. Antibiotics 25:*732-737.

Carrasco, L., and D. Vazquez. 1973a. Differences in eukaryotic ribosomes detected by the selective action of an antibiotic. *Biochim. Biophys. Acta 319:*209-215.

Carrasco, L., and D. Vazquez. 1973b. Ribosomal sites involved in binding of aminoacyl-tRNA and TF-2, mode of action of fusidic acid. *FEBS Lett. 32:*152-156.

Carrasco, L., M. Barbacid, and D. Vazquez. 1973. The trichodermin group of antibiotics, inhibitors of peptide bond formation by eukaryotic ribosomes. *Biochim. Biophys. Acta 312:*368-376.

Casjens, S. R., and A. J. Morris. 1965. The selective inhibition of protein assembly by gougerotin. *Biochim. Biophys. Acta 108:*677-686.

Caskey, C. T., and A. L. Beaudet. 1972. Antibiotic inhibitors of peptide chain termination. in: *Molecular Mechanisms of Antibiotic*

Action on Protein Biosynthesis (E. Munoz, F. Garcia-Ferrendiz, and D. Vazquez, eds.). New York: American Elsevier, pp. 326-336.

Clark, J. M., Jr., and A. Y. Chang. 1965. Inhibitors of the transfer of amino acids from aminoacyl soluble ribonucleic acid to proteins. *J. Biol. Chem. 240:*4734-4739.

Colombo, B., L. Felicetti, and C. Baglioni. 1966. Inhibition of protein synthesis in reticulocytes by antibiotics. I. Effects on polysomes. *Biochim. Biophys. Acta 119:*109-118.

Connamacher, R. H., and H. G. Mandel. 1968. Studies on the intracellular localization of tetracycline in bacteria. *Biochim. Biophys. Acta 166:*475-486.

Crystal, R. G., and W. F. Anderson. 1972. Initiation of hemoglobin synthesis: comparison of model reactions that use artificial templates with those using natural messenger RNA. *Proc. Nat. Acad. Sci.* US *69:*706-711.

Cundliffe, E., M. Cannon, and J. Davies. 1974. Mechanism of inhibition of eukaryotic protein synthesis by trichothecene fungal toxins. *Proc. Nat. Acad. Sci.* US *71:*30-34.

Day, L. E. 1966. Tetracycline inhibition of cell-free protein synthesis. I. Binding of tetracycline to components of the system. *J. Bacteriology 91:*1917-1923.

Dettman, G. L., and W. M. Stanley, Jr. 1973. The ternary complex of initiation factor IF-I, Met-tRNA, and GTP, an aurintricarboxylate-sensitive intermediate in the initiation of eukaryotic protein synthesis. *Biochim. Biophys. Acta 299:*142-147.

Donaldson, G. R., M. R. Atkinson, and A. W. Murry. 1968. Inhibition of protein synthesis in Ehrlich ascites-tumor cells by the phenanthrene alkaloids, tylophorine, tylocrebrine, and cryptopleucine. *Biochem. Biophys. Res. Commun. 31:*104-109.

Erdmann, V. A., S. Fahnestock, K. Higo, and M. Nomura. 1971. Role of 5S RNA in the functions of 50S ribosomal subunits. *Proc. Nat. Acad. Sci.* US *68:*2932-2936.

Franklin, T. J. 1963. The inhibition of incorporation of leucine into protein of cell-free systems from rat liver and *Escherichia coli* by chlorotetracycline. *Biochem. J. 87:*449-453.

Franklin, T. J. 1964. The effect of chlorotetracycline on the transfer of leucine and "transfer" ribonucleic acid to rat liver ribosomes *in vitro*. *Biochem. J. 90:*624-628.

Funatsu, M. 1972. The structure and toxic function of ricin. in: *Proteins: Structure and Function* Vol. II. (M. Funatsu, K. Hiromi, K. Imahori, T. Murachi, and K. Narita, eds.). New York: Wiley. pp. 103-139.

Gellert, E., T. R. Govindachari, M. V. Lakshmikantham, I. S. Ragade, R. Rudzats, and N. Viswanathan. 1962. The alkaloids of tylophora crebriflora: structure and synthesis of tylocrebrine, a new phenanthroindolizidine alkaloid. *J. Chem. Soc.:*1008-1014.

Goldberg, I. H., and K. Mitsugi. 1967a. Sparsomycin inhibition of polypeptide synthesis promoted by synthetic and natural polynucleotides. *Biochemistry* 6:372-382.

Goldberg, I. H., and K. Mitsugi. 1967b. Inhibition by sparsomycin and other antibiotics of the puromycin-induced release of polypeptides from ribosomes. *Biochemistry* 6:383-390.

Goldberg, I. H., M. L. Stewart, M. Ayuso, and L. S. Kappen. 1973. On the mechanisms of inhibition of polypeptide synthesis by the antibiotics sparsomycin and pactamycin. *Fed. Proc.* 32:1688-1697.

Gregg, R. E., and R. L. Heintz. 1972. The inhibition of eukaryotic aminoacyl transferase I by chartreusin. *Arch. Biochem. Biophys.* 152:451-456.

Grollman, A. P. 1967. Inhibitors of protein synthesis. II. Mode of action of anisomycin. *J. Biol. Chem.* 242:3226-3233.

Grollman, A. P. 1968a. Inhibition of messenger ribonucleic acid attachment to ribosomes. II. Proposed mechanism for the design of novel antiviral agents. in: *Antimicrobial Agents and Chemotherapy* (G. L. Hobby, ed.). Bethesda: Am. Soc. Microbiol., pp. 36-40.

Grollman, A. P. 1968b. Inhibitors of protein biosynthesis. V. Effects of emetine on protein and nucleic acid biosynthesis in HeLa cells. *J. Biol. Chem.* 243:4089-4094.

Grollman, A. P. 1975. Cytotoxic inhibitors of protein synthesis. in: *Handbook of Experimental Pharmacology* Vol. 38, Chap. 57 (P. Sartorelli, ed). New York: Springer-Verlag.

Grollman, A. P., C. Grunfeld, C. F. Brewer, and D. Marcus. 1974. Molecular pharmacology of plant lectins: studies on ricin and concanavalin A. *Cancer Chemotherapy Rep.* 58:491-501.

Grollman, A. P., and M.-T. Huang. 1973. Inhibitors of protein synthesis in eukaryotes: tools in cell research. *Fed. Proc.* 32: 1673-1678.

Grollman, A. P., and Z. Jarkovsky. 1974. Emetine and related alkaloids. in: *Antibiotics: Mode of Action* Vol. III. (J. W. Corcoran and F. E. Hahn, eds.). New York: Springer-Verlag. pp. 424-435.

Grollman, A. P., and M. L. Stewart. 1968. Inhibition of the attachment of messenger ribonucleic acid to ribosomes. *Proc. Nat. Acad. Sci.* US 61:719-725.

Grunfeld, C. 1974. Drug receptor interactions: effects of ricin on macromolecular synthesis, Ph.D. dissertation. New York: Albert Einstein College of Medicine, Yeshiva University.

Hansen, B. S., and M. H. Vaughn, Jr. 1973. Effects of a peptide synthesis termination inhibitor, trichodermin, on protein synthesis and polyribosomes in intact human cells. *Fed. Proc.* 32:494 (abstract).

Haselkorn, R., and L. B. Rothman-Denes. 1973. Protein biosynthesis. *Ann. Rev. Biochem. 42*:397-438.

Hettinger, T. P., and L. C. Craig. 1970. Edeine. IV. Structures of the antibiotic peptides edeine A and B. *Biochemistry 9*:1224-1232.

Hoerz, W., and K. S. McCarty. 1969. Evidence for a proposed initiation complex for protein synthesis in reticulocyte polyribosomes profiles. *Proc. Nat. Acad. Sci. US 63*:1206-1213.

Hoerz, W., and K. S. McCarty. 1971. Initiation of protein synthesis in a rabbit reticulocyte lysate system. *Biochim. Biophys. Acta 228*:526-535.

Honjo, T., Y. Nishizuka, O. Hayaishi, and I. Kato. 1968. Diphtheria toxin-dependent adenosine diphosphate ribosylation of aminoacyl transferase II and inhibition of protein synthesis. *J. Biol. Chem. 243*:3553-3555.

Housman, D., M. Jacobs-Lorena, U. L. Rajbhandary, and H. Lodish. 1970. Initiation of haemoglobin synthesis by methionyl-tRNA. *Nature 227*:913-918.

Huang, M.-T., and A. P. Grollman. 1972a. Effects of aurintricarboxylic acid on ribosomes and the biosynthesis of globin in rabbit reticulocytes. *Mol. Pharm. 8*:111-127.

Huang, M.-T., and A. P. Grollman. 1972c. Harringtonine: a novel inhibitor of initiation of protein synthesis. *5th Intern. Congr. Pharm.*

Huang, M.-T., and A. P. Grollman. 1973. Pyrocatechol violet: an inhibitor of initiation of protein synthesis. *Biochem. Biophys. Res. Commun. 53*:1049-1059.

Huang, M.-T. 1975. Harringtonine, an inhibitor of initiation of protein biosynthesis. *Mol. Pharm. 11*:511-519.

Jacobs-Lorena, M., A. Brega, and C. Baglioni. 1971. Inhibition of protein synthesis in reticulocytes by antibiotics. V. Mechanism of action of pederine, an inhibitor of initiation and elongation. *Biochim. Biophys. Acta 240*:263-272.

Johnson, F. 1971. The chemistry of glutarimide antibiotics. in: *Progress in the Chemistry of Organic Natural Products* Vol. 29 (W. Herz, H. Crusebach, and G. W. Kirby, eds.). New York: Springer-Verlag, p. 140.

Kaempfer, R., and J. Kaufman. 1972. Translational control of hemoglobin synthesis by an initiation factor required for recycling of ribosomes and for their binding to messenger RNA. *Proc. Nat. Acad. Sci. US 69*:3317-3321.

Kupchan, S. M., R. W. Britton, M. F. Ziegler, and C. W. Sigel. 1973. Bruceantin, a new potent antileukemic semiroubolide from *Brucea Antidysenterica*. *J. Org. Chem. 38*:178-179.

Laskin, A. I. 1967. Tetracycline. in: *Antibiotics.* I. *Mechanism of Action* (D. Gottlieb and P. I. Shaw, eds.) New York: Springer-Verlag. pp. 331-359.

Liao, L. L., S. B. Horwitz, and A. P. Grollman. 1974a. Gallin [9-(2'-carboxyphenyl)-3,4,5,6-tetrahydroxyxanthene], a new inhibitor of *Escherichia coli* ribonucleic acid polymerase. *Biochemistry 13:* 1331-1335.

Liao, L. L., S. M. Kupchan, and S. B. Horwitz. 1976. Bruceantin: an inhibitor of protein biosynthesis. *Mol. Pharm. 12:*167-176.

Liao, L. L., S. B. Horwitz, D. Steward, J. Martin, M.-T. Huang, and A. P. Grollman. 1975. Triphenylmethane dyes as inhibitors of reverse transcriptase, RNA polymerase and protein synthesis. Structure-activity relationships. *J. Med. Chem. 18:*117-120.

Lin, J.-Y., K. Liu, C.-C. Chen, and T.-C. Tung. 1971. Effect of crystalline ricin on the biosynthesis of protein, RNA, and DNA in experimental tumor cells. *Cancer Res. 31:*921-924.

Lin, J.-Y., C.-C. Pao, S.-T. Ju, and T.-C. Tung. 1972. Polyribosome disaggregation in rat liver following administration of the phytotoxic proteins, abrin and ricin. *Cancer Res. 32:*943-947.

Lin, S. Y., R. D. Mosteller, and B. Hardesty. 1966. The mechanism of sodium fluoride and cycloheximide inhibition of hemoglobin biosynthesis in the cell-free reticulocyte system. *J. Mol. Biol. 21:* 51-69.

Lindenbaum, A., and J. Schubert. 1956. Binding of organic anions by serum albumin. *J. Phys. Chem. 60:*1663-1665.

Lingrel, J. B., and H. Borsook. 1963. A comparison of amino acid incorporation into the hemoglobin and ribosomes of marrow erythroid cells and circulating reticulocytes of severely anemic rabbits. *Biochemistry 2:*309-314.

Lodish, H. F., D. Housman, and M. Jacobsen. 1971. Initiation of hemoglobin synthesis. Specific inhibition by antibiotics and bacteriophage ribonucleic acid. *Biochemistry 10:*2348-2356.

Lucas-Lenard, J., and F. Lipmann. 1971. Protein biosynthesis. *Ann. Rev. Biochem. 40:*409-448.

Lynch, J. E., A. R. English, H. Bauck, and H. Deligianis. 1954. Studies on the *in vitro* activity of anisomycin. *Antibiotics and Chemotherapy 4:*844-848.

Malkin, M., and F. Lipmann. 1969. Fusidic acid: inhibition of factor TF-2 in reticulocyte protein synthesis. *Science 164:*71-72.

Marcus, A., J. D. Bewley, and D. P. Weeks. 1970a. Aurintricarboxylic acid and initiation factors of wheat embryo. *Science 167:* 1135-1136.

Marcus, A., D. P. Weeks, J. P. Leis, and E. B. Keller. 1970b. Protein chain initiation of methionyl-tRNA in wheat embryo. *Proc. Nat. Acad. Sci. US 67:*1681-1687.

Marks, P. A., E. R. Burka, R. Rifkind, and D. Danon. 1963. Polyri-bosomes active in reticulocyte protein synthesis. *Cold Spring Harbor Symp. Quant. Biol. 28*:223-226.

Marks, P. A., E. R. Burka, F. M. Conconi, W. Perl, and R. A. Rifkind. 1965. Polyribosome dissociation and formation in intact reticu-locytes with conservation of messenger ribonucleic acid. *Proc. Nat. Acad. Sci. US 53*:1437-1443.

McKeehan, W., and B. Hardesty. 1969. The mechanism of cycloheximide inhibition of protein synthesis in rabbit reticulocytes. *Biochem. Biophys. Res. Commun. 36*:625-630.

Medappa, K. C., C. McLean, and R. R. Rueckert. 1971. On the struc-ture of rhinovirus 1A. *Virology 44*:259-270.

Miller, D. L. 1972. Elongation factors EF-Tu and EF-G interact at related sites on ribosomes. *Proc. Nat. Acad. Sci. US 69*:752-755.

Modolell, J., and D. Vazquez. 1973. Inhibition by aminoacyl trans-fer ribonucleic acid of elongation factor G-dependent binding of guanosine nucleotide to ribosomes. *J. Biol. Chem. 248*:488-493.

Monro, R. E., and K. A. Marcker. 1967. Ribosome-catalyzed reaction of puromycin with a formylmethionine-containing oligonucleotide. *J. Mol. Biol. 25*:347-350.

Monro, R. E., J. Cerna, and K. A. Macker. 1968. Ribosome-catalyzed peptidyl transfer: substrate specificity at the P site. *Proc. Nat. Acad. Sci. US 61*:1042-1049.

Montanaro, L., S. Sperti, and F. Stirpe. 1973. Inhibition by ricin of protein synthesis *in vitro,* ribosomes as the target of the tox-in. *Biochem. J. 136*:677-683.

Munoz, E., F. Garcia-Ferrandiz, and D. Vazquez. 1972. *Molecular Mechanisms of Antibiotic Action on Protein Biosynthesis and Mem-branes.* New York: Elsevier.

Neth, R., R. E. Monro, G. Heller, E. Battaner, and D. Vazquez. 1970. Catalysis of peptidyl transfer by human tonsil ribosomes and ef-fects of some antibiotics. *FEBS Lett. 6*:198-202.

Obrig, T., J. Irvin, W. Culp, and B. Hardesty. 1971. Inhibition of peptide initiation on reticulocyte ribosomes by edeine. *Eur. J. Biochem. 21*:31-41.

Olsnes, S., and A. Pihl. 1972. Treatment of abrin and ricin with β-mercaptoethanol, opposite effects on their toxicity in mice and their ability to inhibit protein synthesis in a cell-free system. *FEBS Lett. 28*:48-50.

Olsnes, S., and A. Pihl. 1973. Different biological properties of the two constituent peptide chains of ricin, a toxic protein in-hibiting protein synthesis. *Biochemistry 12*:3121-3126.

Olsnes, S., K. Refsnes, and A. Pihl. 1974. Mechanism of action of the toxic lectins, abrin, and ricin. *Nature 249*:627-628.

Openshaw, H. T. 1970. The ipecacuanha alkaloids. in: *Chemistry of the Alkaloids* (S. W. Pelletier, ed.). New York: Van Nostrand Reinhold, p. 85.

Owen, S. P., A. Dietz, and G. W. Camiener. 1962. Sparsomycin, a new antitumor antibiotic. I. Discovery and biological properties. in: *Antimicrobial Agents and Chemotherapy* (J. C. Sylvester, ed.). Ann Arbor: Am. Soc. Microbiol., p. 772-779.

Pappenheimer, A. M., Jr., and D. M. Gill. 1973. Diphtheria. *Science 182:*355-358.

Perani, A., B. Parisi, L. De Carli, and O. Ciferri. 1968. Incorporation of amino acids by a cell-free system prepared from human cells cultured *in vitro. Biochim. Biophys. Acta 161:*223-231.

Pestka, S. 1971. Inhibitors of ribosome functions. *Ann. Rev. Microbiology 25:*487-562.

Pestka, S., H. Rosenfeld, R. Harris, and H. Hintikka. 1972. Studies on transfer ribonucleic acid-ribosome complexes. XXI. Effect of antibiotics on peptidyl-puromycin synthesis by mammalian polyribosomes. *J. Biol. Chem. 247:*6895-6900.

Pestka, S., R. Vince, S. Daluge, and R. Harris. 1973. Effect of puromycin analogues and other agents on peptidyl-puromycin synthesis on polyribosomes. *Antimicrobial Agents and Chemotherapy 4:* 37-43.

Powell, R. G., D. Weisleder, C. R. Smith, Jr., and I. A. Wolff. 1969. Structure of cephalotaxine and related alkaloids. *Tetrahedron Lett. 46:*4081-4084.

Raeburn, S., R. S. Goor, J. A. Schneider, and E. S. Maxwell. 1968. Interaction of aminoacyl transferase II and guanosine triphosphate: inhibition by diphtheria toxin and nicotinamide adenine dinucleotide. *Proc. Nat. Acad. Sci. US 61:*1428-1434.

Rao, S. S., and A. P. Grollman. 1967. Cycloheximide resistance in yeast: a property of the 60S ribosomal subunit. *Biochem. Biophys. Res. Commun. 29:*696-704.

Richman, N., and J. W. Bodley. 1972. Ribosomes cannot interact simultaneously with elongation factors EF-Tu and EF-G. *Proc. Nat. Acad. Sci. US 69:*686-689.

Richter, D. 1973. Competition between the elongation factors 1 and 2 and phenylalanyl-transfer ribonucleic acid for the ribosomal binding sites in a polypeptide-synthesizing system from brain. *J. Biol. Chem. 248:*2853-2857.

Roberts, W. K., and W. H. Coleman. 1971. Polyuridylic acid binding by protein from Ehrlich ascites cell ribosomes and its inhibition by aurintricarboxylic acid. *Biochemistry 10:*4304-4313.

Shigeura, H. T., and C. N. Gordon. 1963. The biological activity of tenuazonic acid. *Biochemistry 2:*1132-1137.

Sinohara, H., and H. H. Sky-Peck. 1965. Effect of gougerotin on the protein synthesis in the mouse liver. *Biochem. Biophys. Res. Commun. 18:*98-102.

Sisler, H. D., and M. R. Siegel. 1967. Cycloheximide and other glutarimide antibiotics. in: *Antibiotics* (D. Gottlieb and P. D. Shaw, ed.) Vol. 1, New York: Springer-Verlag, pp. 283-307.

Stafford, M. E., and C. S. McLaughlin. 1973. Trichodermin, a possible inhibitor of termination process of protein synthesis. *J. Cell Physiol. 82:*121-128.

Stewart, M. L., A. P. Grollman, and M.-T. Huang. 1971. Aurintricarboxylic acid: inhibitor of initiation of protein synthesis. *Proc. Nat. Acad. Sci. US 68:*97-101.

Suhadolnik, R. J. 1970. *Nucleoside Antibiotics*. New York: Wiley.

Szer, W., and Z. Kurylo-Borowska. 1970. Effect of edeine on aminoacyl-tRNA binding to ribosomes and its relationship to ribosomal binding sites. *Biochim. Biophys. Acta 224:*477-486.

Tanaka, N., T. Kinoshita, and H. Masukawa. 1969a. Mechanism of inhibition of protein synthesis by fusidic acid and related steroidal antibiotics. *J. Biochem.* (Tokyo) *65:*459-464.

Tanaka, N., T. Nishimura, T. Kinoshita, and H. Umezawa. 1969b. The effect of fusidic acid on protein synthesis in a mammalian system. *J. Antibiotics 22:*181-182.

Tanaka, N., T. Nishimura, and T. Kinoshita. 1970. Inhibition by fusidic acid of transfer II in reticulocyte protein synthesis. *J. Biochem.* (Tokyo) *67:*459-463.

Ueno, Y., M. Hosoya, Y. Morita, I. Ueno, and T. Taksuno. 1968. Inhibition of protein synthesis in rabbit reticulocyte by nivalenol, a toxic principle isolated from fusarium nivale-growing rice. *J. Biochem.* (Tokyo) *64:*479-485.

Vasquez, D., and E. Battaner. 1972. Site of action of a number of inhibitors acting on ribosomes of the 80S type. in: *Molecular Mechanisms of Antibiotic Action on Protein Biosynthesis and Membranes* (E. Munoz, ed.). New York: Elsevier. pp. 219-241.

Wiley, P. F., H. K. Jahnke, F. MacKellar, R. B. Kelly, and A. D. Agroudelis. 1970. The structure of pactamycin. *J. Org. Chem. 35:* 1420-1425.

Wilkie, D., and B. K. Lee. 1965. Genetic analysis of actidione resistance in *Saccharomyces cerevisiae*. *Genet. Res. Camb. 6:*103-138.

Zasloff, M., and S. Ochoa. 1973. Polypeptide chain initiation in eukaryotes. IV. Purification and properties of supernatant initiation factor from *Artemia salina* embryos. *J. Mol. Biol. 73:*65-76.

Chapter 4
PROTEIN SYNTHESIS IN CHLOROPLASTS

J. Kenneth Hoober

Department of Biochemistry
Temple University School of Medicine
Philadelphia, Pennsylvania

I. INTRODUCTION

The inquiry into the ability of chloroplasts to synthesize proteins
has now spanned about 20 years. During this time the existence of
ribosomes in chloroplasts was demonstrated with a wide variety of
plant cell types, and the features of protein synthesis on these ribosomes were sketched in outline form. Many details remain unstudied,
and in this respect the investigation into protein synthesis in chloroplasts has lagged behind that in other systems. It is assumed that
the mechanism of protein synthesis on chloroplast ribosomes follows
the general pattern determined for other systems, particularly that
for bacterial cells.

However, questions other than those concerning the mechanism of
protein synthesis make the chloroplast a compelling system for study.
There is now no doubt that both the chloroplast and the cytoplasm of
plant cells have the ability to synthesize proteins. The questions
yet to be answered are: How many proteins of the cell are synthesized
on chloroplast ribosomes? And of these, how many are coded by chloroplast DNA? Recent results have presented the possibility that some
messenger RNA molecules synthesized within the cell's nucleus are
translated on chloroplast ribosomes and, conversely, that some of the
messenger RNA molecules transcribed on chloroplast DNA are translated
on cytoplasmic ribosomes. The actual extent of heterologous interaction between DNA and ribosomes of the nuclear-cytoplasmic system and
an organelle system cannot at this time be estimated, but plant cells,
and particularly certain algae, are well suited as a source of material for a study of this type. A more detailed understanding than that
presently available about the mechanism of chloroplast protein synthesis may be necessary to attack these questions, but if exchange
of messenger RNA molecules between chloroplast and cytoplasm occurs,
the mechanism of messenger RNA selectivity becomes a provocative
question.

This chapter summarizes several aspects of protein synthesis in chloroplasts. First, the characteristics of chloroplast protein synthesis at the levels of the intact cell, the isolated organelle, and the extracted components are reviewed. Second, data on the sites of synthesis of chloroplast proteins, gained primarily through the use of selective inhibitors of protein synthesis, are evaluated. And third, evidence on the origins of messenger RNA for chloroplast proteins is discussed. Other reviews on these subjects have appeared (Boulter, 1970; Boulter, Ellis and Yarwood, 1972; Zalik and Jones, 1973).

II. INCORPORATION OF CARBON FROM CO_2 INTO CHLOROPLAST PROTEINS BY WHOLE CELLS

In 1954 evidence was presented that the reactions of photosynthesis (the evolution of oxygen, photophosphorylation, and CO_2 fixation) occur within the chloroplast of plant cells (Arnon et al., 1954). Calvin and his colleagues had observed, in the course of their detailed studies on photosynthetic $^{14}CO_2$ fixation, that several amino acids were among the first compounds labeled (Aronoff et al., 1947; Stepka et al., 1948). Proteins subsequently became labeled as the result of incorporation of the newly synthesized amino acids. The first indication that these amino acids were incorporated into protein in chloroplasts came from experiments reported by Nichiporovich (1955). After intact leaves were labeled with $^{14}CO_2$, protein in a chloroplast fraction prepared from these leaves was labeled and enriched in ^{14}C. Compared to CO_2, carbohydrates from endogenous stores or exogenously supplied hexoses and amino acids were poor precursors of chloroplast proteins (Nichiporovich, 1955; Stephenson et al., 1956).

These observations were extended by Heber (1962) and Parthier (1964), who found that the specific radioactivity of both soluble and particulate chloroplast proteins increased initially at a more rapid rate during photosynthesis in the presence of $^{14}CO_2$ than did the specific radioactivity of cytoplasmic proteins. This pattern of ^{14}C incorporation into proteins of these two cellular compartments suggested

that labeled amino acids, synthesized from CO_2, were incorporated immediately into proteins of the chloroplast but were not incorporated
into cytoplasmic proteins until after a lag of several minutes.

Smith et al. (1961) and Bassham et al. (1964) examined both the
kinetics of labeling of amino acid pools from $^{14}CO_2$ and the subsequent incorporation of the labeled amino acids into proteins by suspensions of *Chlorella* cells. Several amino acids, particularly alanine, aspartic acid, glutamic acid, serine, and glycine, were labeled
rapidly. Within 30-60 min the specific radioactivites of carbon in
the total cellular pools of these amino acids reached plateaus at
various levels, but all were considerably less than the specific radioactivity of the $^{14}CO_2$ presented to the cells. In contrast to the
variety of specific radioactivities of carbon in the free amino acids,
the specific radioactivities of carbon in these same amino acids in
protein, at any point in time, were approximately the same. Most
notably, the specific radioactivity of carbon in protein-bound glycine was similar to that in protein-bound alanine and glutamate at
any time point, even though the specific radioactivites of carbon
in the cellular pools of free alanine and glutamate were about 30
times greater than that of glycine. These results suggested that in
Chlorella there are at least two pools of amino acids, one of which
is rapidly labeled from $^{14}CO_2$ and from which amino acids are incorporated into protein. The differing specific radioactivities of carbon in the free amino acids apparently resulted from variable distributions of the amino acids between the intracellular compartments.
Since the chloroplast is the site of CO_2 fixation, these results im-
plied that labeling of proteins during exposure of cells to $^{14}CO_2$
was the result of synthesis of amino acids *and* protein within the
chloroplast.

III. INCORPORATION OF AMINO ACIDS INTO PROTEIN BY PURIFIED CHLOROPLASTS

The chloroplast can be purified from many plant species, but considerable care must be exercised to obtain the organelle intact and

free of other organelles and bacteria. Stephenson et al. (1956) and
Sissakian and co-workers (see Sissakian, 1958) pioneered in efforts
to find out if this organelle is capable of protein synthesis. But
it was not clear in these initial attempts whether the incorporation
of amino acids into protein was the result of synthetic activity in
the chloroplasts or in contaminating bacteria (App and Jagendorf,
1964).

The achievement of success in the isolation of intact, active
chloroplasts was dependent on the use of a suitable medium. The im-
petus for much of the early work on intact chloroplasts was the development of a medium (Honda's medium), which in addition to sucrose
(0.25 M), tris-HCl, pH 7.8 (25 mM), and magnesium acetate (1-10 mM),
contained Ficoll (2.5%), dextran (5%), and a sulfhydryl compound
(Spencer and Wildman, 1964; Honda et al., 1966). Chloroplasts isolated with this medium from cells of tobacco, bean, and *Euglena* appeared intact as judged by phase-contrast and electron microscopy
(Spencer and Wildman, 1964; Honda et al., 1966; Margulies et al.,
1968; Vasconcelos et al., 1971). Boardman et al. (1965) found that
the amount of protein synthesized in vitro was related to the degree
of intactness of the chloroplasts. When removed from Honda's medium,
the chloroplasts ruptured easily and only a small portion of the original total activity was associated with the chloroplast fragments
recovered by centrifugation. A portion of the original activity was
recovered in the supernatant fraction, suggesting that the protein-
synthesizing activity was largely in the matrix phase of the chloroplast.

A clear demonstration of the advantages of using Honda's medium
were the results of Gnanam et al. (1969), who compared chloroplasts
prepared with a solution containing sucrose, Tricene buffer, $MgCl_2$,
and mercaptoethanol to those prepared with this medium supplemented
with Ficoll and dextran (Honda's medium). Protein synthesis in preparations of "sucrose" chloroplasts showed no sensitivity to ribonuclease and no dependency on added ATP, but were sensitive to inhibitors such as KCN, atebrin, dinitrophenol, and hydrogen peroxide generated by glucose plus glucose oxidase, all inhibitors of bacterial

cells. The opposite effect was found for the chloroplast fraction prepared with the medium containing Ficoll and dextran. This latter medium appeared to preserve chloroplast integrity and activity, but also permitted the preparation of chloroplast fractions containing fewer bacterial cells. The requirement of added ATP for protein synthesis, as well as the ineffectiveness of the inhibitors of oxidative phosphorylation, in preparations made with Ficoll and dextran indicated that the activity was due to chloroplasts rather than to contaminating bacterial cells. Intact chloroplasts as well as bacterial cells are impermeable to ribonuclease (Margulies et al., 1968; Blair and Ellis, 1973), and therefore the lack of an effect of this enzyme on amino acid incorporation cannot be used to indicate the presence of contaminating cells. Boulter et al. (1972) suggested that the best criterion for assessing protein synthesis in preparations of chloroplasts is dependency on an added source of energy, either as exogenously supplied ATP or as light.

Blair and Ellis (1973) abandoned the use of Honda's medium and examined protein synthesis in pea chloroplasts after breaking cells in a medium containing sucrose (0.35 M), EDTA (2 mM), sodium isoascorbate (2 mM), and HEPES buffer (25 mM, pH 7.6) and then suspending the chloroplast fraction in a solution containing KCl (0.2 M), $MgCl_2$ (6.6 mM), and Tricene buffer (66 mM, pH 8.3). The KCl was required for maximal amino acid incorporation and also acted as an osmotic agent. Approximately one-half of the chloroplasts obtained from pea leaves by this procedure were intact. In preparations from spinach leaves, 60-80% of the chloroplasts were intact (Hartley and Ellis, 1973).

The incorporation of amino acids into polypeptides has been demonstrated in chloroplasts purified from a wide variety of plants (see reviews by Smillie and Scott, 1969; Boulter, 1970; Boulter et al., 1972). Activity is high in developing chloroplasts isolated from young seedlings (Spencer, 1965; Bamji and Jagendorf, 1966; Ranalletti et al., 1969; Drumm and Margulies, 1970), which correlates with the observation that chloroplasts from young plants have more ribosomes as polysomes than do chloroplasts from older plants (Had-

ziyev and Zalik, 1970). Protein synthesis in isolated, intact chloroplasts requires, in addition to the usual buffer and salts, only a
source of energy. In the presence of light, photophosphorylation can
support protein synthesis by functioning as an in situ ATP-generating
system (Spencer, 1965; Ramirez et al., 1968; Ellis and Hartley, 1971;
Shephard and Levin, 1972), and under optimal conditions illumination
of an incubation mixture leads to a greater incorporation of amino
acids than does the addition of exogenous ATP (Blair and Ellis, 1973).
Exogenously supplied amino acids are generally not required, since
isolated chloroplasts retain a pool of amino acids (Kirk and Leech,
1972), but the addition of amino acids usually increases activity
(Spencer, 1965; Parenti and Margulies, 1967; Ranalletti et al., 1969).
Wheat chloroplasts showed more activity for protein synthesis at 1
mM than at 20 mM $MgCl_2$, whereas the opposite effect was found for
bean chloroplasts (Ranalletti et al., 1969). The concentration of
$MgCl_2$ in the reaction mixture in most studies ranged from 5 mM (Spencer, 1965; Ranalletti et al., 1969) to 10 mM (Spencer et al., 1971),
but the optimum concentration was not described in these studies.

   Chloroplasts isolated from *Vicia faba* and from *Acetabularia* are
capable of synthesizing amino acids from CO_2 (Kirk and Leech, 1972;
Shephard and Levin, 1972), and those from *Acetabularia* synthesize protein from these amino acids for extended perios of time (Shephard and
Levin, 1972). The amino acid and protein synthesis in the isolated
Acetabularia chloroplasts was dependent on light; ATP added in the
dark did not support these activities. Also, compared to carbon from
CO_2, carbon from exogenously provided amino acids was incorporated
poorly into protein by these chloroplasts.

   The investigations referred to above have provided firm evidence
for the ability of chloroplasts to synthesize protein. However, the
incorporation of amino acids generally stops after about 30 min of
incubation. Actinomycin D, an inhibitor of DNA-dependent RNA synthesis, has very little if any effect on protein synthesis by isolated
chloroplasts (Spencer and Wildman, 1964; Spencer, 1965; Harris and
Eisenstadt, 1971; Harris et al., 1973; Blair and Ellis, 1973). Addition of the four common ribonucleoside triphosphates does not enhance

amino acid incorporation (Ranalletti et al., 1969; Harris et al., 1973), even though it has been amply demonstrated that chloroplasts are capable of synthesizing RNA in vitro (Semal et al., 1964; Kirk, 1964; Schweiger and Berger, 1964; Bottomley et al., 1971b; Harris et al., 1973; Hartley and Ellis, 1973). Goffeau and Brachet (1965) observed that very high concentrations of actinomycin D inhibited protein synthesis in isolated chloroplasts, but their data indicate that, rather than acting via an effect on synthesis of RNA, actinomycin D affected protein synthesis directly. Thus, chloroplasts in the studies referred to above apparently only completed polypeptide chains already initiated in vivo (Harris and Eisenstadt, 1971).

IV. PROPERTIES OF CHLOROPLAST RIBOSOMES

A. Sedimentation Coefficient of Monomers

Lyttleton (1962) reported that isolated spinach chloroplasts contain ribosomes smaller than those in the cytoplasm. Monomers, which had a sedimentation coefficient of 66S (at a concentration of 2 mg/ml), were observed when extracts of isolated chloroplasts were examined with the analytical ultracentrifuge. Lyttleton's discovery of chloroplast ribosomes has subsequently been confirmed by many investigators. Clark (1964) and Clark et al. (1964) found in preparations of chloroplasts from Chinese cabbage that greater than 80% of the ribosomes were 70S particles. Essentially all ribosomes contained in washed preparations of spinach chloroplasts were 70S particles (Spencer, 1965; Spencer and Whitfeld, 1966). Boardman et al. (1966) and Svetailo et al. (1967) also found predominantly 70S ribosomes in chloroplasts isolated from tobacco and pea leaves, respectively. From an analysis of the sedimentation properties of tobacco ribosomes, Boardman et al. (1966) obtained $S^{\circ}_{w,20}$ values of 69.9S for ribosomes obtained from chloroplasts as compared with 82.0S for cytoplasmic ribosomes. Sedimentation coefficients of 69.8S for chloroplast and 79.8S for cytoplasmic ribosomes of pea seedlings were obtained by Svetailo et al. (1967).

Several laboratories have shown that chloroplasts isolated from the alga *Euglena gracilis* contain 70S ribosomes (Rawson and Stutz, 1969; Scott et al., 1970; Vasconcelos et al., 1971). Mihara and Hase (1969) obtained evidence for both 70 and 80S ribosomes in the green alga *Chlorella protothecoides*. Also, Sager and Hamilton (1967), Pakhomova et al. (1968), Hoober and Blobel (1969), and Bourque et al. (1971) observed two populations of ribosomes in extracts of the green alga *Chlamydomonas*; $S^{\circ}_{w,20}$ values of approximately 70 and 82S were determined for *Chlamydomonas* ribosomes (Pakhomova et al., 1968; Sager and Hamilton, 1967; Siersma and Chiang, 1971). However, chloroplasts are difficult to isolate from this organims, and definitive evidence for the intracellular localization of the 70S ribosomes came from studies with mutants and inhibitors. Mixotrophically grown cells, i.e., cells grown in the light on medium containing acetate, of the *ac*-20 and the *cr*-1 strains of *Chlamydomonas reinhardtii* are deficient in 70S particles as analyzed by sucrose gradient centrifugation. Examination of these cells by electron microscopy showed a normal level of cytoplasmic ribosomes but few particles in the chloroplasts (Goodenough and Levine, 1970; Boynton et al., 1970; Boynton et al., 1972). Furthermore, cells treated several days with the drug rifampicin, which inhibits synthesis of the ribosomal RNA of the 70S particles in *Chlamydomonas* cells (Surzycki, 1969; Surzycki and Rochaix, 1971), contained few ribosomal particles in the chloroplast as determined by electron microscopy (Goodenough, 1971).

From these studies with isolated chloroplasts, mutant strains, and rifampicin, it has been established that ribosomes with a sedimentation coefficient of 70S in extracts of plant cells are derived from the chloroplast. Although the contribution of mitochondrial ribosomes to results with extracts of plant cells has largely been ignored, Avadhani and Buetow (1972) have obtained, from *Euglena* mitochondria, ribosomes that are similar in size to those in the chloroplast. However, relatively few particles with sedimentation coefficients about 70S have been found in extracts of nonphotosynthetic species of algae and of mutant strains of *Chlamydomonas* that are de-

ficient in chloroplast ribosomes (Tiboni et al., 1970; Goodenough and
Levine, 1970; Siu et al., 1971; Boynton et al., 1972). Since the mutant cells apparently have normal mitochondria, these results suggest
that mitochondria contribute a minor portion of the 70S ribosomes in
extracts of normal, green cells.

As a general approximation, chloroplast ribosomes are referred to
as 70S particles, and those from the cytoplasm as 82S particles. The
70S particles are indistinguishable in size from bacterial ribosomes,
whereas the 82S plant ribosomes are similar in size to the cytoplasmic ribosomes of animal cells (Stutz and Noll, 1967; Hoober and Blobel, 1969).

B. Sedimentation Coefficients of Subunits

Under a given set of conditions, a higher concentration of Mg^{2+}
is required to maintain chloroplast ribosomes as monomers than is required for cytoplasmic ribosomes (Boardman et al., 1966; Sager and
Hamilton, 1967; Hoober and Blobel, 1969; Hadziyev and Zalik, 1970;
Bourque et al., 1971; Chua et al., 1973a). But since dissociation of
ribosomes into subunits is influenced by several other variables, including temperature, ionic strength of the medium, and the hydrodyna-
mic pressures on the ribosomes during centrifugation (Infante and
Baierlein, 1971; Hauge, 1971; Chua et al., 1973a), analyses of the effects of Mg^{2+} by centrifugation cannot be directly extrapolated to
the in vivo situation. Boardman et al. (1966) and Hadziyev and Zalik (1970) determined the sedimentation coefficients of the subunits
of chloroplast ribosomes from tobacco and wheat, respectively, as 50
and 35S, whereas the values for the subunits of cytoplasmic ribosomes
of higher plants are 58 and 36S (Boardman et al., 1966; Cammarano et
al., 1972). Chua et al. (1973a) have determined similar values for
Chlamydomonas ribosomes; functional subunits derived from chloroplast
ribosomes with 0.5 M KCl have sedimentation coefficients of 50 and
33S as compared to 57 and 37S for cytoplasmic ribosomes. Subunits
of *Euglena* chloroplast ribosomes are reported to have sedimentation
coefficients of 50 and 30S (Rawson and Stutz, 1969; Scott et al.,
1970).

C. Ribosomal RNA Components

Chloroplast ribosomes contain major RNA molecules with sedimentation coefficients of 23 and 16S (Stutz and Noll, 1967; Loening and Ingle, 1967; Mehta et al., 1968; Hoober and Blobel, 1969). The molecular weights of these major chloroplast ribosomal RNA molecules from higher plants and algae are about 1.1×10^6 and 0.56×10^6, respectively (Loening, 1968a,b; Mehta et al., 1968; Ingle, 1968; Surzycki and Rochaix, 1971). In contrast, the 82S cytoplasmic ribosome contains major RNA molecules with sedimentation coefficients of 25 and 18S, corresponding to molecular weights of 1.3×10^6 and 0.69×10^6, respectively (Stutz and Noll, 1967; Loening and Ingle, 1967; Loening, 1968a,b; Hoober and Blobel, 1969; Surzycki and Rochaix, 1971). There is evidence that chloroplast ribosomes also contain a 5S RNA molecule (Dyer and Leech, 1968; Bourque et al., 1971; Payne and Dyer, 1971) but not the 5.8S RNA molecule (Payne and Dyer, 1972) that is found in the large subunit of 80S-type ribosomes (Pène et al., 1968; Sy and McCarty, 1970; Udem et al., 1971). Chloroplast ribosomes therefore resemble those of bacteria with respect to the RNA components they contain (see Kurland, 1972).

Although the ratio of the chloroplast ribosomal 23S RNA to the 16S RNA molecules, on an absorbance basis, was approximately 1 in the initial studies rather than the expected ratio of 2 (Stutz and Noll, 1967; Loening and Ingle, 1967; Loening, 1968a), more recent experiments have produced the expected 2:1 abosrbance ratio (i.e., a 1:1 molar ratio) for the chloroplast ribosomal RNA (Hoober and Blobel, 1969; Leaver and Ingle, 1971; Scott et al., 1970; Cattolico and Jones, 1972).

D. Ribosomal Proteins

Ribosomes in the chloroplast contain a set of ribosomal proteins distinct from those in the cytoplasm. Lyttleton (1968) and Odintsova and Yurina (1969) showed that the electrophoretic patterns of total ribosomal proteins from pea and clover chloroplasts differed considerably from the pattern obtained for cytoplasmic ribosomes. Hoober and

Blobel (1969) showed that the subunits of *Chlamydomonas* 70 and 82S
ribosomes yielded different patterns after electrophoresis of the
proteins. Gualerzi and Cammarano (1969, 1970) and Vasconcelos and
Bogorad (1971) studied proteins of ribosomal subunits from several
plants. The extent of variation in the electrophoretic patterns for
the proteins of 82S ribosomes was correlated with the taxonomic rela-
tedness of the plants. But the patterns for chloroplast ribosomal
proteins from the various plants differed more markedly than did those
of the cytoplasmic ribosomal proteins.

     Gualerzi and Cammarano (1969) found that about 20 fractions were
resolved by gel electrophoresis of protein samples from each of the
subunits of spinach chloroplast ribosomes. In samples of total ribosomal proteins, 8-10 acidic proteins were found that moved toward the
anode at pH 8.3. These results were confirmed by two-dimensional gel
electrophoresis, which resolved approximately 40 different proteins
from spinach chloroplast ribosomes (Gualerzi et al., 1974). Of this
number, 11 or 12 proteins moved toward the anode at pH 8.6. The electrophoretic pattern for cytoplasmic ribosomal proteins bore little
resemblance to that of the proteins from chloroplast ribosomes. Moreover, cytoplasmic ribosomes contain perhaps only one acidic protein
that is located in the small subunit (Gualerzi et al., 1974).

     Jones et al. (1972) reported that, as judged by two-dimensional
electrophoresis, the majority of the proteins of chloroplast ribosomes of wheat leaves are different, with respect to their electrophoretic properties, from those of cytoplasmic ribosomes. A total of
about 75 components was resolved by electrophoresis of preparations
of chloroplast ribosomal proteins, and of these 26 moved toward the
anode at pH 8.6. Some of the proteins were present at low levels,
and it was not established that all 75 components were integral components of the ribosomes. Freyssinet and Schiff (1974) found seven
acidic components in ribosomes of *Euglena* chloroplasts, and they estimated, from patterns obtained from only one-dimensional gel electrophoresis, that these ribosomes contained about 40 basic proteins.

E. Polysomes

Extracts of chloroplasts often contain a portion of the ribosomes as polysomes (Clark, 1964; Boardman et al., 1966; Chen and Wildman, 1967; Stutz and Noll, 1967; Ranalletti et al., 1969; Rawson and Stutz, 1969; Hadziyev and Zalik, 1970), but monomers usually predominate in such preparations. Polyvinyl sulfate was used by Clark (1964) to prevent degradation of polysomes, but this nuclease inhibitor promotes dissociation of ribosomes (Vanyushin and Dunn, 1967). Furthermore, polyvinyl sulfate strongly inhibited protein synthesis and could not be used when studying the activity of ribosomes in vitro (Boardman et al., 1966). A greater proportion of the ribosomes was recovered as polysomes in extracts of chloroplasts isolated from four-day-old wheat seedlings than from seven-day-old plants (Hadziyev and Zalik, 1970). Bentonite was included in the buffers used for preparation of polysomes in this study. Avadhani and Buetow (1972) have developed a procedure that permitted good recovery of polysomes from *Euglena*. Most of the ribosomes in the cytoplasm, mitochondria, and chloroplasts of this alga were recovered as polysomes when 2-mercaptoethanol, heparin, and a cytosolic inhibitor from rat liver were included in all solutions used to prepare the polysomes. As with ribosomes from animal cells (Blobel and Potter, 1966, 1967), the isolation of polysomes from chloroplasts seems to depend on the presence of effective inhibitors of nucleases.

Ribosomes in the form of polysomes become bound to thylakoid membranes in synchronously growing *Chlamydomonas* cells during the early part of the light phase of the cell cycle (Chua et al., 1973b). During this time the amounts of cytochromes in these membranes increase (Schor et al., 1970b). This correlation suggests that some of the thylakoid membrane proteins are synthesized on ribosomes attached to the membrane. The addition of chloramphenicol to the culture medium to "fix" polysomal structures resulted in an almost quantitative recovery of chloroplast ribosomes with isolated thylakoid membranes (Chua et al., 1973b; Margulies and Michaels, 1974). An examination of these membrane fractions by electron microscopy revealed polysomes,

consisting frequently of five to six ribosomes per structure, attached
to exposed, unfused surfaces of the thylakoids. These ribosomes could
be released from the purified membranes with puromycin in a medium of
high ionic strength, conditions similar to those that release bound
ribosomes from the endoplasmic reticulum of rat liver (Adelman et al.,
1973). These results suggest that the polysomes recovered with thylakoid membranes were attached partly via nascent polypeptide chains.
The nature of these polypeptides has not yet been determined.

F. Sensitivity to Inhibitors

Protein synthesis in isolated chloroplasts or on ribosomes extracted therefrom is inhibited by chloramphenicol (Spencer and Wildman, 1964; Spencer, 1965; Bamji and Jagendorf, 1966; Ellis, 1969;
Margulies and Brubaker, 1970; Reger et al., 1972; Avadhani and Buetow,
1972; Schlanger et al., 1972). Only the D-threo isomer of chloramphenicol is effective as an inhibitor (Ellis, 1969). Anderson and
Smillie (1966) reported that chloroplast ribosomes bound chloramphени-
col at a level two- to threefold greater than the amount bound to cytoplasmic ribosomes, but this level of binding was still considerably
less than that obtained with ribosomes of *E. coli*.

Chloroplast ribosomes are also sensitive to inhibition by tetra-
cyclines (Bonotto et al., 1969; Ellis, 1970; Harris et al., 1973),
spectinomycin (Ellis, 1970; Harris et al., 1973), streptomycin (Chua
et al., 1973a), erythromycin, lincomycin (Ellis, 1970; Ellis and Hartley, 1971), carbomycin, and other macrolide antibiotics (Schlanger
et al., 1972). Mets and Bogorad (1971) reported the binding of erythromycin to the large subunit of chloroplast ribosomes of *Chlamydomonas,* and Burton (1972) showed that the small subunit of these ribosomes bind spectinomycin. The effects of these antibiotics on chloroplast ribosomes parallel results with bacterial ribosomes (Pestka,
1971). These inhibitors are not effective against the cytoplasmic
ribosomes of plants.

Chloroplast ribosomes are insensitive to cycloheximide, a commonly used inhibitor of 82S ribosomes (Bonotto et al., 1969; Ellis, 1969;

Hoober and Blobel, 1969; Avadhani and Buetow, 1972; Reger et al., 1972; Harris et al., 1973; Chua et al., 1973a). But since the 82S ribosomes in the cytoplasm of plant cells are sensitive to cycloheximide (Ellis, 1969; Hoober and Blobel, 1969; Reger et al., 1972; Schlanger et al., 1972; Chua et al., 1973a), the ability to selectively inhibit protein synthesis on either the chloroplast or the cytoplasmic ribosomes has provided an opportunity to study the sites of synthesis of chloroplast proteins in whole cells. Most data presently available on the sites of synthesis of plant proteins have been obtained using this approach (see Sect. VI,B).

V. CHARACTERISTICS OF PROTEIN SYNTHESIS ON CHLOROPLAST RIBOSOMES

A. Initiation

Although considerable effort has been given to characterizing the components of chloroplast ribosomes, the mechanism of protein synthesis on these ribosomes has not received detailed examination. Most of the efforts have been aimed at determining if the ribosomes are active and what general requirements are necessary for amino acid incorporation into protein. Because of the size of the ribosomes, and their sensitivity to antibiotics effective against bacterial ribosomes, it has been anticipated, and even assumed, that the mechanism of protein synthesis in chloroplasts is similar to the process in bacterial cells (for a review see Bretscher, 1971).

The initiation of protein synthesis on chloroplast ribosomes is similar to the process on other 70S ribosomes insofar as the initiating aminoacyl-tRNA is concerned. Chloroplasts of bean leaves contain N-formylmethionyl-tRNA (Burkard et al., 1969; Guillemaut et al., 1972), and Schwarz et al. (1967) found N-terminal N-formylmethionine after translation of f_2 viral RNA by extracts of *Euglena* chloroplasts. Leis and Keller (1970) presented evidence that synthesis of polypeptide chains in wheat chloroplasts is initiated with N-formylmethionyl-tRNA. Two major fractions of methionine-accepting tRNA were found in extracts of isolated chloroplasts from wheat and cotton seedlings

(Leis and Keller, 1970; Merrick and Dure, 1971, 1972), but after charging with methionine only one of the two methionyl-tRNA fractions accepted formyl groups from N^{10}-formyl tetrahydrofolate in the presence of the homologous plant transformylase or the enzyme from *E. coli*. Leis and Keller (1971) partially purified the transformylase from wheat, and found that it was specific for the initiating form of the chloroplast methionyl-tRNA. The N-formylmethionyl-tRNA synthesized in vitro was capable of donating N-formylmethionine to puromycin in the presence of ribosomes and initiating factors from *E. coli,* suggesting a role of this aminoacyl-tRNA in initiation of protein synthesis.

Evidence for utilization of N-formylmethionyl-tRNA in initiation of protein synthesis in chloroplasts of *Acetabularia* was presented by Bachmayer (1970). N-Formylmethionylpuromycin was produced by purified chloroplasts when incubated with puromycin, and the amount of N-terminal N-formylmethionine in protein was inversely related to the amount of puromycin added to the reaction mixture.

B. Transfer RNA and Aminoacyl-tRNA Synthesis

During studies of protein synthesis by isolated chloroplasts, it was observed that initially the labeled amino acids were incorporated into a form that was insoluble in cold trichloroacetic acid but that was solubilized by hot acid (Spencer and Wildman, 1964; Parenti and Margulies, 1967; Ranalletti et al., 1969). The amount of amino acids incorporated into this form increased rapidly, but then did not change after the first few minutes of incubation. However, the amount of amino acids incorporated into protein continued to rise. It was by this type of analysis that aminoacyl-tRNA complexes were first discovered (Hoagland et al., 1958).

Sissakian et al. (1965) and Hadziyev and Zalik (1970) isolated a tRNA fraction from pea and wheat chloroplasts, respectively, and also observed enzymic transfer of amino acids to tRNA. Bové and Raacke (1959) and Marcus (1959) had previously determined the presence of amino-acid-activating enzymes in spinach chloroplasts by assaying

amino-acid-dependent pyrophosphate exchange with ATP and by the formation of aminoacyl hydroxamates. Ranalletti et al. (1969) purified
aminoacyl-tRNA complexes after a period of in vitro labeling of chloroplasts. The complexes were precipitable by cold trichloroacetic
acid but were degraded in hot acid, characteristics expected for aminoacyl-tRNA complexes. After lysis of the chloroplasts, aminoacyl-
tRNA synthetase activity was recovered in the supernatant fraction
after high-speed centrifugation.

A considerable degree of specificity resides in the formation of
aminoacyl-tRNA complexes in the chloroplasts of both algae and higher
plants. From recent studies it appears that, for many and perhaps
most amino acids, forms of tRNA reside in the chloroplast that are
different from the isoaccepting forms in the cytoplasmic matrix. Furthermore, the several aminoacyl-tRNA synthetases that have been assayed in chloroplast fractions are specific for the chloroplast forms
of tRNA.

In *Euglena* cells Reger et al. (1970) found two tRNA species specific for isoleucine. One fraction was found only in light-grown
cells, and was recovered in purified chloroplasts. Two isoleucyl-
tRNA synthetases were also found, a constitutive form and a light-
inducible form. The inducible form of the enzyme, found in the chloroplast fraction, added isoleucine only to the light-inducible, chloroplast tRNA^{Ile}. Kislev et al. (1972) reported the separation of
five types of tRNA^{Ile} in *Euglena,* one of which was located in chloroplasts and acylated only by the chloroplast enzyme. Another species
of tRNA^{Ile} was found in the mitochondrial fraction, while the remaining three species were recovered in the cytosol fraction (Kislev et
al., 1972). Reger et al. (1970) also found three phenylalanyl-tRNA
synthetases in *Euglena,* one of which was recovered in the chloroplast
fraction of the cells. Although constitutive, this chloroplast enzyme was specific for a light-inducible tRNA^{Phe}, one of two major
forms of this tRNA in *Euglena* (Barnett et al., 1969). The cytoplasmic form of tRNA^{Phe} contains a fluorescent base as does the cytoplasmic tRNA^{Phe} of other eukaryotic organisms including higher plants
(Yoshikami and Keller, 1971), but the chloroplast tRNA^{Phe} is not fluo-

rescent and in this respect is similar to the tRNA^{Phe} of bacteria (Fairfield and Barnett, 1971). *Euglena* chloroplasts also apparently contain specific aminoacyl-tRNA synthetases and tRNA molecules for aspartic acid, glutamic acid, leucine, proline, valine, and possibly other amino acids (Kislev et al., 1972).

Examples of chloroplast-specific forms of tRNA and aminoacyl-tRNA synthetases have also emerged from work with higher plants. During their studies on initiation of protein synthesis in chloroplasts, Leis and Keller (1970) separated five methionine-isoaccepting species in extracts of wheat seedlings, two of which were recovered in the chloroplast fraction. Whereas the initiating form of chloroplast methionyl-tRNA was formylated by wheat transformylase (see Sect. V, A above), the cytoplasmic methionyl-tRNA species were not formylated (Leis and Keller, 1970, 1971). Similar results were obtained by Merrick and Dure (1971, 1972), who found at least three isoaccepting species of tRNA^{Met} in cotton seedlings, two of which were found in purified chloroplasts. In cotton plants, the only species of methionyl-tRNA that accepted formyl groups was again one of the two found in chloroplasts.

Aliev and Filippovich (1968) demonstrated by cell fractionation and chromatography that the forms of tRNA^{Arg} and arginyl-tRNA synthetases found in the chloroplasts of pea leaves are different from their cytoplasmic counterparts. Burkard et al. (1970) compared tRNA molecules and aminoacyl-tRNA synthetases of the chloroplast and cytoplasmic matrix of bean leaves. The chloroplast fraction contained five tRNA species for leucine, and two of these were found after cell fractionation in the cytoplasmic matrix fraction. All five tRNA^{Leu} species were recognized by the leucyl-tRNA synthetase in the chloroplast fraction, but only the two types of tRNA^{Leu} found in both the cytoplasmic and the chloroplast fractions could be charged by the cytoplasmic enzyme. In contrast to these findings with bean leaves, Guderian et al. (1972) resolved six species of tRNA^{Leu} from extracts of tobacco leaves, but only two of these were recovered in purified chloroplasts. Two of the remaining four species were recovered in purified mitochondria, and presumably the other two were located in the cytoplasmic matrix. The tobacco chloroplast leucyl-tRNA synthetase acylated the two chloroplast tRNA^{Leu} species but not the mitochondrial or the cytoplasmic tRNA^{Leu}. Since Guderian et al. (1972) observed that the mitochondrial enzyme would recognize the four tRNA^{Leu} species exclusive of the chloroplast species, the findings of Burkard et al. (1970), that the leucyl-tRNA synthetase recovered in the chloroplast fraction would recognize the five species of tRNA^{Leu} found in bean, may have been the result of the presence of mitochondria in the chloroplast fraction.

Merrick and Dure (1972) observed six isoaccepting species of tRNA^{Leu} in cotton cotyledons and suggested that three of these species reside in the chloroplast on the basis of increases in amount during development of chloroplasts upon seed germination. No evidence was presented on whether the amounts of mitochondrial tRNA^{Leu} species remain relatively constant during germination or increase along with chloroplast tRNA levels. However, chloroplasts purified from cotton seedlings contained the same three species of tRNA that increased in amount during the germination of seeds and consequent development of chloroplasts (Merrick and Dure, 1972).

Burkard et al. (1970) separated four fractions of valine-specific tRNA from bean leaves. Three species were recovered in the chloroplast fraction, and three were also found in the cytoplasmic fraction, with two forms common to both fractions. The form found only in the chloroplast fraction was recognized only by the chloroplast valyl-tRNA synthetase, whereas the form unique to the cytoplasm was recognized only by the cytoplasmic enzyme. The forms in common to both the chloroplast and cytoplasmic fractions were acylated by both enzymes.

Using the increases in tRNA species that accompany seed germination as the basis for identifying chloroplast tRNA, Merrick and Dure (1972) have shown that chloroplasts of cotton seedlings contain, in addition to those already mentioned above, specific tRNA species for histidine, isoleucine, lysine, phenylalanine, tryptophan, and valine. Differences between cytoplasmic and chloroplast tRNA species for other amino acids were not detected by their procedures.

C. Elongation

Although reports on studies of elongation factors in chloroplasts
are few in number, they indicate that the elongation factors in the
chloroplast are active in protein synthesis with 70S ribosomes but
are not active with 82S ribosomes. Eisenstadt and Brawerman (1966)
found that heterologous mixtures of enzymes and ribosomes from *E. coli* and *Euglena* chloroplasts had protein synthetic activity comparable
to the homologous systems. But although amino acid incorporation into protein in both homologous systems was stimulated by the addition
of f_2 viral RNA, the heterologous mixtures were not. Their results
suggested that elongation factors are not species specific, whereas
one or several initiation factors may be species specific. Ciferri
and Parisi (1970) described experiments with heterologous mixtures of
ribosomes from *E. coli* or yeast and elongation factors from several
green and blue-green algae. Factors extracted from cells that contain
both 70 and 82S ribosomes were active in protein synthesis when tested
with either *E. coli* 70S or yeast 82S ribosomes. Factors from *Nostoc
muscorum, Chlorella vulgaris,* and *Euglena gracilis* were active in protein synthesis with *E. coli* ribosomes, since all three species contain 70S ribosomes. But only extracts of *Chlorella* and *Euglena* were
active with 82S yeast ribosomes. Extracts of *Nostoc* were inactive
with yeast ribosomes, as blue-green algae contain only 70S ribosomes
(Taylor and Storck, 1964). Elongation factors from either light-
grown, etiolated, or permanently bleached cells of *Euglena* were active with the yeast ribosomes, but only extracts of light-grown cells
that contained developed chloroplasts functioned with *E. coli* ribosomes. The activity observed with extracts of green *Euglena* and *Chlorella,* cells with both types of ribosomes, was apparently the result
of the combined presence of both chloroplast and cytoplasmic elongation factors in the extracts.

These results suggest that green algae, and probably higher plants
as well, contain specific factors for 70 and 82S ribosomes. This
suggestion was tested by Tiboni et al. (1970), who fractionated the
elongation factors from the nonphotosynthetic alga *Prototheca zopfii.*

Although these cells lack chloroplast ribosomes, two forms of elongation factor I and two forms of elongation factor II were resolved by chromatography of cell extracts. One of the forms of each factor was specific for 70S ribosomes, whereas the other form of each was specific for 82S ribosomes. However, recently Ciferri and Tiboni (1973) found that the factor II, which is specific for 70S ribosomes in *P. zopfii,* is a mitochondrial rather than a chloroplast protein. The chloroplast elongation factor II (EF-G$_{chl}$ by their nomenclature) could not be detected in dark-grown cells of *Chlorella vulgaris* or in an apochlorotic mutant strain of this alga. Extracts of light-grown, wild-type *Chlorella* cells contained two elongation factor-II proteins specific for 70S ribosomes, and cell fractionation studies indicated that one form was in the chloroplasts and the other in mitochondria. Ciferri and Tiboni (1973) suggested that a developed and functional chloroplast is required for the presence of the chloroplast-specific elongation factors. Similar results were also found for *Euglena,* since etiolated and permanently bleached mutant cells have no combined elongation factor activity specific for 70S ribosomes. In *Euglena,* synthesis of the chloroplast form of elongation factor I, in particular, appears to be induced by light; it may possibly be synthesized on chloroplast ribosomes (Ciferri and Parisi, 1970).

The requirement of Mg^{2+} for elongation on chloroplast ribosomes differs from that on cytoplasmic ribosomes. With endogenous messenger RNA, Boardman et al. (1966) found that chloroplast ribosomes required about 15 mM Mg^{2+} for optimal activity, whereas cytoplasmic ribosomes had greatest activity at 4-5 mM Mg^{2+}. Purified chloroplast ribosomes from *Chlamydomonas reinhardtii* required about 25 mM Mg^{2+} for optimal activity in poly(U)-directed synthesis of polyphenylalanine, while ribosomes from the cytoplasm of these cells had optimal activity in the same system at 10-12 mM Mg^{2+} (Schlanger et al., 1972; Chua et al., 1973a).

VI. SITES OF SYNTHESIS OF CHLOROPLAST PROTEINS

A. Identification of Products of in vitro
    Chloroplast Protein Synthesis

The work on protein synthesis with isolated chloroplasts has had
a twofold goal. The first has been to establish that these organelles
synthesize protein, and to learn the mechanism by which this synthesis occurs. In general terms, this goal has been achieved. The second aim has been to identify the products of protein synthesis, a
goal that has been particularly difficult to attain. Not only has it
been difficult to provide conditions that facilitate synthesis of complete polypeptides, but once achieved, the minuscule amounts of protein synthesized have made identification difficult. However, the
theoretical soundness of this approach has prompted a number of investigators to attempt this goal of identifying the proteins synthesized by chloroplasts in vitro.

Isolated chloroplasts incorporate labeled amino acids into membrane-bound as well as soluble polypeptides, but the radioactivity
recovered in soluble proteins after an in vitro incubation accounts
for only 25-50% of the total (Margulies and Parenti, 1968; Ranalletti
et al., 1969; Goffeau, 1969; Spencer et al., 1971; Harris et al.,
1973). Ranalletti et al. (1969) examined the incorporation of labeled
amino acids into the membrane-bound Ca^{2+}-dependent adenosine triphosphatase, and Margulies (1970) and Harris et al. (1973) have studied
the synthesis in vitro of the soluble enzyme ribulose 1,5-diphosphate
carboxylase. Although a small amount of radioactivity remained with
the most purified fractions in each study, the characterization of
these fractions was not sufficient to prove unequivocally that the
labeled amino acids were incorporated into these enzymes. Both enzymes contain subunit polypeptides of different molecular weights
(Rutner, 1970; Kawashima and Wildman, 1970; Nelson et al., 1972b),
but incorporation of labeled amino acids into the different subunits
was not determined. Spencer et al. (1971) demonstrated synthesis of
both high- and low-molecular-weight polypeptides in isolated spinach
chloroplasts, but they were unable to identify the polypeptides. The

large polypeptides synthesized were approximately the size of the
large subunits of ribulose 1,5-diphosphate carboxylase, whereas the
small polypeptides were similar in size to the small subunit of this
enzyme.

Blair and Ellis (1973) examined the polypeptides (synthesized in
chloroplasts from pea seedlings) by electrophoresis on polyacrylamide
gels in the presence of sodium dodecylsulfate. The radioactivity in
soluble polypeptides, which accounted for 25% of the total [^{14}C]leu-
cine incorporated into protein by the preparation, was recovered after electrophoresis in a single polypeptide fraction. The labeled
component was identified as the large subunit of ribulose 1,5-diphos-
phate carboxylase by its size and from an analysis of peptides produced by digestion with trypsin. The absence of significant radioactivity in other polypeptides may indicate that chloroplasts synthesize few, and perhaps only one, soluble protein. Incorporation of
labeled amino acids into the small subunit of ribulose 1,5-diphosphate
was not observed (Blair and Ellis, 1973).

In a continuation of this work, Eaglesham and Ellis (1974) determined the number of labeled polypeptides recovered with the chloroplast membranes after a period of in vitro protein synthesis. Six
or seven labeled polypeptide fractions, whose molecular weights ranged
from 85,000 to 18,000, were observed after electrophoresis in the
presence of sodium dodecylsulfate. These polypeptides could not be
equated with any of the thylakoid proteins already characterized, and
so their identity remains unknown. Yet these studies have provided
the most definitive results so far on in vitro protein synthesis in
chloroplasts, and perhaps have as much value with regard to those
polypeptides not labeled as to those that were.

B. Studies with Antibiotics and Ribosome-deficient Mutant Strains

Lacking the means to easily identify proteins synthesized by isolated chloroplasts, most investigators have attempted to determine
what proteins are synthesized within the organelle by studying the
effects of antibiotics. The approach has been to determine which

proteins are localized within the chloroplast, and then to test the effects of inhibitors of 70S ribosomes on the synthesis of these proteins. Since intact cells synthesize net amounts of protein, the effects of inhibitors can be determined by assaying the functional activity of a protein rather than attempting identification by chemical means. Although problems of interpretation of such data may arise as the result of secondary effects of the antibiotics (Ellis and Hartley, 1971), this approach has provided most of the information available on the sites of synthesis of chloroplast proteins. Because of the greater fidelity of translation in vivo, the effects of inhibitors in vivo have provided the basis by which results with isolated chloroplasts can be evaluated.

A second approach toward determining which proteins are synthesized within chloroplasts in situ is the use of mutants deficient in chloroplast ribosomes. Two mutant strains of *Chlamydomonas reinhardtii, ac*-20 and *cr*-1, have been used for this purpose (see Goodenough and Levine, 1970; Boynton et al., 1972). These strains have advantages over some other types of mutants, e.g., the W₃BUL strain of *Euglena,* which lack chloroplast DNA (Edelman et al., 1965). The mutations in the *Chlamydomonas* strains reside in nuclear DNA, and apparently specifically affect the cells' ability to make chloroplast ribosomes, resulting in deficiencies in the proteins translated by these ribosomes. Generally, the studies on the ribosome-deficient strains have complemented those with antibiotics. A more detailed review of these studies follows.

1. Aminoacyl-tRNA Synthetases

In Sect. V,B it was noted that synthesis of several forms of tRNA and aminoacyl-tRNA synthetases was induced in *Euglena* when etiolated cells were transferred to light. This observation suggests that these components may be synthesized within the chloroplast. Furthermore, the light-inducible form of isoleucyl-tRNA synthetase in *Euglena* was not found in the W₃BUL strain, which lacks chloroplast DNA (Reger et al., 1970). Since this lack of chloroplast DNA results in a lack of chloroplast ribosomes, it is not clear if this enzyme is coded by chloroplast DNA, or if the messenger RNA for this enzyme is only translated in the chloroplast. But in contrast to the chloroplast isoleucyl-tRNA synthetase, the chloroplast form of phenylalanyl-tRNA synthetase in *Euglena* is present in dark-grown and in W$_3$BUL cells.

These findings were extended by Hecker et al. (1974). Streptomycin did not inhibit synthesis of the phenylalanyl-tRNA or the valyl-tRNA synthetases during development of chloroplasts in *Euglena*, but cycloheximide strongly inhibited the increases in these activities. Cells of the W$_3$BUL strain showed activities of these enzymes comparable to the level in dark-grown, wild-type cells. Although the activities of these enzymes increase concomitantly with chloroplast development, the results of this study demonstrate that these two chloroplast enzymes are neither coded for nor synthesized within the chloroplast.

Parthier et al. (1972) attempted to distinguish chloroplast from cytoplasmic forms of aminoacyl-tRNA synthetases of *Euglena* by their activity with tRNA molecules prepared from the blue-green alga *Anacystis nidulans*. During greening of etiolated *Euglena* cells, the activities of several enzymes, in particular those specific for arginine, aspartic acid, leucine, lysine, methionine, phenylalanine, tyrosine, and valine, increased about 10-fold during a 48-hr period. After chromatographic resolution of the synthetases for a given amino acid, the light-induced form was considerably more active with tRNA from *Anacystis* than with tRNA from *Euglena* cells. The effect of cycloheximide on the increases in these activities was difficult to interpret. Chloramphenicol, at concentrations sufficiently high to prevent chloroplast development, did not prevent an increase in any of these activities, although the extent of the increases was about one-half that found with no inhibitor.

2. *Ribosomal Proteins*

The sites of synthesis of proteins of chloroplast ribosomes have received less attention that those of mitochondrial ribosomes. How-

ever, sufficient data are available to suggest that chloroplast ribosomal proteins are synthesized outside the organelle on cytoplasmic
ribosomes, as is the case for mitochondrial ribosomal proteins (Lizardi and Luck, 1972). Long-term growth of *Chlamydomonas reinhardtii*
cells in the presence of spectinomycin or chloramphenicol did not im-
pair the cells' ability to form chloroplast ribosomes, although these
drugs inhibited protein synthesis on these ribosomes (Goodenough,
1971). Nor did chloramphenicol affect the four- to fivefold increase
in the amount of chloroplast ribosomes that occurred after cells of
Chlamydomonas reinhardtii ac-20 were transferred from mixotrophic to
phototrophic growth conditions (Goodenough and Levine, 1971). Since
mixotrophically grown *ac*-20 cells are deficient in chloroplast ribosomes and, as a result, in proteins synthesized on these ribosomes,
these data suggest that chloroplast ribosomal proteins are synthesized
on cytoplasmic ribosomes.

Margulies (1971) reported that cycloheximide but not chloramphenicol inhibited incorporation in vivo of labeled amino acids into proteins of 70S ribosomes of *Chlamydomonas*. Since in these experiments
the chloroplast ribosomal proteins were not separated to determine
the effect of the inhibitors on the labeling of each protein, one can
conclude only that the bulk of chloroplast ribosomal proteins may be
synthesized on cytoplasmic ribosomes in this alga. This conclusion
was advanced by Honeycutt and Margulies (1973), who studied the kinetics of labeling of chloroplast ribosomal proteins by [³H]arginine.
A lag of several minutes was observed before significant labeling of
these proteins occurred, whereas cytoplasmic ribosomal proteins became labeled immediately. Cycloheximide, but not chloramphenicol,
inhibited the labeling of chloroplast ribosomal proteins.

Ellis and Hartley (1971) reached a different conclusion from
their studies of the effects of lincomycin, also an inhibitor of protein synthesis on chloroplast ribosomes, on the greening of etiolated
pea seedlings. Lincomycin did not inhibit incorporation of [³²P]or-
thophosphate into chloroplast ribosomal RNA, but it inhibited the
accumulation of chloroplast ribosomes over a two-day period during
greening. These authors suggest that the lack of accumulation of

chloroplast ribosomes under these conditions resulted from the inhibition by lincomycin of the synthesis of at least some ribosomal proteins. However, Ingle et al. (1971) observed that chloramphenicol
inhibited the incorporation of $[^{32}P]$orthophosphate into chloroplast
ribosomal RNA in cotyledons of radish seedlings over a 6-hr period.
It seems unwarranted to conclude from the probably indirect effects
of the inhibitors on RNA synthesis that synthesis of ribosomal proteins was inhibited.

3. Calvin Cycle Enzymes

During greening of etiolated plant cells, synthesis of many chloroplast enzymes involved in the photosynthetic carbon-reduction cycle
is stimulated. The effect of inhibitors of protein synthesis on the
increase in activity of each enzyme has been used as an indication
of their site of synthesis. By this criterion, most enzymes of this
cycle, even though perhaps localized almost exclusively in the chloroplast (Smillie and Scott, 1969; Boulter et al., 1972), are synthesized on cytoplasmic ribosomes. Synthesis of this group of enzymes,
which includes ribose 5-phosphate isomerase, 3-phosphoglycerate kinase, triose phosphate isomerase, fructose 1,6-diphosphate aldolase,
and transketolase, is unaffected by treating leaves with chloramphenicol (Ireland and Bradbeer, 1971; Ellis and Hartley, 1971). However,
studies on other enzymes have provided variable results. Ireland
and Bradbeer (1971) reported that chloramphenicol strongly inhibited
development of the activity of phosphoribulose kinase in greening
bean leaves, but other investigators found with pea leaves (Ellis and
Hartley, 1971) and with *Chlamydomonas* (Armstrong et al., 1971) that
inhibitors of chloroplast protein synthesis had no effect on the synthesis of this enzyme. Smillie et al. (1967) reported that, in *Euglena*, chloramphenicol but not cycloheximide inhibited the increase
in the amount of NADP-dependent glyceraldehyde 3-phosphate dehydrogenase. However, Margulies (1964) and Ellis and Hartley (1971) found
that chloramphenicol and lincomycin, respectively, did not inhibit
synthesis of the NADP-dependent glyceraldehyde 3-phosphate dehydrogenase in bean and pea leaves. Also, Schiff (1971) observed no sig-

nificant inhibition of the synthesis of this enzyme in Euglena cells
treated with streptomycin; and light-grown cells of the W$_3$BUL strain
of Euglena, a permanently bleached strain that lacks chloroplast DNA
and therefore also chloroplast ribosomes, contained an amount of this
enzyme comparable to that in dark-grown, wild-type cells. Togasaki
and Levine (1970) measured the activities of these chloroplast enzymes in the ac-20 strain of Chlamydomonas reinhardtii under conditions in which the cells were deficient in chloroplast ribosomes, and
found them all similar to the activities in wild-type cells. The
data from the ribosome-deficient strain complement the findings in
higher plants with antibiotics and indicate that these soluble chloroplast enzymes of the Calvin cycle are synthesized outside the organelle.

4. Ribulose 1,5-Diphosphate Carboxylase

Ribulose 1,5-diphosphate carboxylase (fraction I protein) is the
major protein of plant cells, accounting for approximately 10% of the
total cellular protein of Chlamydomonas (Boynton et al., 1972) and
for about 40% of the soluble protein of tobacco chloroplasts (Wildman, 1971). Because of its functional importance and physical pro-
minence in plant cells, this enzyme is discussed separately. The
enzyme has a molecular weight of 557,000 (Paulsen and Lane, 1966) and
is composed of two types of subunits with molecular weights of about
55,000 for the large and either 12,000, as determined by gel electrophoresis, or 24,500, as determined by Sephadex gel filtration, for
the small subunit (Rutner, 1970; Sugiyama and Akazawa, 1970; Kawashima and Wildman, 1970). Data from many laboratories have provided
evidence that chloroplast ribosomes are involved in the synthesis of
this enzyme (see Boulter et al., 1972, for a review). The question
that has not been adequately resolved is whether both subunits of
this enzyme are synthesized within the organelle or if one is synthesized on cytoplasmic ribosomes. Smillie et al. (1967, 1971) found
that chloramphenicol, but not cycloheximide, inhibited the increase
of the carboxylase activity during a 72-hr period of illumination of

etiolated *Euglena* cells. However, in many other plant cells both
chloramphenicol and cycloheximide inhibited synthesis of the enzyme
(Kleinkopf et al., 1970; Criddle et al., 1970; Givan and Criddle,
1972; Armstrong et al., 1971; Oshio and Hase, 1972). Since the possibility exists that synthesis of each subunit is affected differently
by the antibiotics, only those experiments in which incorporation of
labeled amino acids into each subunit has been examined can provide
data on the question of the sites of synthesis. Criddle et al. (1970)
reported that synthesis of the large subunit of the enzyme in barley
leaves was more sensitive to chloramphenicol than synthesis of small
subunits by comparing the $^{14}C/^3H$ ratios for each subunit produced by
the incorporation of $[^{14}C]$leucine (in chloramphenicol-treated cells)
with that of $[^3H]$leucine (in control cells). Cycloheximide, on the
other hand, inhibited incorporation into small subunits slightly more
than the incorporation into large subunits. In *Chlamydomonas,* however, no differential sensitivity was observed (Givan and Criddle,
1972), and both inhibitors were equally effective against synthesis
of either subunit. These investigators suggested that the subunits
are synthesized at separate sites in the cell, and that a tight control exists by which the presence of one subunit regulates the synthesis of the other. Since the complete enzyme is found within the
chloroplast, the implication of these results is that the rate of
entrance of the small subunit into the chloroplast would control the
rate of synthesis of large subunits on chloroplast ribosomes. Conversely, it is not clear in this regard how inhibition of the synthesis of large subunits could affect synthesis of small subunits.

Unfortunately, on close examination, these experiments have provided equivocal results with respect to the sites of synthesis of
ribulose 1,5-diphosphate carboxylase. Criddle et al. (1970) isolated
the complete enzyme before determining the $^{14}C/^3H$ ratios, and it does
not seem reasonable that the protein would assemble without stoichiometric amounts of both subunits present. Preferential labeling of
one subunit might occur if free subunits exchange with the complete
enzyme, but whether this occurs in vivo is not known.

The other inhibitor, cycloheximide, strongly inhibited the synthesis of both subunits of ribulose 1,5-diphosphate carboxylase (Criddle et al., 1970; Givan and Criddle, 1972), even though the large
subunit is apparently synthesized within the chloroplast (Blair and
Ellis, 1973). In experiments with cycloheximide, there are no soluble proteins known whose synthesis in the chloroplast is not affected
by the drug and which could act as controls. Therefore, inhibition
of synthesis of the carboxylase by cycloheximide cannot be interpreted. The cause of this inhibition of chloroplast as well as cytoplasmic protein synthesis by cycloheximide in vivo is not known, but
nearly complete inhibition of [^{3}H]arginine incorporation into the
large subunit of ribulose 1,5-diphosphate carboxylase can be observed
in *Chlamydomonas* after treating cells for a time period as short as
15 min (Stegeman and Hoober, unpublished experiments). Because chloroplast ribosomes are insensitive to cycloheximide in vitro (Blair
and Ellis, 1973), the inhibition of synthesis of this enzyme in vivo
is probably a secondary effect and not the result of direct interaction with chloroplast ribosomes. Moreover, cycloheximide does not
inhibit translation on mitochondrial ribosomes in *Euglena* cells
(Avadhani and Buetow, 1972).

Kawashima (1970) made an interesting observation on the incorporation of carbon from $^{14}CO_2$ into both subunits of ribulose 1,5-diphos-
phate carboxylase. After intact leaves of tobacco were labeled for
15 min, several amino acids in the small subunit had specific radioactivities lower than those for the corresponding amino acids in the
large subunit. Kawashima's suggestion from these data, that small
subunits are synthesized in the cytoplasm, is compatible with experiments described in Sect. II, which showed that chloroplast proteins
are labeled sooner than cytoplasmic proteins by $^{14}CO_2$. But these
data have several noteworthy features. The ratio of the specific
radioactivity of a given amino acid in the small subunit to that of
the same amino acid in the large subunit was approximately the same
for all the amino acids that were labeled. Moreover, the relationship of the specific radioactivity of any amino acid to that of others
in the same subunit were similar for both polypeptides. Such uni-

formity would not be expected if the polypeptides were synthesized from different amino acid pools, unless rapid equilibration of the radioactivity between the chloroplast and cytoplasmic amino acid pools occurred. If such rapid equilibration did occur, then it is not clear how the specific radioactivity of the small subunit could be less than that of the large subunit unless a significant pool of small subunits existed in the cells. Since only the complete enzyme was examined in these experiments and the labeling period was 15 min, a pool of small subunits would lower the specific radioactivity of this subunit by dilution. If this is true, a conclusion on the sites of synthesis of the subunits from these experiments cannot be made. This question might be resolved if the kinetics of incorporation of $^{14}CO_2$ into the total amount of each subunit, in the complete enzyme as well as free subunits, were measured.

Hoober (1972) found that synthesis of both large and small subunits of ribulose 1,5-diphosphate carboxylase in *Chlamydomonas* was inhibited by chloramphenicol when the incorporation of [^3H]arginine into total cellular protein was examined by electrophoresis. Synthesis of both subunits also stopped when cells were transferred from light to the dark. The latter experiment involved no antibiotics, and since general synthesis of cytoplasmic proteins was not affected by the transfer from light to the dark, it is not known how synthesis of the small subunit could be inhibited by this condition if not also synthesized within the chloroplast. However, the effects of chloramphenicol, and of transfer to the dark, on the synthesis of these polypeptides could involve processes not evident, such as translational control or rapid degradation of one subunit if not complexed with the other.

Holdsworth (1971) presented evidence that ribulose 1,5-diphosphate carboxylase is concentrated in pyrenoid structures of *Eremosphaera viridis*. Schiff (1973) has also described results that indicate the pyrenoid of *Euglena* chloroplasts consists of a highly condensed form of this enzyme. These findings are.supported by the atrophy that occurs in pyrenoid structures of mixotrophically grown cells of *Chlamydomonas reinhardtii ac-20*, which are deficient in

chloroplast ribosomes and synthesize little of the carboxylase (Togasaki and Levine, 1970; Goodenough and Levine, 1970; Boynton et al.,
1972). A similar decrease in size of the pyrenoid body was observed
in wild-type cells treated over several days with inhibitors of chloroplast protein synthesis (Goodenough, 1971). In the ac-20 strain of
Chlamydomonas reinhardtii, transfer to phototrophic conditions enables
recovery of ribosomes, synthesis of the enzyme, and regeneration of
the pyrenoid structures. The increase in size of the pyrenoid structure was not affected by a low level (1 µg/ml) of cycloheximide, but
was inhibited by chloramphenicol (Goodenough and Levine, 1970).

On the basis of the experiments discussed above, a definite conclusion cannot be reached on the site or sites of synthesis of ribulose 1,5-diphosphate carboxylase. The inhibitory action of cycloheximide cannot be interpreted for reasons that for the moment are
obscure. However, there is agreement, from the effect of chloramphenicol and in vitro protein synthesis (Blair and Ellis, 1973), that
at least the large subunit of this enzyme is synthesized within the
chloroplast. Also, the studies with Euglena (Smillie et al., 1967,
1971) and Chlamydomonas (Hoober, 1972) suggest that the small subunit
may also be synthesized within the chloroplast of these algae.

Several groups of investigators have attacked directly the problem
of the sites of synthesis of the carboxylase subunits by separating
chloroplast from cytoplasmic ribosomes and testing for the presence
of nascent carboxylase polypeptides with specific antibodies. Gooding et al. (1973) treated ribosomes from wheat leaves with [3H]puro-
mycin and studied the ability of antiserum prepared against purified
carboxylase subunits to precipitate labeled peptides. Although antiserum to the large carboxylase subunit precipitated peptides released
from both 70 and 82S ribosomes, antiserum to the small subunit exhibited a preference for peptides released from 82S ribosomes. The conclusion from this study, that the small subunit is synthesized on cytoplasmic ribosomes whereas the large subunit is synthesized on chloroplast ribosomes, must be moderated by noting that nonspecific, back-
ground precipitation was often equal to or greater than specific an-

tibody precipitation and that the techniques employed did not completely resolve the two types of ribosomes.

More convincing evidence for the synthesis of the small subunit
of ribulose 1,5-diphosphate carboxylase on cytoplasmic ribosomes was
obtained by Gray and Kekwick (1973). Cytoplasmic polysomes were prepared from greening bean leaves and allowed to complete synthesis of
nascent polypeptides in the presence of labeled amino acids. Antiserum against carboxylase small subunits precipitated 30% of the labeled polypeptides released from the ribosomes, and most of the radioactivity in the precipitated fraction was eluted from Sephadex
G-100 in the position of carboxylase small subunits.

These latter studies, coupled with the absence of a labeled polypeptide corresponding to the small subunit of the carboxylase after
in vitro protein synthesis with isolated chloroplasts (Blair and Ellis, 1973), have provided the most direct evidence supporting the
cytoplasmic ribosomes as the site of synthesis of the small subunit
of this enzyme.

5. *Proteins of Thylakoid Membranes*

The sites of synthesis of thylakoid membrane components have been
studied by the use of inhibitors of protein synthesis, and to a limited extent with the ac-20 strain of *Chlamydomonas reinhardtii*. Relatively low levels of chloramphenicol inhibit the development of
photosynthetic activity of these membranes in plants and algae (Hudock
et al., 1964; Hoober et al., 1969; Armstrong et al., 1971; Nadler et
al., 1972; Eytan and Ohad, 1972a; Wallach et al., 1972), whereas high
concentrations of the drug also inhibit strongly but not completely
formation of the membranes (Margulies, 1966; Smillie et al., 1967;
Eytan and Ohad, 1970; Hoober, 1972; Smith-Johannsen and Gibbs, 1972).
The mechanism by which the formation of membranes is inhibited by
chloramphenicol is not known, but may be the result of secondary effects on chlorophyll synthesis. In some cases the inhibition of
chlorophyll synthesis can be partially overcome by adding δ-aminolevulinic acid to the cultures (Gassman and Bogorad, 1967; Kirk, 1968;

Nadler and Granick, 1970; Nadler et al., 1972). Furthermore, chloramphenicol is less inhibitory when added to cells actively synthesizing chlorophyll than when added to etiolated cells at the time of illumination (Armstrong et al., 1971; Hase, 1971; Hoober, 1972). This
effect suggests that chloramphenicol may affect the synthesis of enzymes producing precursors of chlorophyll. The results with low
levels of chloramphenicol are less problematic, since membranes form
at nearly the control rate but are deficient in proteins necessary
for photosynthetic function (Hoober et al., 1969).

On the other hand, cycloheximide completely inhibits chlorophyll
synthesis and formation of thylakoid membranes (Kirk, 1968; Hoober
et al., 1969; Eytan and Ohad, 1970; Armstrong et al., 1971). Perl
(1972) reported that low concentrations of cycloheximide stimulate
chlorophyll synthesis in greening *Euglena,* but similar results have
not been obtained with *Chlamydomonas* (Hoober et al., 1969; Hoober,
unpublished results). The inhibition of synthesis of chlorophyll by
cycloheximide can be partially overcome by feeding treated barley
leaves δ-aminolevulinic acid (Nadler et al., 1972), but such feeding
does not overcome this inhibition by cycloheximide in *Euglena* (Kirk,
1968). In greening *Euglena* and *Chlamydomonas,* the rapidity of the
effect, both when cycloheximide is added and when it is removed, suggests that the effect is not directly on the synthesis of enzymes involved on the pathway of chlorophyll synthesis, but that proteins
synthesized on cytoplasmic ribosomes are required in stoichiometric
amounts for the accumulation of chlorophyll and assembly of the thylakoid membranes (Kirk, 1968; Hoober et al., 1969). Moreover, Smillie et al. (1968) reported that chlorophyll synthesis during illumination of etiolated *Euglena* cells, which were previously treated in
the dark with chloramphenicol, was insensitive to cycloheximide.
These results suggest that during the treatment with chloramphenicol
in the dark, proteins of cytoplasmic origin accumulated, which enabled
chlorophyll synthesis to proceed after cycloheximide was added (see
also Eytan and Ohad, 1970; Hoober and Stegeman, 1973).

The inhibition of chloroplast protein synthesis does no permanent
damage to the thylakoid membranes made in the presence of chloram-

phenicol. Transferring chloramphenicol-treated cells to drug-free
medium permitted the recovery of chloroplast protein synthesis and
the development of photosynthetic activities (Hoober et al., 1969;
Wallach et al., 1972; Eytan and Ohad, 1972a). The presence of cycloheximide, at concentrations sufficient to completely inhibit chlorophyll synthesis and membrane formation, did not prevent repair of the
membranes (Hoober et al., 1969; Wallach et al., 1972). Thus, proteins of chloroplast origin are necessary for complete, functional
activity of the membranes, and these proteins can be incorporated into already existing, deficient membranes. The normal assembly of the
membranes may also follow a similar stepwise course, although not as
separated in time (Eytan and Ohad, 1972a,b). Activation of thylakoid
membranes has also been observed (in cells not treated with antibiotics) during the initial 4 hr of the light phase of synchronous culture before chlorophyll synthesis and membrane formation proceed
(Schor et al., 1970b). Since the usual assays for function measure
systems rather than individual reactions, the activity of these systems may also require proteins synthesized on cytoplasmic ribosomes.
This possibility cannot be determined directly, because of the inhibitory effect of cycloheximide on membrane assembly.

Hoober (1970) and Eytan and Ohad (1970) studied the sites of synthesis of thylakoid membrane polypeptides by taking advantage of the
observation that *Chlamydomonas* cells treated with chloramphenicol
formed membranes deficient in polypeptides synthesized in the chloroplast. Subsequent transfer of such cells to cycloheximide-containing
medium inhibited synthesis of cytoplasmic proteins and further assembly of membranes, but permitted synthesis and incorporation of proteins made in the chloroplast into the membranes. Polypeptides synthesized in the presence of each inhibitor were labeled with either
$[^{14}C]$- or $[^3H]$arginine. Then the membranes were purified and the
polypeptides separated by electrophoresis in buffers containing sodium dodecylsulfate. The electrophoretic pattern of the protein stain
consisted of two major bands of protein stain and numerous minor
bands (Hoober, 1970, 1972). The patterns of radioactivity revealed
that some polypeptides present in low amounts are synthesized in the

chloroplast, others in the cytoplasm. However, the results of this
approach indicated that the two major polypeptides of the thylakoid
membranes of Chlamydomonas are synthesized on cytoplasmic ribosomes
(Hoober, 1970, 1972).

Additional evidence for synthesis of the two major thylakoid polypeptides on cytoplasmic ribosomes in Chlamydomonas reinhardtii was
obtained in a study of the incorporation of [^{3}H]leucine into protein
of these cells (Stegeman and Hoober, 1974). At low extracellular
concentrations of the amino acid, no labeling of polypeptides synthesized on cytoplasmic or chloroplast ribosomes occurred. But at a
several hundred-fold greater concentration (2 x 10^{-5} M) in the medium,
polypeptides synthesized on cytoplasmic ribosomes were labeled. Electrophoresis of total cellular protein revealed two prominent radioactive peaks, corresponding to the two major thylakoid polypeptides.
However, even at the higher concentration, labeling of the principal
product of chloroplast ribosomes, the large subunit of ribulose 1,5-
diphosphate carboxylase, was barely detectable.

The ac-20 and cr-1 strains of Chlamydomonas reinhardtii, which
are deficient in chloroplast ribosomes, form thylakoid membranes similar to those formed in chloramphenicol-treated cells, in that the
photochemical activities of these membranes are low (Goodenough and
Levine, 1970; Boynton et al., 1972). However, the fact that thylakoid membranes are present in these mutant cells further supports the
conclusion that the major proteins required for assembly of the membrane structure are synthesized on cytoplasmic ribosomes.

Thylakoid membranes in Vicia faba leaves treated with tetracycline
or chloramphenicol lacked several minor polypeptides but still contained the major membrane polypeptide (Machold, 1971). Machold and
Aurich (1972) extended these observations and found that chloramphenicol inhibited the incorporation of [^{14}C]leucine into most of the minor
polypeptide components of the chloroplast membranes. Chloramphenicol
strongly inhibited synthesis of the protein involved in the photosystem I protein-chlorophyll complex. However, synthesis of the major
polypeptide of the membranes was not affected by chloramphenicol,
but was strongly inhibited by cycloheximide. Eaglesham and Ellis

(1974) also found that the major thylakoid polypeptide of pea chloroplasts was not labeled during in vitro protein synthesis with isolated
chloroplasts, confirming the in vivo studies that this protein is synthesized on cytoplasmic ribosomes.

It is interesting that *Chlamydomonas* chloroplast membranes contain
two major polypeptides of molecular weights about 24,000 and 28,000
(Hoober, 1970; Levine and Duram, 1973), whereas higher plant chloroplast membranes contain only one major polypeptide of approximately
25,000 molecular weight (Menke and Ruppel, 1971; Lagoutte and Duranton, 1971; Herrmann, 1972; Levine et al., 1972; Machold and Aurich,
1972; Anderson and Levine, 1974). However, in both systems the major
polypeptides are synthesized on cytoplasmic ribosomes. These polypeptides also appear to be involved with the photosystem II complex
(Machold et al., 1971; Kung and Thornber, 1971; Levine et al., 1972).
Since the isolated photosystem II complex comprises the major portion
of thylakoid membranes of higher plants (Kung and Thornber, 1971),
and the development of the activity of photosystem II precedes that
of photosystem I during greening of etiolated cells (Alberte et al.,
1972, 1973), the major polypeptides may be required for the initial
assembly or growth of the membrane. A similar suggestion was made
earlier on the basis of the rapid inhibition of chlorophyll synthesis
and membrane formation observed in the presence of cycloheximide
(Kirk, 1968; Hoober et al., 1969).

Margulies (1962, 1964) observed that etiolated seedlings treated
with chloramphenicol failed to develop photoreductive activities when
exposed to light. Over the last few years it has been observed that
this drug inhibits the development, during greening of etiolated plant
cells, of many activities of the thylakoid membranes including photosystems I and II (Margulies, 1964; Hudock et al., 1964; Hoober et al.,
1969), oxygen evolution (Armstrong et al., 1971; Nadler et al., 1972),
and photophosphorylation (Wallach et al., 1972). These observations
have been interpreted to mean that chloramphenicol, by inhibiting
protein synthesis in the chloroplasts, inhibits the synthesis of proteins necessary for these activities. In some cases, assaying individual components has supported this view.

Smillie et al. (1967) observed that both chloramphenicol and cycloheximide inhibited synthesis in *Euglena* of cytochromes-552 and
-561, but the inhibitory effect of chloramphenicol was greater than
that of cycloheximide on the synthesis of each of these components.
Synthesis of cytochrome-552 was also diminished in *Euglena* cells
treated with streptomycin (Schiff, 1971) and 5-fluorouracil (Evans
and Smillie, 1971), both of which also inhibited chloroplast development. In *Chlamydomonas,* chloramphenicol and spectinomycin inhibited
completely the synthesis of cytochrome-553 and -563 (Armstrong et al.,
1971). However, in mixotrophically grown cells of *Chlamydomonas rein-*
hardtii ac-20, which are deficient in chloroplast ribosomes, significant amounts of these two cytochromes were present, but no cytochrome-
559 was detected (Levine and Paszewski, 1970). This anomaly between
the results with antibiotics and with the *ac*-20 mutant strain prompted
a reinvestigation of the levels of the cytochromes in *ac*-20 (Levine
and Armstrong, 1972). Mixotrophically grown *ac*-20 cells contain about
10% of the chloroplast ribosomes and about 20% of the amount of cyto-
chrome-553 and -563 found in wild-type cells. Phototrophically grown
cells contain about 25% of the wild-type level of chloroplast ribosomes and 40% of the amounts of the two cytochromes found in wild-type
cells. After transfer of mixotrophically grown cells to phototrophic
growth conditions, synthesis of the cytochromes increased, but by 13
hr of phototrophic growth the amounts of the cytochromes had increased
only 20 to 40% over the original level. In contrast, the phototrophic
level of chloroplast ribosomes, which was a three to fourfold increase
over the amount in mixotrophically grown cells, was reached by 6-8 hr
(Goodenough and Levine, 1970).

However, further support for the chloroplast as the site of synthesis of chloroplast cytochromes in *Chlamydomonas* was the observation
that during synchronous growth of the cells, the level of the cytochromes fell during the dark phase but increased rapidly when the
cells entered the light phase of the growth cycle (Schor et al.,
1970b). This increase in the synthesis of cytochromes preceded the
synthesis of chlorophyll and the formation of new membranes. During

the dark phase, protein synthesis in the chloroplast appears to be inactive (Schor et al., 1970a; Hoober, 1972).

Ferredoxin-NADP reductase and ferredoxin are synthesized on cytoplasmic ribosomes in *Chlamydomonas*. Cycloheximide completely inhibited the synthesis of these membrane-associated proteins, whereas neither chloramphenicol nor spectinomycin had any effect (Armstrong et al., 1971). The levels of these proteins were not depressed in mixotrophically grown cells of the *ac-20* strain (Levine and Paszewski, 1970). In contrast to these results, Smillie et al. (1967) found that chloramphenicol strongly inhibited the synthesis of ferredoxin-NADP reductase in *Euglena*. Whether this is a difference between the two algae remains to be determined.

Wallach et al. (1972) have shown that the activity of photophosphorylation increases shortly after etiolated cells of *Chlamydomonas reinhardtii* y-1 are exposed to light and that the increase in this activity can be prevented with chloramphenicol. A coupling factor necessary for photophosphorylation has been purified from chloroplasts, which when appropriately activated has the activity of an adenosine triphosphatase (see Nelson et al., 1972a). Horak and Hill (1972) have studied the increase in this adenosine triphosphatase activity extracted from chloroplasts during greening of etiolated bean seedlings. The activity increased fivefold during a 72-hr light period, but in leaves treated with chloramphenicol the increase in activity was completely blocked. Cycloheximide was also inhibitory, although less so than chloramphenicol, and may have had toxic effects on the seedlings during the extended period of time of the experiment. These results indicated that the coupling factor is synthesized within the chloroplast, confirming the suggestion by Ranalletti et al. (1969) that this protein was synthesized by purified chloroplasts incubated in vitro.

6. Other Enzymes

Data are available that suggest both the DNA polymerase and RNA polymerase of chloroplasts are synthesized outside the organelle.

Surzycki (1969) reported that the presence of rifampicin, which inhibited synthesis of ribosomal RNA, had no effect on the synthesis of
DNA over four generations of growth of *Chlamydomonas*. These results
imply that chloroplast ribosomes are not required for synthesis of
the DNA polymerase. The RNA polymerase apparently is also synthesized
on cytoplasmic ribosomes, since the recovery of chloroplast ribosomes
that occurred when cells of the *ac-20* strain of *Chlamydomonas reinhardtii* were transferred from medium containing acetate to minimal
medium was not affected by chloramphenicol (Goodenough and Levine,
1971). This conclusion presumes that a normal level of RNA polymerase was maintained by cytoplasmic protein synthesis under conditions
in which the cells were deficient in chloroplast ribosomes. However,
the activity of RNA polymerase was not measured directly in these experiments. Ellis and Hartley (1971) observed a threefold increase
in the activity of the chloroplast RNA polymerase during greening of
etiolated pea seedlings. The increase in activity of this enzyme was
not inhibited by lincomycin, which indicates that synthesis of this
enzyme occurs on cytoplasmic ribosomes.

Sirevåg and Levine (1972) have studied the site of synthesis of
an acyl carrier protein-dependent fatty acid synthetase in *Chlamydomonas reinhardtii*. During synchronous growth of the algal cells, the
activity of the enzyme increased during the light phase of the cycle.
This increase was not affected by cycloheximide, but was partially inhibited by spectinomycin. In *Euglena gracilis,* two fatty acid synthetases were found, one dependent on acyl-carrier protein for activity,
and one independent of this protein (Delo et al., 1971). The dependent form of the enzyme was found in green but not in etiolated cells,
suggesting its location in the chloroplast. Furthermore, synthesis
of the acyl-carrier protein-dependent form of fatty acid synthetase,
upon illumination of etiolated cells, was not affected by cycloheximide but was completely inhibited by chloramphenicol (Ernst-Fonberg
and Bloch, 1971). These results suggest that the acyl-carrier protein-dependent fatty acyl synthetase is made within the chloroplast
in both *Chlamydomonas* and *Euglena*.

Nitrite reductase has been shown to be localized within the chloroplasts (Ritenous et al., 1967; Dalling et al., 1972). Nitrate reductase in these same experiments was recovered in the cytoplasmic matrix fraction. Synthesis of nitrite reductase, which is induced upon illumination of etiolated leaves, was abolished by chloramphenicol, whereas synthesis of nitrate reductase was not affected (Schrader et al., 1967). These enzymes, therefore, remain in the compartment in which they are synthesized.

VII. ORIGINS OF MESSENGER RNA FOR CHLOROPLAST PROTEINS

A. Properties of Chloroplast DNA

The informational content of chloroplast DNA remains a central problem in a consideration of the origins of chloroplast components. On the basis of current information, chloroplasts contain an appreciable amount of DNA, but may play a limited role in coding for chloroplast proteins. The amount of DNA in the single chloroplast of a *Chlamydomonas reinhardtii* cell is $7.2-8.6 \times 10^{-15}$g ($4-5 \times 10^9$ daltons) in gametes and twice this amount in vegetative cells (Sueoka et al., 1967; Wells and Sager, 1971; Bastia et al., 1971a). A chloroplast from tobacco leaves contains about 9×10^{-15}g (5×10^9 daltons) of DNA (Tewari and Wildman, 1970; Wildman, 1971), similar to the amount in a chloroplast from *Euglena* (Brawerman and Eisenstadt, 1964; Edelman et al., 1964; Scott, 1973). For comparison, some typical bacteria contain $3-5 \times 10^{-15}$ g ($2-3 \times 10^9$ daltons) of DNA per cell (Gillis et al., 1970). But several lines of evidence suggest that the informational content of chloroplast DNA is considerably less than might be expected from the amount of DNA present.

First, chloroplast DNA has been isolated from several sources as intact, circular molecules, and the molecular weight, as calculated from contour lengths, of the DNA structures obtained from *Euglena*, spinach, corn, and pea is approximately 9×10^7 (Manning et al., 1971; Manning et al., 1972; Kolodner and Tewari, 1972). Possibly, the chloroplast DNA in *Chlamydomonas reinhardtii* also exists as cir-

cular molecules, since an analysis of the genetic markers on the cytoplasmic genome suggests a functionally circular DNA (Sager, 1972).

Second, the kinetics of renaturation of DNA have been used to estimate the lengths of reiterated segments and consequently the genomic
size (Wetmur and Davidson, 1968; Gillis et al., 1970). This technique
has been applied to chloroplast DNA. A value of approximately 19 x
10^7 daltons was calculated for the genomic size of chloroplast DNA
from the algae *Euglena* and *Chlamydomonas* (Stutz, 1970; Wells and Sager, 1971; Bastia et al., 1971a) when bacteriophage T4 DNA was used
as a standard. In these same experiments the kinetic complexity of
T4 DNA provided a value of about 20 x 10^7 daltons. Since the molecular weight of T4 DNA is 10.6 x 10^7 (Dubin et al., 1970), or one-half
the kinetic complexity, the actual molecular weight of chloroplast
DNA might also be approximately one-half the kinetic complexity or
near 9 x 10^7 (Kolodner and Tewari, 1972). This suggestion is borne
out by the finding that the molecular weight of *Euglena* chloroplast
DNA, as estimated from contour lengths on electron micrographs (Manning et al., 1971), is about one-half the value determined from the
kinetic complexity (Stutz, 1970). Wells and Sager (1971) detected a
fraction of the chloroplast DNA of *Chlamydomonas*, amounting to about
10% of the DNA, that renatured more rapidly than the bulk of the DNA.
However, Bastia et al. (1971a), working with the same alga, could not
detect such a rapidly renaturing component within the limitations of
their experimental procedure.

The genomic size of chloroplast DNA from lettuce, tobacco, and
pea leaves was estimated from the kinetics of renaturation. Wells
and Birnstiel (1969) observed, with DNA from lettuce, a rapidly renaturing component accounting for 24% and a more slowly renaturing component accounting for the remaining 76% of the DNA. The slow fraction comprising the bulk of the DNA was estimated to have a size of
12 x 10^7 daltons on the basis of a molecular weight of 13 x 10^7 for
T4 DNA. Correcting this value for the revised molecular weight of
T4 DNA provides a value of about 9 x 10^7 for the major component of
lettuce DNA. Tewari and Wildman (1970) also detected two fractions
in tobacco chloroplast DNA when renaturation of heated DNA was exa-

mined, which was the procedure used by Wells and Birnstiel (1969). However, when alkaline-denatured DNA was allowed to renature, only a single component was found. This component had an estimated mass of about 11×10^7 daltons when compared with $E. coli$ DNA as the standard. Kolodner and Tewari (1972) carefully analyzed the DNA of pea chloroplasts and found that the molecular weight determined from the kinetics of renaturation of alkaline-denatured chloroplast DNA, based on a value of 10.6×10^7 for the standard T4 DNA, was in close agreement with the molecular weight determined by measuring the contour lengths of the circular DNA. The kinetics of renaturation yielded a value of 9.5×10^7, whereas measurements of contour lengths of the circular DNA provided a value of 9.1×10^7 for the molecular weight. Thus, chloroplast DNA from at least several plants and algae is of similar molecular weight, about 9×10^7, and circular.

Third, although the kinetics of renaturation of alkaline-denatured DNA suggested no repeating sequences in pea and tobacco chloroplast DNA smaller than the genomic size of 9×10^7 daltons, there may be minor fractions of the DNA that contain tandemly repeated sequences, which renature too rapidly to be detected by these procedures. A suggestion for repeating sequences in chloroplast DNA of Chlamydomonas was obtained from the formation of circular structures by annealing exonuclease-treated fragments of the DNA (Rochaix, 1972). The contour lengths of the circles formed in these experiments suggest the presence of repeated sequences on the order of $1-2 \times 10^6$ daltons, which is in the range of the size of the rapidly renaturing portion of heated DNA (Wells and Birnstiel, 1969; Wells and Sager, 1971).

These data indicate that chloroplast DNA may be highly redundant, at least in terms of the number of copies present within a chloroplast. A consideration of the amounts of DNA per chloroplast and the molecular weight of the DNA determined by Manning et al. (1971, 1972) and Kolodner and Tewari (1972) leads to the conclusion that a gamete cell of Chlamydomonas reinhardtii may have approximately 50 molecules of DNA per chloroplast, while a vegetative cell of this alga may have twice this number of DNA molecules per chloroplast. Likewise, each chloroplast in Euglena and higher plants may contain 40-80 copies of a single species of DNA. Direct evidence for the presence of multiple copies of DNA within chloroplasts of beet leaves was the observation that DNA is distributed among several regions of the chloroplast, each entirely separated from other regions by thylakoid membranes (Kowallik and Herrmann, 1972). The number of the DNA-containing regions in the beet chloroplasts varied with the size of the chloroplast, and, moreover, the amount of DNA per region varied.

It is not known if all molecules of chloroplast DNA are identical. A suggestion that they are similar in pea chloroplasts is the close agreement between the molecular weight determined for circular DNA from measurements of the contour lengths, and the genomic size as calculated from the kinetics of renaturation (Kolodner and Tewari, 1972).

B. Products of Chloroplast DNA

1. Ribosomal RNA

a. Hybridization studies. Several laboratories have shown that chloroplast ribosomal RNA molecules hybridize with chloroplast DNA. From these studies it was concluded that chloroplast DNA is the site of synthesis of these RNA molecules (Tewari and Wildman, 1968, 1970; Ingle et al., 1971; Bastia et al., 1971b; Scott et al., 1971). Although chloroplast ribosomal RNA molecules also hybridize with nuclear DNA (Tewari and Wildman, 1970; Ingle et al., 1971; Scott et al., 1971), such hybrids dissociate with a relatively low and broad melting profile, indicative of weak, perhaps nonspecific interactions (Stutz, 1971; Bastia et al., 1971b; Scott, 1973). Furthermore, Ingle et al. (1971) have found that chloroplast and cytoplasmic ribosomal RNA molecules compete for the same sites on nuclear DNA. This evidence suggests that no specific, functional genes for chloroplast ribosomal RNA reside in nuclear DNA. An explanation for the fact that chloroplast ribosomal RNA does hybridize with nuclear DNA, albeit at low matching frequencies, may reside in the evolutionary history of the organelle (Uzzell and Spolsky, 1974).

The level of hybridization generally found indicates that 1.5-2% of the chloroplast DNA is specific for chloroplast ribosomal RNA.

Rawson and Haselkorn (1973), by preparing *Euglena* chloroplast DNA
that was neither enriched nor deficient in any component, found that
1.9% of the DNA hybridized with chloroplast ribosomal RNA. This value is identical to the theoretical value calculated on the basis of
one gene for each ribosomal RNA per chloroplast DNA molecule of 9.2
x 10^7 daltons. As shown by these investigators, the purification of
chloroplast DNA must be carefully controlled to prevent enrichment
or loss of ribosomal genes.

Thomas (1972) reported that 4.4% of pea chloroplast DNA hybridized
with chloroplast ribosomal RNA, and this level of hybridization, considerably higher than those mentioned above, would indicate at least
two cistrons for each ribosomal RNA on a molecule of DNA. Scott
(1973) calculated from results of hybridization experiments that 6.2%
of chloroplast DNA from *Euglena* anneals with chloroplast ribosomal
RNA, a value sufficient to account for three cistrons per DNA molecule for each type of ribosomal RNA. A recalculation of the data reported in this paper provided a value of approximately one-tenth the
stated value. Also, equilibrium density profiles of the chloroplast
DNA fraction revealed that nearly one-half the DNA had a density of
1.707 g/cm^3, the density of nuclear DNA. Material of this density
was absent from samples of DNA prepared from purified chloroplasts by
Rawson and Haselkorn (1973).

Thus, hybridization experiments have provided conflicting results.
Most investigators obtained levels of hybridization that indicated
only about 2% of chloroplast DNA to be complementary to chloroplast
ribosomal RNA, indicating that one cistron for each ribosomal RNA
species is the correct value. These cistrons are part of segments
of DNA that have a slightly greater density (1.696-1.700 g/cm^3) than
does the bulk of the chloroplast DNA (1.686 g/cm^3) (Rawson and Haselkorn, 1973; Scott, 1973). A complicating factor, generally not considered, is the possible presence of mitochondrial DNA in the preparations of chloroplast DNA. *Euglena* mitochondrial DNA has a density of 1.691-1.692 g/cm^3 (Edelman et al., 1966; Stolarsky et al.,
1973) and is present in cells at amounts comparable to chloroplast
DNA (Schiff, 1973). Although Rawson and Haselkorn (1973) state that

their preparations of chloroplast DNA were free of mitochondrial DNA,
the purification procedure was based on an earlier study (Brown and
Haselkorn, 1972) that does not support this statement. No means is
presently available to ascertain the purity of chloroplast DNA samples, but the fact that a minimal value, expected on the basis of one
cistron for each ribosomal RNA species, was obtained by hybridization
experiments suggests that the conclusion of Rawson and Haselkorn
(1973) concerning purity of their DNA samples is valid.

    b. Studies with rifampicin. Surzycki (1969) reported that ri-
fampicin specifically inhibits DNA-dependent RNA polymerase in chloroplasts of *Chlamydomonas reinhardtii*. Incorporation of labeled adenine and guanine into 23 and 16S ribosomal RNA was inhibited by this
drug, whereas that into 25 and 18S cytoplasmic ribosomal RNA was not
affected (Surzycki and Rochaix, 1971). Surzycki and Rochaix (1971)
compared their experimental data on the inhibition of chloroplast
ribosomal RNA synthesis by rifampicin and actinomycin D with mathe-
matical analyses of various model situations and concluded that transcriptional units for ribosomal RNA on chloroplast DNA of *Chlamydomonas* consist of 2 to 3 tandem pairs of cistrons for 23 and 16S chloroplast ribosomal RNA. These data are consistent with the level of
hybridization reported by Thomas (1972) and Scott (1973), but do not
agree with the results of Rawson and Haselkorn (1973).

    Bogorad and Woodcock (1971) and more recently Munsche and Woll-
giehn (1973) observed that rifampicin inhibited the incorporation of
[32P]orthophosphate into chloroplast ribosomal RNA in excised leaves
of maize and tobacco, respectively. However, these experiments covered extended periods of time (22 hr), and a direct effect of rifampicin on chloroplast RNA synthesis cannot unequivocally be concluded
from these results. Bottomley et al. (1971b) could find no effect of
rifampicin on in vivo incorporation of [32P]orthophosphate or [3H]-
orotic acid into chloroplast ribosomal RNA in excised pea, corn, and
radish leaves. The effects of rifampicin on the synthesis of ribosomal RNA in higher plants, therefore, remains uncertain.

    Further evidence that chloroplast DNA is the template for chloro-
plast ribosomal RNA was the finding by Munsche and Wollgiehn (1973)

of highly labeled RNA components associated with chloroplast membranes after they fed tobacco leaves with [^{32}P]orthophosphate. These RNA components (1.3 x 10^6 and 0.7 x 10^6 daltons) were slightly larger than chloroplast ribosomal RNA and may have been precursor RNA molecules. These results have been confirmed by Hartley and Ellis (1973), who also found precursor molecules (1.2 x 10^6 and 0.65 x 10^6 daltons) of ribosomal RNA in chloroplasts after in vitro incubations with [^{3}H]-uridine or [^{32}P]phosphate. In addition, a rapidly labeled component of 2.7 x 10^6 daltons was observed. The function of this component has not been determined.

2. Transfer RNA

a. *Hybridization studies.* Results from hybridization studies suggest that transcription of some tRNA molecules occurs on chloroplast DNA. Williams and Williams (1970) found that bean chloroplast DNA would hybridize with a portion of the tRNA^{Leu} purified from bean leaves. Tewari and Wildman (1970) observed hybridization of tobacco chloroplast tRNA with chloroplast DNA and, from the extent of hybridization, they suggested that there may be 20-30 cistrons for tRNA per unit molecular weight of chloroplast DNA. However, these experiments provided no information on the numbers of different tRNA molecules these cistrons represent.

b. *Induction of synthesis by light.* The induction of several types of tRNA by light in etiolated *Euglena* (mentioned in Sect. V, B) suggests that these tRNA molecules are synthesized within the chloroplast. Furthermore, the light-inducible forms of tRNA^{Ile}, tRNA^{Phe}, and tRNA^{Glu} found in wild-type cells are not present in light-grown cells of the mutant W$_3$BUL (Reger et al., 1970), a permanently bleached strain of *Euglena* that lacks chloroplast DNA (Edelman et al., 1965). However, light induction does not seem to be a general characteristic of the synthesis of chloroplast tRNA species, since no differences in the relative amounts of isoaccepting species were observed when tRNA^{Arg} or tRNA^{Leu} from light- and dark-grown *Euglena* cells were compared. Also, Merrick and Dure (1972) observed no requirement of

light for the synthesis of chloroplast tRNA species during the development of cotton plants.

3. *Messenger RNA*

After considering the amount of DNA involved in hybridization
with ribosomal RNA and tRNA, at least 95% of the chloroplast DNA remains unaccounted for. Numerous mutant strains of higher plants and
algae have been isolated in which the traits are inherited in a uniparental fashion, a property considered characteristic of the chloroplast genetic system (Sager and Lane, 1972; see also reviews by Kirk
and Tilney-Bassett, 1967; Preer, 1971; Sager, 1972). The map of this
genetic system, the cytoplasmic linkage group, in *Chlamydomonas reinhardtii* shows a distribution of markers for several phenotypes, including a requirement of acetate for growth, temperature sensitivity
of growth, and antibiotic resistance (Sager, 1972). It is unclear
what changes are brought by the mutations in chloroplast DNA, but it
has been assumed from the existence of the uniparental mutant strains
that messenger RNA molecules for some proteins are transcribed on this
DNA. Support for the suggestion that chloroplast DNA contains structural genes for at least a few proteins is slowly emerging.

a. *Uniparentally inherited mutations affecting chloroplast ribosomes*. Mutant strains of *Chlamydomonas reinhardtii* resistant to
antibiotics have been isolated either after spontaneous mutational
events or after treatment of cells with various mutagens (Sager, 1954;
Gillham, 1965; Surzycki and Gillham, 1971; Mets and Bogorad, 1971;
Gillham et al., 1970; Schlanger et al., 1972; Burton, 1972; Sager,
1972). Crossing sensitive and resistant strains revealed that many
of the mutant traits are inherited in a uniparental fashion. The
components altered in these strains are largely unknown, but on the
basis of recent studies with bacterial cells, it is clear that one
cannot simply assume that mutations conferring resistance to antibiotics result only in changes in ribosomal proteins. Although in bacteria the antibiotic resistance is often the result of amino acid sub-
stitutions in ribosomal proteins (Nomura, 1970; Funatsu and Wittmann,

1972; Funatsu et al., 1972a,b; Davies and Nomura, 1972), changes in ribosomal RNA also can confer antibiotic resistance. Ribosomal RNA is involved in binding erythromycin to bacterial ribosomes (Mao and Putterman, 1969), and free 16S ribosomal RNA binds streptomycin with the same stoichiometry as do intact ribosomes (Biswas and Gorini, 1972). In a mutant strain of *Staphylococcus aureus,* which is resistant to a group of antibiotics including erythromycin, spiramycin, and lincomycin (Weisblum et al., 1971; Lai et al., 1973b), the 23S ribosomal RNA contains N^6,N^6-dimethyladenine ($m_2{}^6A$), whereas the 23S RNA in the sensitive strain does not (Lai and Weisblum, 1971). The dimethyladenine arises as the result of methylation of an adenine residue in the sequence $A_pA_pA_pG$ (Lai et al., 1973a), and this change in the 23S RNA appears to be the cause of resistance to the antibiotics (Lai et al., 1973). In a strain of *E. coli* resistant to kasugamycin, the oligonucleotide sequence $m_2{}^6A_pm_2{}^6A_pC_pC_pU_pG$ found near the 3' end of normal 16S RNA is replaced by the unmethylated sequence (Helser et al., 1971). Reconstitution experiments with 16S RNA and ribosomal proteins from resistant and sensitive strains have proved that the altered 16S RNA is the cause of resistance to the antibiotic (Helser et al., 1971; Zimmerman et al., 1973). It is interesting in this regard that in *E. coli* the precursor of the 16S RNA contains the unmethylated sequence $A_pA_pC_pC_pU_pG$ (Lowry and Dahlberg, 1971; Hayes et al., 1971). The defect in this kasugamycin-resistant strain of *E. coli* has been traced to a deficiency in a methylating enzyme (Helser et al., 1972). In *S. aureus,* the resistance to erythromycin may be the result of induction of the synthesis of a methylase (Lai et al., 1973a). The apparent specificity of these methylating enzymes suggests that a multiplicity of these activities are present in cells, each specific for certain sites in ribosomal RNA or tRNA.

By analogy to the results with bacteria, antibiotic resistance caused by mutations in chloroplast DNA of *Chlamydomonas* could result in altered ribosomal proteins, methylating enzymes, or even ribosomal RNA. At least two uniparental mutant strains are resistant to a group of macrolide antibiotics. Markers of resistance to erythromycin, spiramycin, oleandomycin, and carbomycin are tightly linked and may represent mutations in the same gene (Schlanger et al., 1972). Schlan ger et al. (1972) have shown that the resistance to carbomycin is the result of a change in chloroplast ribosomes of *Chlamydomonas*. But the similarity of these strains to the multiantibiotic-resistant strain of *S. aureus* suggests the possibility that changes in modification of chloroplast ribosomal RNA occurred in these mutant algal strains rather than changes in ribosomal proteins. Rijven and Zwar (1973) reported that isolated cotyledons of fenugreek incorporated methyl groups from methionine into chloroplast and cytoplasmic ribosomal RNA molecules. Interestingly, methyl groups were incorporated predominantly onto the bases of chloroplast ribosomal RNA (nearly 80% of the methyl groups incorporated), whereas the addition of methyl groups to cytoplasmic ribosomal RNA occurred principally (nearly 80%) onto the ribose moieties.

Burton (1972) observed that in a strain of *Chlamydomonas reinhardtii,* rendered resistant to spectinomycin by a uniparentally inherited mutation, binding of the drug to the small subunit of chloroplast ribosomes did not occur. By analogy to the spectinomycin-resistant strain of *E. coli,* in which an amino acid substitution occurred in one of the proteins of the small ribosomal subunit (Funatsu et al., 1972b), Burton (1972) suggested that a mutation in chloroplast DNA altered the structural gene for a chloroplast ribosomal protein. Mets and Bogorad (1972) have shown that in an erythromycin-resistant strain of *Chlamydomonas reinhardtii,* one of the proteins of the large subunit of 70S ribosomes has an altered electrophoretic mobility. Since in this strain the resistance to erythromycin is inherited in a uniparental fashion, these results provide the first direct evidence linking chloroplast DNA with a chloroplast ribosomal protein.

Chloroplast ribosomes of several strains in which antibiotic resistance is inherited uniparentally have altered sedimentation characteristics (Gillham et al., 1970; Burton, 1972). The chloroplast monomer sediments as a 66S rather than a 70S particle, and in some strains an excess of large subunits is present. The results suggest that in these strains the structure of the ribosomal subunits is altered by the loss of, or a change in, a ribosomal component. Whe-

ther the change in these altered ribosomes is in RNA or protein must
await an analysis of the ribosomes from the mutant strains.

b. *Mendelian mutations affecting chloroplast ribosomes*. Chloroplast ribosomes are affected by mutations in nuclear as well as in
chloroplast DNA (Boynton et al., 1970; Goodenough and Levine, 1970;
Mets and Bogorad, 1971, 1972), but again in most cases the primary
effects of such mutations are not known. Burton (1972) has shown that
a nuclear mutation in *Chlamydomonas reinhardtii*, which confers resistance to spectinomycin, does not change the ability of chloroplast
ribosomes to bind the drug. This mutation apparently affects the
permeability of the cell to spectinomycin rather than causing a change
in a ribosomal component. However, Mets and Bogorad (1971) reported
that a nuclear mutation conferring resistance of *Chlamydomonas reinhardtii* to erythromycin abolished the binding of this antibiotic by
chloroplast ribosomes, suggesting that a chloroplast ribosomal protein may have been altered by the mutation and therefore may be en-
coded in nuclear DNA. These investigators confirmed this suggestion
by showing that in a Mendelian mutant strain resistant to erythromycin,
the electrophoretic property of one of the proteins of the large subunit of chloroplast ribosomes is altered (Mets and Bogorad, 1972).
This protein, altered by a mutation in nuclear DNA, is a different
protein than the one changed by a mutation in chloroplast DNA, although both mutations confer resistance to erythromycin.

Bourque and Wildman (1973) have taken advantage of differences in
the electrophoretic mobilities of several proteins of the large subunit of chloroplast ribosomes in two species of tobacco to test the
proteins' mode of inheritance. The two species were crossed, and the
electrophoretic patterns for proteins of large ribosomal subunits,
obtained from the hybrids, were compared with those for the ribosomal
proteins of the parental strains. From their results, these authors
concluded that the genes for two proteins are inherited in a Mendelian fashion and therefore reside in nuclear DNA.

The *ac-20* and the *cr-1* strains of *Chlamydomonas reinhardtii*, both
containing mutations inherited in a Mendelian fashion, cause a conditional deficiency of chloroplast ribosomes (Goodenough and Levine,

1970; Boynton et al., 1970, 1972). Mixotrophically grown cells of
the ac-20 strain contain a low level of 70S ribosomes and few subunits,
whereas those of the cr-1 strain accumulate large subunits of chloroplast ribosomes (Boynton et al., 1972). Phototrophic growth conditions allow partial recovery of the chloroplast ribosomes, but cells
grown under these conditions still contain fewer chloroplast ribosomes than do wild-type cells (Goodenough and Levine, 1970; Boynton
et al., 1972). It has not been determined if the mutations resulted
in the conditional inactivation of proteins required for assembly of
one or both of the subunits or if synthesis of ribosomal RNA is affected in some manner by the mutations (Goodenough and Levine, 1971).

From these studies it seems that the messenger RNA molecules for
ribosomal proteins originate from both chloroplast and nuclear DNA.
But because of the possibility of mutations affecting ribosomal RNA,
a known product of chloroplast DNA, or enzymes that catalyze RNA modifying reactions, the existence of a mutation is not a sufficient reason for assuming that a change occurred in a ribosomal protein. The
altered component must be tracked down, and such detective work has
barely begun with chloroplast ribosomes.

c. *Uniparentally inherited mutations affecting chloroplast membrane components.* Several strains of *Chlamydomonas reinhardtii* were
isolated in which mutations inherited uniparentally caused a requirement of acetate for growth (Sager and Ramanis, 1973). The defect in
these strains has not been determined. Chloroplast mutants that are
not active in photosynthesis have also been found in higher plants.
The thylakoid membranes of two such mutant strains of *Antirrhinum*
have been analyzed by electrophoresis (Herrmann and Bauer-Stäb, 1969;
Herrmann, 1972). The electrophoretic patterns of these membranes revealed deficiencies in several minor proteins. Among the chloroplast
mutants in *Oenothera,* one group of mutants is deficient in photosystem II activity, while another group has low photosystem I activity
(Fork and Heber, 1968; Hallier, 1968). These mutant strains, in particular those deficient in a component of photosystem II, also show
disrupted and vesiculated thylakoid membranes, with each mutant dis-

playing a characteristic morphology (Dolzmann, 1968). Although vesiculated membranes are observed in chloramphenicol-treated cells (Margulies, 1966; Goodenough, 1971), the morphological lesions in these
Oenothera mutant strains are not the result of the lack of protein
synthesis in the chloroplast, since high levels of ribulose 1,5-di-
phosphate carboxylase activity are present in these cells (Hallier
et al., 1968).

The underlying mechanisms for the profound morphological effects
found in photosystem II deficient mutants of Oenothera are not known,
but a possibility is that a protein necessary for structure of thylakoid membranes is lacking in the chloroplast mutant strains. On the
basis of membrane fractionation studies, the major polypeptides of
thylakoid membranes appear to be associated with photosystem II particles (Remy, 1971; Levine et al., 1972; Machold and Aurich, 1972).
Hoober and Stegeman (1973) have studied the control of synthesis of
a major polypeptide of thylakoid membranes in Chlamydomonas. Both
chlorophyll synthesis and chloramphenicol-sensitive protein synthesis
appear to be involved in the control mechanism. Since these events
occur in the chloroplast, whereas synthesis of the membrane polypeptide occurs on cytoplasmic ribosomes, these authors have suggested
that chloroplast DNA may contain the gene for this polypeptide, and
that regulation of the synthesis of this polypeptide occurs at the
transcriptional level. If in fact the cistron for a major membrane
polypeptide resides in chloroplast DNA, a mutation there might result in cytological changes.

An interesting plastid mutant has been found in tobacco, in which
normal and mutant plastids coexist within the same cells (Wildman,
1971; von Wettstein et al., 1971). The mutant plastids are apparently unable to form thylakoid membranes, but contain DNA, ribosomes,
and at least some products of chloroplast protein synthesis. It is
not known if the defective plastids lack the ability to synthesize
membrane proteins, lipids, or both.

   d. Mendelian mutations affecting chloroplast membrane components.
The information for many of the functional components of thylakoid

membranes resides in nuclear DNA. A number of Mendelian mutants of
Chlamydomonas have been isolated that are unable to carry out photosynthesis (reviewed by Levine, 1968, 1969; Levine and Goodenough,
1970). These strains generally lack a functional component of the
photosynthetic electron transport system (Levine, 1968). Such mutants,
unable to grow on CO_2, require acetate for growth. Disorders in membrane structure occur in some of these mutant strains of Chlamydomonas, but these abnormalities are generally found as changes in the
degree of stacking and not in significant changes in the appearance
of the thylakoid membranes themselves (Goodenough and Levine, 1969).
In two of these strains the polypeptides affected by mutations are
normally present in the membranes in low amounts relative to the major
polypeptides of the membranes (Levine et al., 1972).

Kung et al. (1972) reported that the gene for a major polypeptide
associated with photosystem II in tobacco thylakoid membranes resides
in nuclear DNA. This conclusion was reached from studies on the mode
of inheritance of a peptide derived from the membrane protein by digestion with trypsin. The inheritance of the composition of this peptide was independent of the maternal strain and thus was concluded to
be Mendelian. Apel and Schweiger (1972) have shown that several of
the thylakoid membrane polypeptides in Acetabularia are species specific, and that the presence of these polypeptides is determined by
the nucleus of the cell. Transplanting the nucleus of one species
into another enucleated species caused a change in part of the electrophoretic pattern of the thylakoid membrane polypeptides. No information is available on the origin of messenger RNA molecules for the
other polypeptides.

These data indicate that cistrons in both nuclear and chloroplast
DNA are involved in coding for membrane polypeptides. Although certain specific membrane proteins can be linked directly to nuclear DNA
as the result of genetic studies, the link between chloroplast DNA
and membrane proteins is only indirect. It is hoped that procedures
will soon be developed that will allow pursuit of the question of
what membrane polypeptides are coded by chloroplast DNA.

e. *Inheritance of ribulose 1,5-diphosphate carboxylase*. Andersen et al. (1970) observed changes in the properties of ribulose 1,5-diphosphate carboxylase in several strains of tomato in which the mutations affecting this enzyme are inherited in a Mendelian fashion. Although it was not determined in this study which of the subunits is altered in each strain, the mutations were mapped on two different nuclear linkage groups, suggesting that both subunits are coded by nuclear DNA. Contradicting results have been obtained by Wildman and his colleagues. The latter investigators initiated studies on the mode of inheritance of the carboxylase by examining the peptides produced by trypsin digestion of purified subunits. Species of tobacco were selected that showed different peptide patterns for the subunits, and preparations of the enzyme were then examined from hybrids produced by crossing the species. The differences, observed in peptides of the small subunit, were inherited in a Mendelian fashion, independent of the maternal species, and therefore it was concluded that this subunit is coded by nuclear DNA (Kawashima and Wildman, 1972). However, Chan and Wildman (1972) found by the same procedure that in tobacco the differences in the peptide patterns for the large subunit are inherited in a non-Mendelian fashion, dependent on DNA from the maternal species. These results suggest that information for this polypeptide resides in chloroplast DNA. It would be desirable to have confirmation of this conclusion by another approach, and an examination of the uniparental nonphotosynthetic mutants of *Chlamydomonas* that require acetate for growth (Sager and Ramanis, 1963) may possibly provide this information.

f. *Studies with rifampicin*. Surzycki (1969) suggested that, since a high concentration of rifampicin inhibited synthesis of chloroplast ribosomal RNA in *Chlamydomonas,* the drug inhibits all RNA synthesis in the chloroplast. However, there is no evidence that the drug inhibits synthesis of messenger RNA. RNA synthesis in chloroplasts isolated from corn, peas, and spinach leaves was not affected by rifampicin (Bottomley et al., 1971b), nor was the activity of partially purified preparations of RNA polymerase of corn and wheat

chloroplasts inhibited by the drug (Bottomley et al., 1971a; Polya and
Jagendorf, 1971). However, Harris et al. (1973) found that even re-
latively low concentrations (e.g., 5 µg/ml) of rifampicin inhibited
RNA synthesis in isolated chloroplasts of *Euglena*.

Rifampicin is effective as an inhibitor of bacterial RNA polymerases only if present prior to initiation of the synthesis of RNA
chains and is most effective if the polymerase is dissociated from
DNA (Sippel and Hartmann, 1968; Hinkle et al., 1972). To test if the
chloroplast RNA polymerases are sensitive to rifampicin, conditions
must prevail under which the polymerases are not engaged in RNA synthesis at the time the drug is added. These conditions apparently
occur in vivo for the synthesis of chloroplast ribosomal RNA (Surzycki and Rochaix, 1971). During synthesis of messenger RNA in vivo,
initiation steps should also occur continuously, and synthesis of the
RNA should eventually become blocked by rifampicin, if indeed the
polymerase is sensitive to this drug. But no conveniently measurable
proteins have been found that are definitely products of chloroplast
DNA, and without such markers the effects of rifampicin on the synthesis of messenger RNA cannot be adequately tested.

However, Levine and his colleagues (Armstrong et al., 1971; Sirevåg and Levine, 1972) have attempted to use rifampicin as a means
toward identifying products of the chloroplast genome. The drug was
added near the end of a dark period to cultures synchronized by light-
dark cycles, and its effect on the subsequent increases of various
chloroplast components was determined throughout the light period.
An assumption made in these experiments is that the primary effect of
rifampicin is not on the levels of ribosomes or tRNA during the experiment. This assumption has not been adequately tested, but from
the data it appears to be valid, since synthesis of ribulose 1,5-di-
phosphate carboxylase was not affected. Rifampicin did not inhibit
chlorophyll synthesis but did prevent the increase in photosynthetic
activity. The nearly complete inhibition of the expected increases
in the amounts of cytochromes-553 and -563 may be the reason for the
effect of the drug on the photosynthetic activity. These results
suggest that chloroplast DNA may be involved in the synthesis of these

cytochromes. But it was previously reported by Gorman and Levine (1966) that cytochrome-553 was not detected in a photosynthetically deficient strain of Chlamydomonas reinhardtii, ac-206, that contains a mutation inherited in a Mendelian fashion. Furthermore, chloroplast cytochrome-559 was not detected in three other Mendelian mutant strains, ac-115, ac-141, and F-34 (Levine and Gorman, 1966; Levine, 1968). The mode of inheritance of cytochrome-563 has not been reported. The conflict between the results with rifampicin and those with the mutant strains, therefore, does not permit a firm conclusion as to the site of transcription of the messenger RNA for the cyto-chrome proteins. It is worth noting that for those components studied so far, rifampicin inhibits the synthesis of few, if any, chloroplast components in Chlamydomonas other than chloroplast ribosomal RNA. The unaffected components include the major product of chloroplast protein synthesis, ribulose 1,5-diphosphate carboxylase.

The suggestion of uniparental inheritance of the gene for the large subunit of the carboxylase in tobacco (Chan and Wildman, 1972) casts doubt on the effectiveness of rifampicin in inhibiting messen-ger RNA transcription on chloroplast DNA. It should be noted again that the specificity, or lack of it, of rifampicin in inhibiting the synthesis of various types of chloroplast RNA molecules has not been worked out. Although cytological changes occur within the chloroplast during long-term exposure of the cells to the drug, these changes may be secondary to the effects of the drug on the synthesis of chloro-plast ribosomal RNA (Goodenough, 1971).

Jennings and Ohad (1972) have used rifampicin to test for the origin of messenger RNA translated by chloroplast ribosomes in Chlamy-domonas. In the presence of high concentrations of rifampicin, the synthesis of ribosomal RNA was inhibited, but [^{3}H]uracil was still incorporated into the 70S ribosomes. These investigators suggested that this labeling of chloroplast ribosomes resulted from the presence of RNA transcribed from nuclear DNA. However, from the discussion above, this conclusion cannot be reached without evidence that rifam-picin does inhibit transcription of messenger RNA in chloroplasts.

The difficulty in determining what structural genes reside in chloroplast DNA, and in interpreting the effects of inhibitors, is the lack of information on what proteins (which could be used as markers) the chloroplast DNA codes for. Rifampicin, streptolydigin, acriflavine, and ethidium bromide, inhibitors of bacterial-type RNA polymerase and circular DNA functions, have no selective effect on synthesis of the major thylakoid membrane polypeptides in *Chlamydomonas* (Hoober and Stegeman, 1973; Stegeman and Hoober, 1974). These inhibitors also have no effect on the synthesis in this alga of the large subunit of ribulose 1,5-diphosphate carboxylase (Stegeman and Hoober, 1974), a polypeptide that, according to available evidence, is coded by chloroplast DNA (see Sect. VII,B,3,e). It remains to be resolved whether this polypeptide is not coded by chloroplast DNA, or if chloroplast DNA and RNA polymerase are refractory to the action of the aforementioned inhibitors.

VIII. CONCLUSIONS

The mechanism, as much as is known of it, of protein synthesis in chloroplasts resembles the process in prokaryotic cells. The ribosomes are of similar size, N-formylmethionyl-tRNA is involved in initiation of synthesis of polypeptide chains, and several of the soluble factors in chloroplasts resemble those in prokaryotic cells in terms of their specificity for 70S ribosomes.

The sources of the messenger RNA molecules translated by chloroplast ribosomes remain largely unknown. The genetic information in chloroplast DNA may be limited to that which can be accommodated by a circular molecule of 9×10^7 daltons in molecular weight. Chloroplasts synthesize RNA, including the RNA of chloroplast ribosomes. And from the existence of uniparentally inherited mutations, particularly to antibiotic resistance, it is considered that some messenger RNA molecules also are transcribed on chloroplast DNA. But how many of these messenger RNA molecules are translated within the chloroplast is not known.

So, the current situation can be formulated as follows. The capability of chloroplasts to provide the genetic information and machinery for the synthesis of proteins has been extensively documented. But the capacity of the chloroplast in terms of the number of proteins it provides for itself, genetically and synthetically, remains unclear.

Data on the sites of synthesis of chloroplast proteins have largely been obtained from the use of inhibitors of protein synthesis. These results indicate that many of the chloroplast proteins, including most of the enzymes of the Calvin cycle and the major thylakoid polypeptides, are synthesized on cytoplasmic ribosomes. But at least several proteins, in particular the singly most prominent polypeptide in plant cells, large subunits of ribulose 1,5-diphosphate carboxylase, are synthesized within the chloroplast. However, it is highly desirable to identify proteins synthesized by the chloroplast by studying the products of protein synthesis in isolated chloroplasts. But problems remain that presently preclude much success with this approach. Conditions have not yet been developed that permit net synthesis of protein in isolated chloroplasts. This problem may be the result of not providing certain required factors for the incubation medium, or it may be the result of chloroplast ribosomes not translating much of the messenger RNA produced by the organelle. Transcription and translation of the messenger RNA for the large subunit ribulose 1,5-diphosphate carboxylase should provide an excellent test system for such a study.

It is not known how much of the protein synthesized within the chloroplast is the result of translation of messenger RNA molecules transcribed on nuclear DNA. If there indeed is a flux of messenger RNA across the chloroplast envelope, then a question to be solved is what mechanism determines the selection of messenger RNA by cytoplasmic and chloroplast ribosomes.

REFERENCES

Adelman, M. R., D. D. Sabatini, and G. Blobel. 1973. Ribosome-membrane interaction. Nondestructive disassembly of rat liver microsomes into ribosomal and membranous components. *J. Cell Biol. 56:* 206-229.

Alberte, R. S., J. P. Thornber, and A. W. Naylor. 1972. Time of appearance of photosystems I and II in chloroplasts of greening jack bean leaves. *J. Exptl. Bot. 23:*1060-1069.

Alberte, R. S., J. P. Thornber, and A. W. Naylor. 1973. Biosynthesis of the photosystem I chlorophyll-protein complex in greening leaves of higher plants. *Proc. Nat. Acad. Sci. US 70:*134-137.

Aliev, K. A., and I. I. Filippovich. 1968. Differences in tRNAs and aminoacyl tRNA synthetases from chloroplasts and the cytoplasm of pea seedlings. *Mol. Biol. 2:*364-373.

Andersen, W. R., G. F. Wildner, and R. S. Criddle. 1970. Ribulose diphosphate carboxylase, III. Altered forms of ribulose diphosphate carboxylase from mutant tomato plants. *Arch. Biochem. Biophys. 137:*84-90.

Anderson, J. M., and R. P. Levine. 1974. Membrane polypeptides of some higher plant chloroplasts. *Biochim. Biophys. Acta 333:*378-387.

Anderson, L. A., and R. M. Smillie. 1966. Binding of chloramphenicol by ribosomes from chloroplasts. *Biochem. Biophys. Res. Commun. 23:*535-539.

Apel, K., and H.-G. Schweiger. 1972. Nuclear dependency of chloroplast proteins in *Acetabularia*. *Eur. J. Biochem. 25:*229-238.

App, A. A., and A. T. Jagendorf. 1964. ^{14}C-Amino acid incorporation by spinach chloroplast preparations. *Plant Physiol. 39:*772-776.

Armstrong, J. J., S. J. Surzycki, B. Moll, and R. P. Levine. 1971. Genetic transcription and translation specifying chloroplast components in *Chlamydomonas reinhardi*. *Biochemistry 10:*692-701.

Arnon, D. I., M. B. Allen, and F. R. Whatley. 1954. Photosynthesis by isolated chloroplasts. *Nature 174:*394-396.

Aronoff, S., A. Benson, W. Z. Hassid, and M. Calvin. 1947. Distribution of ^{14}C in photosynthesizing barley seedlings. *Science 105:* 664-665.

Avadhani, N. G., and D. E. Buetow. 1972. Isolation of active polyribosomes from cytoplasm, mitochondria, and chloroplasts of *Euglena gracilis*. *Biochem. J. 128:*353-365.

Bachmayer, H. 1970. Initiation of protein synthesis in intact cells and in isolated chloroplasts of *Acetabularia mediterranea*. *Biochim. Biophys. Acta 209:*584-586.

Bamji, M. S., and A. T. Jagendorf. 1966. Amino acid incorporation by wheat chloroplasts. *Plant Physiol. 41:*764-770.

Barnett, W. E., C. J. Pennington, Jr., and S. A. Fairfield. 1969. Induction of *Euglena* transfer RNA's by light. *Proc. Nat. Acad. Sci. US 63:*1261-1268.

Bassham, J. A., B. Morawiecka, and M. Kirk. 1964. Protein synthesis during photosynthesis. *Biochim. Biophys. Acta 90*:542-552.

Bastia, D., K.-S. Chiang, H. Swift, and P. Siersma. 1971a. Heterogeneity, complexity, and repetition of the chloroplast DNA of *Chlamydomonas reinhardtii*. *Proc. Nat. Acad. Sci. US 68*:1157-1161.

Bastia, D., K.-S. Chiang, and H. Swift. 1971b. Studies on the ribosomal RNA cistrons of chloroplast and nucleus in *Chlamydomonas reinhardtii*. *Absts. Eleventh Ann. Meeting, Am. Soc. Cell Biol., New Orleans*:25

Biswas, D. K., and L. Gorini. 1972. The attachment site of streptomycin to the 30 S ribosomal subunit. *Proc. Nat. Acad. Sci. US 69*:2141-2144.

Blair, G. E., and R. J. Ellis. 1973. Protein synthesis in chloroplasts. I. Light-driven synthesis of the large subunit of Fraction I protein by isolated pea chloroplasts. *Biochim. Biophys. Acta 319*:223-234.

Blobel, G., and V. R. Potter. 1966. Relation of ribonuclease and ribonuclease inhibitor to the isolation of polysomes from rat liver. *Proc. Nat. Acad. Sci. US 55*:1283-1288.

Blobel, G., and V. R. Potter. 1967. Studies on free and membrane-bound ribosomes in rat liver. I. Distribution as related to total cellular RNA. *J. Mol. Biol. 26*:279-292.

Boardman, N. K., R. I. B. Francki, and S. G. Wildman. 1965. Protein synthesis by cell-free extracts from tobacco leaves. II. Association of activity with chloroplast ribosomes. *Biochemistry 4*: 872-876.

Boardman, N. K., R. I. B. Francki, and S. G. Wildman. 1966. Protein synthesis by cell-free extracts of tobacco leaves. III. Comparison of the physical properties and protein synthesizing activities of 70 S chloroplast and 80 S cytoplasmic ribosomes. *J. Mol. Biol. 17*:470-489.

Bogorad, L., and C. L. F. Woodcock. 1971. Rifamycins: the inhibition of plastid RNA synthesis *in vivo* and *in vitro* and variable effects on chlorophyll formation in maize leaves. in *Autonomy and Biogenesis of Mitochondria and Chloroplasts* (N. K. Boardman, A. W. Linnane, and R. M. Smillie, eds). Amsterdam: North-Holland. pp. 92-97.

Bonotto, S., A. Goffeau, M. Janowski, T. Vanden Driessche, and J. Brachet. 1969. Effects of various inhibitors of protein synthesis on *Acetabularia mediterranea*. *Biochim. Biophys. Acta 174*: 704-712.

Bottomley, W., H. J. Smith, and L. Bogorad. 1971a. RNA polymerase of maize: partial purification and properties of the chloroplast enzyme. *Proc. Nat. Acad. Sci. US 68*:2412-2416.

Bottomley, W., D. Spencer, A. M. Wheeler, and P. R. Whitfeld. 1971b. The effect of a range of RNA polymerase inhibitors on RNA synthesis in higher plant chloroplasts and nuclei. *Arch. Biochem. Biophys. 143:*269-275.

Boulter, D. 1970. Protein synthesis in plants. *Ann. Rev. Plant Physiol. 21:*91-114.

Boulter, D., R. J. Ellis, and A. Yarwood. 1972. Biochemistry of protein synthesis in plants. *Biol. Rev. 47:*113-175.

Bourque, D. P., J. E. Boynton, and N. W. Gillham. 1971. Studies on the structure and cellular location of various ribosome and ribosomal RNA species in the green alga *Chlamydomonas reinhardi. J. Cell Sci. 8:*153-183.

Bourque, D. P., and S. G. Wildman. 1973. Evidence that nuclear genes code for several chloroplast ribosomal proteins. *Biochem. Biophys. Res. Commun. 50:*532-537.

Bové, J., and I. D. Raacke. 1959. Amino acid-activating enzymes in isolated chloroplasts from spinach leaves. *Arch. Biochem. Biophys. 85:*521-531.

Boynton, J. E., N. W. Gillham, and B. Burkholder. 1970. Mutations altering chloroplast ribosome phenotype in *Chlamydomonas*. II. A new Mendelian mutation. *Proc. Nat. Acad. Sci. US 67:*1505-1512.

Boynton, J. E., N. W. Gillham, and J. F. Chabot. 1972. Chloroplast ribosome deficient mutants in the green alga *Chlamydomonas reinhardi* and the question of chloroplast ribosome function. *J. Cell Sci. 10:*267-305.

Brawerman, G., and J. M. Eisenstadt. 1964. Deoxyribonucleic acid from the chloroplasts of *Euglena gracilis. Biochim. Biophys. Acta 91:*477-485.

Bretscher, M. S. 1971. Initiation and termination of the polypeptide chain. in *Protein Synthesis, A Series of Advances* (E. McConkey, ed.). Vol. I. New York: Marcel Dekker. pp. 89-120.

Brown, R. D., and R. Haselkorn. 1972. The isolation of *Euglena gracilis* chloroplasts uncontaminated by nuclear DNA. *Biochim. Biophys. Acta 259:*1-4.

Burkard, G., B. Eclancher, and J. H. Weil. 1969. Presence of N-formylmethionyl-transfer RNA in bean chloroplasts. *FEBS Lett. 4:*285-287.

Burkard, G., P. Guillemaut, and J. H. Weil. 1970. Comparative studies on the tRNA's and the amino acyl-tRNA synthetases from the cytoplasm and the chloroplasts of *Phaseolus vulgaris. Biochim. Biophys. Acta 224:*184-198.

Burton, W. G. 1972. Dihydrospectinomycin binding to chloroplast ribosomes from antibiotic-sensitive and -resistant strains of *Chlamydomonas reinhardtii. Biochim. Biophys. Acta 272:*305-311.

Cammarano, P., S. Pons, A. Romeo, M. Galdieri, and C. Gualerzi. 1972. Characterization of unfolded and compact ribosomal subunits from plants and their relationship to those of lower and higher animals: evidence for physiochemical heterogeneity among eucaryotic ribosomes. *Biochim. Biophys. Acta 281*:571-596.

Cattolico, R. A., and R. F. Jones. 1972. Isolation of stable ribosomal RNA from whole cells of *Chlamydomonas reinhardtii*. *Biochim. Biophys. Acta 269*:259-264.

Chan, P.-H., and S. G. Wildman. 1972. Chloroplast DNA codes for the primary structure of the large subunit of Fraction I protein. *Biochim. Biophys. Acta 277*:677-680.

Chen, J. L., and S. G. Wildman. 1967. Functional chloroplast ribosomes from tobacco leaves. *Science 155*:1271-1273.

Chiang, K.-S., and N. Sueoka. 1967. Replication of chloroplast DNA in *Chlamydomonas reinhardi* during vegetative cell cycle: its mode and regulation. *Proc. Nat. Acad. Sci. US 57*:1506-1513.

Chua, N.-H., G. Blobel, and P. Siekevitz. 1973a. Isolation of cytoplasmic and chloroplast ribosomes and their dissociation into active subunits from *Chlamydomonas reinhardtii*. *J. Cell Biol. 57*: 798-814.

Chua, N.-H., G. Blobel, P. Siekevitz, and G. E. Palade. 1973b. Attachment of chloroplast polysomes to thylakoid membranes in *Chlamydomonas reinhardtii*. *Proc. Nat. Acad. Sci. US 70*:1554-1558.

Ciferri, O., and B. Parisi. 1970. Ribosome specificity of protein synthesis *in vitro*. *Prog. Nucleic Acid Res. Mol. Biol. 10*:121-144.

Ciferri, O., and O. Tiboni. 1973. Elongation factors for chloroplast and mitochondrial protein synthesis in *Chlorella vulgaris*. *Nature New Biol. 245*:209-211.

Clark, M. F. 1964. Polyribosomes from chloroplasts. *Biochim. Biophys. Acta 91*:671-674.

Clark, M. F., R. E. F. Matthews, and R. K. Ralph. 1964. Ribosomes and polysomes in *Brassica pekinesis*. *Biochim. Biophys. Acta 91*: 289-304.

Criddle, R. S., B. Dau, G. E. Kleinkopf, and R. C. Huffaker. 1970. Differential synthesis of ribulose diphosphate carboxylase subunits. *Biochem. Biophys. Res. Commun. 41*:621-627.

Dalling, M. J., N. E. Tolbert, and R. H. Hageman. 1972. Intracellular location of nitrate reductase and nitrite reductase, I. Spinach and tobacco leaves. *Biochim. Biophys. Acta 283*:505-512.

Davies, J., and M. Nomura. 1972. The genetics of bacterial ribosomes. *Ann. Rev. Genetics 6*:203-234.

Delo, J., M. L. Ernst-Fonberg, and K. Bloch. 1971. Fatty acid synthetases from *Euglena gracilis*. *Arch. Biochem. Biophys. 143*:384-391.

Dolzmann, P. 1968. Photosynthese-Reacktionen einiger Plastom-Mutanten von *Oenothera*, III. Strukturelle Aspekte. *Z. Pflanzenphysiol. 58*:300-309.

Drumm, H. E., and M. M. Margulies. 1970. *In vitro* protein synthesis by plastids of *Phaseolus vulgaris*. IV. Amino acid incorporation by etioplasts and effect of illumination of leaves on incorporation by plastids. *Plant Physiol.* 45:435-442.

Dubin, S. B., G. B. Benedek, F. C. Bancroft, and D. Freifelder. 1970. Molecular weights of coliphages and coliphage DNA. II. Measurement of diffusion coefficients using optical mixing spectroscopy, and measurement of sedimentation coefficients. *J. Mol. Biol.* 54:547-556.

Dyer, T. A., and R. M. Leech. 1968. Chloroplast and cytoplasmic low-molecular-weight ribonucleic acid components of the leaf of *Vicia faba* L. *Biochem. J.* 106:689-698.

Eaglesham, A. R. J., and R. J. Ellis. 1974. Protein synthesis in chloroplasts. II. Light-driven synthesis of membrane proteins by isolated pea chloroplasts. *Biochim. Biophys. Acta* 335:396-407.

Edelman, M., C. A. Cowan, H. T. Epstein, and J. A. Schiff. 1964. Studies of chloroplast development in *Euglena*. VIII. Chloroplast-associated DNA. *Proc. Nat. Acad. Sci. US* 52:1214-1219.

Edelman, M., J. A. Schiff, and H. T. Epstein. 1965. Studies of chloroplast development in *Euglena*. XII. Two types of satellite DNA. *J. Mol. Biol.* 11:769-774.

Edelman, M., H. T. Epstein, and J. A. Schiff. 1966. Isolation and characterization of DNA from the mitochondrial fraction of *Euglena*. *J. Mol. Biol.* 17:463-469.

Eisenstadt, J. M., and G. Brawerman. 1966. A factor from *Escherichia coli* concerned with the stimulation of cell-free polypeptide synthesis by exogenous ribonucleic acid. I. Evidence for the occurrence of a stimulation factor. *Biochemistry* 5:2777-2783.

Ellis, R. J. 1969. Chloroplast ribosomes. Stereospecificity of inhibition by chloramphenicol. *Science* 163:477-478.

Ellis, R. J. 1970. Further similarities between chloroplast and bacterial ribosomes. *Planta* 91:329-335.

Ellis, R. J., and R. M. Hartley. 1971. Sites of synthesis of chloroplast proteins. *Nature New Biol.* 233:193-196.

Ernst-Fonberg, M. L., and K. Bloch. 1971. A chloroplast-associated fatty acid synthetase system in *Euglena*. *Arch. Biochem. Biophys.* 143:392-400.

Evans, W. R., and R. M. Smillie. 1971. Inhibition of light-dependent development of chloroplasts by 5-fluorouracil. *J. Exptl. Bot.* 22:371-381.

Eytan, G., and I. Ohad. 1970. Biogenesis of chloroplast membranes. VI. Cooperation between cytoplasmic and chloroplast ribosomes in the synthesis of photosynthetic lamellar proteins during the greening process in a mutant of *Chlamydomonas reinhardi*, y-1. *J. Biol. Chem.* 245:4297-4307.

Eytan, G., and I. Ohad. 1972a. Biogenesis of chloroplast membranes. VII. The preservation of membrane homogeneity during development of the photosynthetic lamellar system in an algal mutant (*Chlamydomonas reinhardi* y-1). *J. Biol. Chem. 247*:112-121.

Eytan, G., and I. Ohad. 1972b. Biogenesis of chloroplast membranes. VIII. Modulation of chloroplast lamellae composition and function induced by discontinuous illumination and inhibition of ribonucleic acid and protein synthesis during greening of *Chlamydomonas reinhardi* y-1 mutant cells. *J. Biol. Chem. 247*:122-129.

Fairfield, S. A., and W. E. Barnett. 1971. On the similarity between the tRNA's of organelles and prokaryotes. *Proc. Nat. Acad. Sci. US 68*:2972-2976.

Fork, D. C., and U. W. Heber. 1968. Studies on the electron-transport reactions of photosynthesis in plastom mutants of *Oenothera*. *Plant Physiol. 43*:606-612.

Francki, R. I. B., N. K. Boardman, and S. G. Wildman. 1965. Protein synthesis by cell-free extracts from tobacco leaves. I. Amino acid incorporating activity of chloroplasts in relation to their structure. *Biochemistry 4*:865-872.

Freyssinet, G., and J. A. Schiff. 1974. The chloroplast and cytoplasmic ribosomes of *Euglena*. II. Characterization of ribosomal proteins. *Plant Physiol. 53*:543-554.

Funatsu, G., and H. G. Wittmann. 1972. Ribosomal proteins. XXXIII. Location of amino acid replacements in protein S12 isolated from *Escherichia coli* mutants resistant to streptomycin. *J. Mol. Biol. 68*:547-550.

Funatsu, G., K. Nierhaus, and H. G. Wittmann. 1972a. Ribosomal proteins. XXXVII. Determination of allele types and amino acid exchanges in protein S12 of three streptomycin-resistant mutants of *Escherichia coli*. *Biochim. Biophys. Acta 287*:282-291.

Funatsu, G., K. Nierhaus, and B. Wittmann-Liebold. 1972b. Ribosomal proteins. XXII. Studies on the altered protein S5 from a spectinomycin-resistant mutant of *Escherichia coli*. *J. Mol. Biol. 64*:201-209.

Gassman, M., and L. Bogorad. 1967. Studies on the regeneration of protochlorophyllide after brief illumination of etiolated bean leaves. *Plant Physiol. 42*:781-784.

Gillham, N. W. 1965. Induction of chromosomal and non-chromosomal mutations in *Chlamydomonas reinhardi* with N-methyl-N'-nitro-N-nitrosoguanidine. *Genetics 52*:529-537.

Gillham, N. W., J. E. Boynton, and B. Burkholder. 1970. Mutations altering chloroplast ribosome phenotype in *Chlamydomonas*. I. Non-Mendelian mutations. *Proc. Nat. Acad. Sci. US 67*:1026-1033.

Gillis, M., J. De Ley, and M. De Cleene. 1970. The determination of molecular weight of bacterial genome DNA from renaturation rates. *Eur. J. Biochem. 12*:143-153.

Givan, A. L., and R. S. Criddle. 1972. Ribulose diphosphate carboxylase from Chlamydomonas reinhardi: purification, properties, and its mode of synthesis in the cell. Arch. Biochem. Biophys. 149: 153-163.

Gnanam, A., A. T. Jagendorf, and M. Ranalletti. 1969. Chloroplasts and bacterial amino acid incorporation: a further comment. Biochim. Biophys. Acta 186:205-213.

Goffeau, A. 1969. Incorporation of amino acids into the soluble and membrane-bound proteins of chloroplasts isolated from enucleated Acetabularia. Biochim. Biophys. Acta 174:340-350.

Goffeau, A., and J. Brachet. 1965. Deoxyribonucleic acid-dependent incorporation of amino acids into proteins of chloroplasts isolated from anucleate Acetabularia fragments. Biochim. Biophys. Acta 95:302-313.

Goodenough, U. W. 1971. The effects of inhibitors of RNA and protein synthesis on chloroplast structure and function in wild-type Chlamydomonas reinhardi. J. Cell Biol. 50:35-49.

Goodenough, U. W., and R. P. Levine. 1969. Chloroplast ultrastructure in mutant strains of Chlamydomonas reinhardi lacking components of the photosynthetic apparatus. Plant Physiol. 44:990-1000.

Goodenough, U. W., and R. P. Levine. 1970. Chloroplast structure and function in ac-20, a mutant strain of Chlamydomonas reinhardi. III. Chloroplast ribosomes and membrane organization. J. Cell Biol. 44:547-562.

Goodenough, U. W., and R. P. Levine. 1971. The effects of inhibitors of RNA and protein synthesis on the recovery of chloroplast ribosomes, membrane organization, and photosynthetic electron transport in the ac-20 strain of Chlamydomonas reinhardi. J. Cell Biol. 50:50-62.

Gooding, L. R., H. Roy, and A. T. Jagendorf. 1973. Immunological identification of nascent subunits of wheat ribulose diphosphate carboxylase on ribosomes of both chloroplast and cytoplasmic origin. Arch. Biochem. Biophys. 159:324-335.

Gorman, D. S., and R. P. Levine. 1966. Photosynthetic electron transport chain of Chlamydomonas reinhardi. VI. Electron transport in mutant strains lacking either cytochrome-553 or plastocyanin. Plant Physiol. 41:1648-1656.

Gray, J. C., and R. G. O. Kekwick. 1973. Synthesis of the small subunit of ribulose 1,5-diphosphate carboxylase on cytoplasmic ribosomes from greening bean leaves. FEBS Lett. 38:67-69.

Gualerzi, C., and P. Cammarano. 1969. Comparative electrophoretic studies on the proteins of chloroplast and cytoplasmic ribosomes of spinach leaves. Biochim. Biophys. Acta 190:170-186.

Gualerzi, C., and P. Cammarano. 1970. Species specificity of ribosomal proteins from chloroplast and cytoplasmic ribosomes of higher plants: electrophoretic studies. Biochim. Biophys. Acta 199:203-213.

Gualerzi, C., H. G. Janda, H. Passow, and G. Stöffler. 1974. Studies on the protein moiety of plant ribosomes. Enumeration of the proteins of the ribosomal subunits and determination of the degree of evolutionary conservation by electrophoretic and immunochemical methods. *J. Biol. Chem. 249:*3347-3355.

Guderian, R. H., R. L. Pulliam, and M. P. Gordon. 1972. Characterization and fractionation of tobacco leaf transfer RNA. *Biochim. Biophys. Acta 262:*50-65.

Guillemaut, P., G. Burkard, and J. H. Weil. 1972. Characterization of N-formylmethionyl-tRNA in bean mitochondria and etioplasts. *Phytochemistry 11:*2217-2219.

Hadziyev, D., and S. Zalik. 1970. Amino acid incorporation by ribosomes and polysomes from wheat chloroplasts. *Biochem. J. 116:*111-124.

Hallier, U. W. 1968. Photosynthese-Reaktionen einiger Plastom-Mutanten von *Oenothera*. II. Die Bilding von ATP and NADPH. *Z. Pflanzenphysiol. 58:*289-299.

Hallier, U. W., U. Heber, and W. Stubbe. 1968. Photosynthese-Reaktionen einiger Plastom-Mutanten von *Oenothera*. I. Der reduktive Pentosephosphatzyklus. *Z. Pflanzenphysiol. 58:*222-239.

Harris, E. H., and J. M. Eisenstadt. 1971. Initiation of polysome formation in chloroplasts isolated from *Euglena gracilis*. *Biochim. Biophys. Acta 232:*167-170.

Harris, E. H., J. F. Preston, and J. M. Eisenstadt. 1973. Amino acid incorporation and products of protein synthesis in isolated chloroplasts of *Euglena gracilis*. *Biochemistry 12:*1227-1234.

Hartley, M. R., and R. J. Ellis. 1973. Ribonucleic acid synthesis in chloroplasts. *Biochem. J. 134:*249-262.

Hase, E. 1971. Studies on the metabolism of nucleic acid and protein associated with the processes of de- and re-generation of chloroplasts in *Chlorella protothecoides*. in *Autonomy and Biogenesis of Mitochondria and Chloroplasts* (N. K. Boardman, A. W. Linnane, and R. M. Smillie, eds.). Amsterdam: North-Holland. pp. 434-446.

Hauge, J. G. 1971. Pressure-induced dissociation of ribosomes during ultracentrifugation. *FEBS Lett. 17:*168-172.

Hayes, F., D. Hayes, P. Fellner, and C. Ehresmann. 1971. Additional nucleotide sequences in precursor 16 S ribosomal RNA from *Escherichia coli*. *Nature New Biol. 232:*54-55.

Heber, U. 1962. Protein synthesis in chloroplasts during photosynthesis. *Nature 195:*91-92.

Hecker, L. I., J. Egan, R. J. Reynolds, C. E. Nix, J. A. Schiff, and W. E. Barnett. 1974. The sites of transcription and translation for *Euglena* chloroplastic aminoacyl-tRNA synthetases. *Proc. Nat. Acad. Sci. US 71:*1910-1914.

Helser, T. L., J. E. Davies, and J. E. Dahlberg. 1971. Change in methylation of 16 S ribosomal RNA associated with mutation to ka-sugamycin resistance in *Escherichia coli*. *Nature New Biol.* *233:* 12-14.

Helser, T. L., J. E. Davies, and J. E. Dahlberg. 1972. Mechanism of kasugamycin resistance in *Escherichia coli*. *Nature New Biol.* *235:*6-9.

Herrmann, F. 1972. Chloroplast lamellar proteins of the plastid mutant en: viridis-1 of *Antirrhinum majus* having impaired photo-system II. *Exptl. Cell Res.* *70:*452-453.

Herrmann, F., and G. Bauer-Stäb. 1969. Lamellarproteins mutierter Plastiden der Plastommutante albomaculata-1 von *Antirrhinum majus* L. *Flora* *160:*391-393.

Hinkle, D. C., W. F. Mangel, and M. J. Chamberlin. 1972. Studies on the binding of *Escherichia coli* RNA polymerase to DNA. IV. The effect of rifampicin on binding and on RNA chain initiation. *J. Mol. Biol.* *70:*209-220.

Hoagland, M. B., M. L. Stephenson, J. F. Scott, L. I. Hecht, and P. C. Zamecnik. 1958. A soluble ribonucleic acid intermediate in pro-tein synthesis. *J. Biol. Chem.* *231:*241-257.

Holdsworth, R. H. 1971. The isolation and partial characterization of the pyrenoid protein of *Eremosphaera viridis*. *J. Cell Biol.* *51:*499-513.

Honda, S. I., T. Hongladarom, and G. G. Laties. 1966. A new isola-tion medium for plant organelles. *J. Exptl. Bot.* *17:*460-472.

Honeycutt, R. C., and M. M. Margulies. 1973. Protein synthesis in *Chlamydomonas reinhardi*: evidence for synthesis of proteins of chloroplastic ribosomes on cytoplasmic ribosomes. *J. Biol. Chem.* *248:*6145-6153.

Hoober, J. K., 1970. Sites of synthesis of chloroplast membrane po-lypeptides in *Chlamydomonas reinhardi* y-1. *J. Biol. Chem.* *245:* 4327-4334.

Hoober, J. K. 1972. A major polypeptide of chloroplast membranes of *Chlamydomonas reinhardi*: evidence for synthesis in the cyto-plasm as a soluble component. *J. Cell Biol.* *52:*84-96.

Hoober, J. K., and G. Blobel. 1969. Characterization of the chloro-plastic and cytoplasmic ribosomes of *Chlamydomonas reinhardi*. *J. Mol. Biol.* *41:*121-138.

Hoober, J. K., and W. J. Stegeman. 1973. Control of the synthesis of a major polypeptide of chloroplast membranes in *Chlamydomonas reinhardi*. *J. Cell Biol.* *56:*1-12.

Hoober, J. K., P. Siekevitz, and G. E. Palade. 1969. Formation of chloroplast membranes in *Chlamydomonas reinhardi* y-1: effects of inhibitors of protein synthesis. *J. Biol. Chem.* *244:*2621-2631.

Horak, A., and R. D. Hill. 1972. Adenosine triphosphatase of bean plastids. Its properties and site of formation. *Plant Physiol.* *49*:365-370.

Hudock, G. A., G. C. McLeod, J. Moravkova-Kiely, and R. P. Levine. 1964. The relation of oxygen evolution to chlorophyll and protein synthesis in a mutant strain of *Chlamydomonas reinhardi*. *Plant Physiol*. *39*:898-903.

Infante, A. A., and R. Baierlein. 1971. Pressure-induced dissociation of sedimenting ribosomes: effect on sedimentation patterns. *Proc. Nat. Acad. Sci. US 68*:1780-1785.

Ingle, J. 1968. Synthesis and stability of chloroplast ribosomal RNAs. *Plant Physiol.* *43*:1448-1454.

Ingle, J., R. Wells, J. V. Possingham, and C. J. Leaver. 1971. The origins of chloroplast ribosomal-RNA. in *Autonomy and Biogenesis of Mitochondria and Chloroplasts* (N. K. Boardman, A. W. Linnane, and R. M. Smillie, eds.). Amsterdam: North-Holland. pp. 393-401.

Ireland, H. M. M., and J. W. Bradbeer. 1971. Plastid development in primary leaves of *Phaseolus vulgaris*. The effects of *D-threo* and *L-threo* chloramphenicol on the light-induced formation of enzymes of the photosynthetic carbon pathway. *Plant 96*:254-261.

Jennings, R. C., and I. Ohad. 1972. Biogenesis of chloroplast membranes. XI. Evidence for the translation of extrachloroplast RNA on chloroplast ribosomes in a mutant of *Chlamydomonas reinhardi, y-1*. *Arch. Biochem. Biophys. 153*:79-87.

Jones, B. L., N. Nagabhushan, A. Gulyas, and S. Zalik. 1972. Two dimensional acrylamide gel electrophoresis of wheat leaf cytoplasmic and chloroplast ribosomal proteins. *FEBS Lett. 23*:167-170.

Kawashima, N. 1970. Non-synchronous incorporation of $C^{14}O_2$ into amino acids of the two subunits of Fraction I protein. *Biochem. Biophys. Res. Commun. 38*:119-124.

Kawashima, N., and S. G. Wildman. 1970. A model of the subunit structure of Fraction I protein. *Biochem. Biophys. Res. Commun. 41*:1463-1468.

Kawashima, N., and S. G. Wildman. 1972. Studies on Fraction I protein. IV. Mode of inheritance of primary structure in relation to whether chloroplast or nuclear DNA contains the code for a chloroplast protein. *Biochim. Biophys. Acta 262*:42-49.

Kirk, J. T. O. 1964. Studies on RNA synthesis in chloroplast preparations. *Biochem. Biophys. Res. Commun. 16*:233-238.

Kirk, J. T. O. 1968. Studies on the dependence of chlorophyll synthesis on protein synthesis in *Euglena gracilis,* together with a nomogram for determination of chlorophyll concentration. *Planta 78*:200-207.

Kirk, J. T. O., and R. A. E. Tilney-Bassett. 1967. *The Plastids*. London: W. H. Freeman.

Kirk, P. R., and R. M. Leech. 1972. Amino acid biosynthesis by isolated chloroplasts during photosynthesis. *Plant Physiol.* *50*:228-234.

Kislev, N., M. I. Selsky, C. Norton, and J. M. Eisenstadt. 1972. tRNA and tRNA aminoacyl synthetases of chloroplasts, mitochondria and cytoplasm from *Euglena gracilis*. *Biochim. Biophys. Acta 287*: 256-269.

Kleinkopf, G. E., R. C. Huffaker, and A. Matheson. 1970. Light-induced *de novo* synthesis of ribulose 1,5-diphosphate carboxylase in greening leaves of barley. *Plant Physiol.* *46*:416-418.

Kolodner, R., and K. K. Tewari. 1972. Molecular size and conformation of chloroplast deoxyribonucleic acid from pea leaves. *J. Biol. Chem. 247*:6355-6364.

Kowallik, K. V., and R. G. Herrmann. 1972. Variable amounts of DNA related to the size of chloroplasts. IV. Three-dimensional arrangement of DNA in fully differentiated chloroplasts of *Beta vulgaris* L. *J. Cell Sci. 11*:357-377.

Kung, S. D., and J. P. Thornber. 1971. Photosystem I and II chlorophyll-protein complexes of higher plant chloroplasts. *Biochim. Biophys. Acta 253*:285-289.

Kung, S. D., J. P. Thornber, and S. G. Wildman. 1972. Nuclear DNA codes for photosystem II chlorophyll-protein of chloroplast membranes. *FEBS Lett. 24*:185-188.

Kurland, C. G. 1972. Structure and function of the bacterial ribosome. *Ann. Rev. Biochem. 41*:377-408.

Lagoutte, B., and J. Duranton. 1971. Physiochemical study of structural proteins of chloroplast from *Zea mays* L. *Biochim. Biophys. Acta 253*:232-239.

Lai, C.-J., and B. Weisblum. 1971. Altered methylation of ribosomal RNA in an erythromycin-resistant strain of *Staphylococcus aureus*. *Proc. Nat. Acad. Sci. US 68*:856-860.

Lai, C.-J., J. E. Dahlberg, and B. Weisblum. 1973a. Structure of an inducibly methylatable nucleotide sequence in 23 S ribosomal ribonucleic acid from erythromycin-resistant *Staphylococcus aureus*. *Biochemistry 12*:457-460.

Lai, C.-J., B. Weisblum, S. R. Fahnestock, and M. Nomura. 1973b. Alteration of 23 S ribosomal RNA and erythromycin-induced resistance to lincomycin and spiramycin in *Staphylococcus aureus*. *J. Mol. Biol. 74*:67-72.

Leaver, C. J., and J. Ingle. 1971. The molecular integrity of chloroplast ribosomal ribonucleic acid. *Biochem. J. 123*:235-243.

Leis, J. P., and E. B. Keller. 1970. Protein chain-initiating methionine tRNA's in chloroplasts and cytoplasm of wheat leaves. *Proc. Nat. Acad. Sci. US 67*:1593-1599.

Leis, J. P., and E. B. Keller. 1971. N-Formylmethionyl-tRNA$_f$ of wheat chloroplasts. Its synthesis by a wheat transformylase. *Biochemistry 10*:889-894.

Levine, R. P. 1968. Genetic dissection of photosynthesis. *Science 162*:768-771.

Levine, R. P. 1969. The analysis of photosynthesis using mutant strains of algae and higher plants. *Ann. Rev. Plant Physiol. 20*: 523-540.

Levine, R. P., and J. Armstrong. 1972. The site of synthesis of two chloroplast cytochromes in *Chlamydomonas reinhardi*. *Plant Physiol. 49*:661-662.

Levine, R. P., and H. A. Duram. 1973. The polypeptides of stacked and unstacked *Chlamydomonas reinhardi* chloroplast membranes and their relation to photosystem II activity. *Biochim. Biophys. Acta 325*:565-572.

Levine, R. P., and U. W. Goodenough. 1970. The genetics of photosynthesis and of the chloroplast in *Chlamydomonas reinhardi*. *Ann. Rev. Genetics 4*:397-408.

Levine, R. P., and D. S. Gorman. 1966. Photosynthetic electron transport chain of *Chlamydomonas reinhardi*. III. Light-induced absorbance changes in chloroplast fragments of the wild-type and mutant strains. *Plant Physiol. 41*:1293-1300.

Levine, R. P., and A. Paszewski. 1970. Chloroplast structure and function in *ac-20*, a mutant strain of *Chlamydomonas reinhardi*. II. Photosynthetic electron transport. *J. Cell Biol. 44*:540-546.

Levine, R. P., W. G. Burton, and H. A. Duram. 1972. Membrane polypeptides associated with photochemical systems. *Nature New Biol. 237*:176-177.

Lizardi, P. M., and D. J. L. Luck. 1972. The intracellular site of synthesis of mitochondrial ribosomal proteins in *Neurospora crassa*. *J. Cell Biol. 54*:56-74.

Loening, U. E. 1968a. Molecular weights of ribosomal RNA in relation to evolution. *J. Mol. Biol. 38*:355-365.

Loening, U. E. 1968b. RNA structure and metabolism. *Ann. Rev. Plant Physiol. 19*:37-70.

Loening, U. E., and J. Ingle. 1967. Diversity of RNA components in green plant tissues. *Nature 215*:363-367.

Lowry, C. V., and J. E. Dahlberg. 1971. Structural differences between the 16 S ribosomal RNA of *E. coli* and its precursors. *Nature New Biol. 232*:52-54.

Lyttleton, J. W. 1962. Isolation of ribosomes from spinach chloroplasts. *Exptl. Cell Res. 26*:312-317.

Lyttleton, J. W. 1968. Protein constituents of plant ribosomes. *Biochim. Biophys. Acta 154*:145-149.

Machold, O. 1971. Lamellar proteins of green and chlorotic chloroplasts as affected by iron deficiency and antibiotics. *Biochim.
Biophys. Acta* 238:324-331.

Machold, O., and O. Aurich. 1972. Sites of synthesis of chloroplast
lamellar proteins in *Vicia faba*. *Biochim. Biophys. Acta* 281:103-
112.

Machold, O., A. Meister, and K. Adler. 1971. Spektroskopische Eigen-
schaften von elektrophoretisch getrennten Chlorophyll-Protein-Kom-
plexen der Photosysteme I und II aus *Vicia faba* und *Chlorella py-
renoidosa*. *Photosynthetica* 5:160-165.

Manning, J. E., D. R. Wolstenholme, R. S. Ryan, J. A. Hunter, and O.
C. Richards. 1971. Circular chloroplast DNA from *Euglena gracilis*. *Proc. Nat. Acad. Sci. US* 68:1169-1173.

Manning, J. E., D. R. Wolstenholme, and O. C. Richards. 1972. Circular DNA molecules associated with chloroplasts of spinach, *Spi-
nacia oleracea*. *J. Cell Biol.* 53:594-601.

Mao, J. C.-H., and M. Putterman. 1969. The intermolecular complex
of erythromycin and ribosome. *J. Mol. Biol.* 44:347-361.

Marcus, A. 1959. Amino acid dependent exchange between pyrophosphate
and adenosine triphosphate in spinach preparations. *J. Biol. Chem.*
234:1238-1240.

Margulies, M. M. 1962. Effect of chloramphenicol on light-dependent
development of seedlings of *Phaseolus vulgaris* var. Black Valen-
tine, with particular reference to development of photosynthetic
activity. *Plant Physiol.* 37:473-480.

Margulies, M. M. 1964. Effect of chloramphenicol on light-dependent
synthesis of proteins and enzymes of leaves and chloroplasts of
Phaseolus vulgaris. *Plant Physiol.* 39:579-585.

Margulies, M. M. 1966. Effect of chloramphenicol on formation of
chloroplast structure and protein during greening of etiolated
leaves of *Phaseolus vulgaris*. *Plant Physiol.* 41:992-1003.

Margulies, M. M. 1970. *In vitro* protein synthesis by plastids of
Phaseolus vulgaris. V. Incorporation of ^{14}C-leucine into a pro-
tein fraction containing ribulose diphosphate carboxylase. *Plant
Physiol.* 46:136-141.

Margulies, M. M. 1971. Concerning the sites of synthesis of proteins of chloroplast ribosomes and of Fraction I protein (ribulose
1,5-diphosphate carboxylase). *Biochem. Biophys. Res. Commun.* 44:
539-545.

Margulies, M. M., and C. Brubaker. 1970. Effect of chloramphenicol
on amino acid incorporation by chloroplasts and comparison with
the effect of chloramphenicol on chloroplast development *in vivo*.
Plant Physiol. 45:632-633.

Margulies, M. M., and A. Michaels. 1974. Ribosomes bound to chloroplast membranes in *Chlamydomonas reinhardtii*. *J. Cell Biol.* 60:
65-77.

Margulies, M. M., and F. Parenti. 1968. *In vitro* protein synthesis by plastids of *Phaseolus vulgaris*. III. Formation of lamellar and soluble chloroplast proteins. *Plant Physiol. 43*:504-514.

Margulies, M. M., E. Gantt, and F. Parenti. 1968. *In vitro* protein synthesis by plastids of *Phaseolus vulgaris*. II. The probable relation between ribonuclease insensitive amino acid incorporation and the presence of intact chloroplasts. *Plant Physiol. 43*:495-503.

Mehta, S. L., D. Hadziyev, and S. Zalik. 1968. Chloroplast and cytoplasmic polysomes and ribosomal RNA from wheat. *Biochim. Biophys. Acta 169*:381-386.

Menke, W., and H.-G. Ruppel. 1971. Molekulargewicht, Grösse and Gestalt von Proteinen der Thylakoidmembran. *z. Naturforsch. 26(b)*: 825-831.

Merrick, W. C., and L. S. Dure III. 1971. Specific transformylation of one methionyl-tRNA from cotton seedling chloroplasts by endogenous and *Escherichia coli* transformylases. *Proc. Nat. Acad. Sci. US 68*:641-644.

Merrick, W. C., and L. S. Dure III. 1972. The developmental biochemistry of cotton seed embryogenesis and germination. IV. Levels of cytoplasmic and chloroplastic transfer ribonucleic acid species. *J. Biol. Chem. 247*:7988-7999.

Mets, L. J., and L. Bogorad. 1971. Mendelian and uniparental alterations in erythromycin binding by plastid ribosomes. *Science 174*: 707-709.

Mets, L., and L. Bogorad. 1972. Altered chloroplast ribosomal proteins associated with erythromycin-resistant mutants in two genetic systems of *Chlamydomonas reinhardi*. *Proc. Nat. Acad. Sci. US 69*:3779-3783.

Mihara, S., and E. Hase. 1969. A note on ribosomes in cells of *Chlorella protothecoides*. *Plant Cell Physiol. 10*:465-470.

Munsche, D., and R. Wolligiehn. 1973. Die Synthese von ribosomaler RNA in Chloroplasten von *Nicotiana rustica*. *Biochim. Biophys. Acta 294*:106-117.

Nadler, K., and S. Granick. 1970. Controls on chlorophyll synthesis in barley. *Plant Physiol. 46*:240-246.

Nadler, K. D., H. A. Herron, and S. Granick. 1972. Development of chlorophyll and Hill activity. *Plant Physiol. 49*:388-392.

Nelson, N., H. Nelson, and E. Racker. 1972a. Partial resolution of the enzymes catalyzing photophosphorylation. XI. Magnesium-adenosine triphosphatase properties of the heat-activated coupling factor from chloroplasts. *J. Biol. Chem. 247*:6506-6510.

Nelson, N., H. Nelson, and E. Racker. 1972b. Partial resolution of the enzymes catalyzing photophosphorylation. XII. Purification and properties of an inhibitor isolated from chloroplast coupling factor I. *J. Biol. Chem. 247*:7657-7662.

Nichiporovich, A. A. 1955. Tracer atoms used to study the products of photosynthesis depending on the conditions under which the process takes place. *Proceedings of the International Conference on Peaceful Uses of Atomic Energy, Geneva 12*:340-346.

Nomura, M. 1970. Bacterial ribosome. *Bacterial Rev. 34*:228-277.

Odintsova, M. S., and N. P. Yurina. 1969. Proteins of chloroplast and cytoplasmic ribosomes. *J. Mol. Biol. 40*:503-506.

Oshio, Y., and E. Hase. 1972. Changes of ribulose 1,5-diphosphate carboxylase level during the processes of degeneration and regeneration of chloroplasts in *Chlorella prototheocoides*. *Plant Cell Physiol. 13*:955-963.

Pakhomova, M. V., G. N. Zaitseva, and S. Yu. Stepanenko. 1968. Ribosomes from chloroplasts and cytoplasm of the unicellular green alga *Chlamydomonas globosa*. *Biokhimiya* (English transl.) *33*:878-882.

Parenti, F., and M. M. Margulies. 1967. *In vitro* protein synthesis by plastids of *Phaseolus vulgaris*. I. Localization of activity in the chloroplasts of a chloroplast-containing fraction from developing leaves. *Plant Physiol. 42*:1179-1186.

Parthier, B. 1964. Proteinsynthese in grünen Blättern. I. Die Proteinsynthese in verschiedenen subzellulären Fracktionen nach $^{14}C_2$-Photosynthese. *Z. Naturforsch. 19(b)*:235-248.

Parthier, B., R. Krauspe, and S. Samtleben. 1972. Light-stimulated synthesis of aminoacyl-tRNA synthetases in greening *Euglena gracilis*. *Biochim. Biophys. Acta 277*:335-341.

Paulsen, J. M., and M. D. Lane. 1966. Spinach ribulose diphosphate carboxylase. I. Purification and properties of the enzyme. *Biochemistry 5*:2350-2357.

Payne, P. I., and T. A. Dyer. 1971. Characterization of cytoplasmic and chloroplast 5 S ribosomal ribonucleic acid from broad-bean leaves. *Biochem. J. 124*:83-89.

Payne, P. I., and T. A. Dyer. 1972. Plant 5.8 S RNA is a component of 80 S but not 70 S ribosomes. *Nature New Biol. 235*:145-147.

Pène, J., E. Knight, and J. E. Darnell. 1968. Characterization of a new low molecular weight RNA in HeLa cell ribosomes. *J. Mol. Biol. 33*:609-623.

Perl, M. 1972. A possible ribosomal-directed regulatory system in *Euglena gracilis*. Chlorophyll synthesis. *Biochem. J. 130*:813-818.

Pestka, S. 1971. Inhibitors of ribosome function. *Ann. Rev. Microbiol. 25*:487-562.

Polya, G. M., and A. T. Jagendorf. 1971. Wheat leaf RNA polymerase. I. Partial purification and characterization of nuclear, chloroplast and soluble DNA-dependent enzymes. *Arch. Biochem. Biophys. 146*:635-648.

Preer, J. R., Jr. 1971. Extrachromosomal inheritance: hereditary symbionts, mitochondria, chloroplasts. *Ann. Rev. Genetics* 5:361-406.

Ramirez, J. M., F. F. del Campo, and D. I. Arnon. 1968. Photosynthetic phosphorylation as energy source for protein synthesis and carbon dioxide assimilation by chloroplasts. *Proc. Nat. Acad. Sci. US* 59:606-612.

Ranalletti, M., A. Gnanam, and A. T. Jagendorf. 1969. Amino acid incorporation by isolated chloroplasts. *Biochim. Biophys. Acta* 186:192-204.

Rawson, J. R. Y., and R. Haselkorn. 1973. Chloroplast ribosomal RNA genes in the chloroplast DNA of *Euglena gracilis*. *J. Mol. Biol.* 77:125-132.

Rawson, J. R., and E. Stutz. 1969. Isolation and characterization of *Euglena gracilis* cytoplasmic and chloroplast ribosomes and their ribosomal RNA components. *Biochim. Biophys. Acta* 190:368-380.

Reger, B. J., S. A. Fairfield, J. L. Epler, and W. E. Barnett. 1970. Identification and origin of some chloroplast aminoacyl-tRNA synthetases and tRNAs. *Proc. Nat. Acad. Sci. US* 67:1207-1213.

Reger, B. J., R. M. Smillie, and R. C. Fuller. 1972. Protein synthesis by isolated etioplasts and chloroplasts from pea and wheat and the effects of chloramphenicol and cycloheximide. *Plant Physiol.* 50:19-23.

Remy, R. 1971. Resolution of chloroplast lamellar proteins by electrophoresis in polyacrylamide gels. Different patterns obtained with fractions enriched in either chlorophyll *a* or chlorophyll *b*. *FEBS Lett.* 13:313-317.

Rijven, A. H. G. C., and J. A. Zwar. 1973. Methylation patterns of ribonucleic acids from chloroplasts and cytoplasm of fenugreek (*Trigonella foenumgraecum* L.) cotyledons. *Biochim. Biophys. Acta* 299:564-567.

Ritenour, G. L., K. W. Joy, J. Bunning, and R. H. Hageman. 1967. Intracellular localization of nitrate reductase, nitrite reductase, and glutamic acid dehydrogenase in green leaf tissue. *Plant Physiol.* 42:233-237.

Rochaix, J. D. 1972. Cyclization of chloroplast DNA fragments of *Chlamydomonas reinhardi*. *Nature New Biol.* 238:76-78.

Rutner, A. C. 1970. Estimation of the molecular weight of ribulose diphosphate carboxylase sub-units. *Biochem. Biophys. Res. Commun.* 39:923-929.

Sager, R. 1954. Mendelian and non-Mendelian inheritance of streptomycin resistance in *Chlamydomonas reinhardi*. *Proc. Nat. Acad. Sci. US* 40:356-363.

Sager, R. 1972. *Cytoplasmic Genes and Organelles*. New York: Academic Press.

Sager, R., and M. G. Hamilton. 1967. Cytoplasmic and chloroplast
    ribosomes of *Chlamydomonas*: ultracentrifugal characterization.
    Science 157:709-711.

Sager, R., and D. Lane. 1972. Molecular basis of maternal inheri-
    tance. *Proc. Nat. Acad. Sci. US 69*:2410-2413.

Sager, R., and Z. Ramanis. 1963. The particulate nature of nonchro-
    mosomal genes in *Chlamydomonas*. *Proc. Nat. Acad. Sci. US 50*:260-
    268.

Sager, R., and Z. Ramanis. 1970. A genetic map of non-Mendelian
    genes in *Chlamydomonas*. *Proc. Nat. Acad. Sci. US 65*:593-600.

Schiff, J. A. 1971. Developmental interactions among cellular com-
    partments in *Euglena*. in *Autonomy and Biogenesis of Mitochondria
    and Chloroplasts* (N. K. Boardman, A. W. Linnane, and R. M. Smillie,
    eds.). Amsterdam: North-Holland. pp. 98-118.

Schiff, J. A. 1973. The development, inheritance, and origin of the
    plastid in *Euglena*. *Adv. Morphogen. 10*:265-312.

Schlanger, G., R. Sager, and Z. Ramanis. 1972. Mutation of a cyto-
    plasmic gene in *Chlamydomonas* alters chloroplast ribosome function.
    Proc. Nat. Acad. Sci. US 69:3551-3555.

Schor, S. L., P. Siekevitz, and G. E. Palade. 1970a. Cyclic varia-
    tions in the protein synthetic activity of the chloroplast and cy-
    toplasm in synchronized *Chlamydomonas reinhardi*. *J. Cell Biol.
    47*:182a-183a.

Schor, S., P. Siekevitz, and G. E. Palade. 1970b. Cyclic changes in
    thylakoid membranes of synchronized *Chlamydomonas reinhardi*. *Proc.
    Nat. Acad. Sci. US 66*:174-180.

Schrader, L. E., L. Beevers, and R. H. Hageman. 1967. Differential
    effects of chloramphenicol on the induction of nitrate and nitrite
    reductase in green leaf tissue. *Biochem. Biophys. Res. Commun.
    26*:14-17.

Schwartz, J. H., R. Meyer, J. M. Eisenstadt, and G. Brawerman. 1967.
    Involvement of N-formylmethionine in initiation of protein synthe-
    sis in cell-free extracts of *Euglena gracilis*. *J. Mol. Biol. 25*:
    571-574.

Schweiger, H. G., and S. Berger. 1964. DNA-dependent RNA synthesis
    in chloroplasts of *Acetabularia*. *Biochim. Biophys. Acta 87*:533-
    535.

Scott, N. S. 1973. Ribosomal RNA cistrons in *Euglena gracilis*. *J.
    Mol. Biol. 81*:327-336.

Scott, N. S., R. Munns, and R. M. Smillie. 1970. Chloroplast and
    cytoplasmic ribosomes in *Euglena gracilis*. *FEBS Lett. 10*:149-152.

Scott, N. S., R. Munns, D. Graham, and R. M. Smillie. 1971. Origin
    and synthesis of chloroplast ribosomal RNA and photoregulation dur-
    ing chloroplast biogenesis. in *Autonomy and Biogenesis of Mito-

chondria and Chloroplasts (N. K. Boardman, A. W. Linnane, and R. M. Smillie, eds). Amsterdam: North-Holland. pp. 383-392.

Semal, J., D. Spencer, Y. T. Kim, and S. G. Wildman. 1964. Properties of a ribonucleic acid synthesizing system in cell-free extracts of tobacco leaves. *Biochim. Biophys. Acta 91:*205-216.

Shephard, D. C., and W. B. Levin. 1972. Biosynthesis in isolated *Acetabularia* chloroplasts. I. Protein amino acids. *J. Cell Biol. 54:*279-294.

Siersma, P. W., and K.-S. Chiang. 1971. Conservation and degradation of cytoplasmic and chloroplast ribosomes in *Chlamydomonas reinhardtii*. *J. Mol. Biol. 58:*167-185.

Sippel, A., and G. Hartmann. 1968. Mode of action of rifamycin on the RNA polymerase reaction. *Biochim. Biophys. Acta 157:*218-219.

Sirevåg, R., and R. P. Levine. 1972. Fatty acid synthetase from *Chlamydomonas reinhardi*. Sites of transcription and translation. *J. Biol. Chem. 247:*2586-2591.

Sissakian, N. M. 1958. Enzymology of the plastids. *Advances in Enzymology and Related Subjects of Biochemistry 20:*201-236.

Sissakian, N. M., I. I. Filippovich, E. N. Svetailo, and K. A. Aliyev. 1965. On the protein-synthesizing system of chloroplasts. *Biochim. Biophys. Acta 95:*474-485.

Siu, C. H., K.-S. Chiang, and H. Swift. 1971. Ribosomal cistrons and cell particulates in *Polytoma obtusum*. *Absts. Eleventh Ann. Meeting Am. Soc. Cell Biol., New Orleans:*277.

Smillie, R. M., D. Graham, M. R. Dwyer, A. Grieve, and N. F. Tobin. 1967. Evidence for the synthesis *in vivo* of proteins of the Calvin cycle and of the photosynthetic electron-transfer pathway on chloroplast ribosomes. *Biochem. Biophys. Res. Commun. 28:*604-610.

Smillie, R. M., N. S. Scott, and D. Graham. 1968. Biogenesis of chloroplasts: roles of chloroplast DNA and chloroplast ribosomes. in *Comparative Biochemistry and Biophysics of Photosynthesis* (K. Shibata, A. Takamiya, A. T. Jagendorf, and R. C. Fuller, eds.). Tokyo: University of Tokyo Press. pp. 332-353.

Smillie, R. M., and N. S. Scott. 1969. Organelle biosynthesis: the chloroplast. *Progress in Molecular and Subcellular Biology 1:*136-202.

Smillie, R. M., D. G. Bishop, G. C. Gibbons, D. Graham, A. M. Grieve, J. K. Raison, and B. J. Reger. 1971. Determination of the sites of synthesis of proteins and lipids of the chloroplast using chloramphenicol and cycloheximide. in *Autonomy and Biogenesis of Mitochondria and Chloroplasts* (N. K. Boardman, A. W. Linnane, and R. M. Smillie, eds.). Amsterdam: North-Holland. pp. 422-433.

Smith, D. C., J. A. Bassham, and M. Kirk. 1961. Dynamics of the photosynthesis of carbon compounds. II. Amino acid synthesis. *Biochim. Biophys. Acta 48:*299-313.

Smith-Johannsen, H., and S. P. Gibbs. 1972. Effects of chloramphenicol on chloroplast and mitochondrial ultrastructure in Ochromonas danica. J. Cell Biol. 52:598-614.

Spencer, D. 1965. Protein synthesis by isolated spinach chloroplasts. Arch. Biochem. Biophys. 111:381-390.

Spencer, D., and P. R. Whitfeld. 1966. The nature of the ribonucleic acid of isolated chloroplasts. Arch. Biochem. Biophys. 117: 337-346.

Spencer, D., and S. G. Wildman. 1964. The incorporation of amino acids into protein by cell-free extracts from tobacco leaves. Biochemistry 3:954-959.

Spencer, D., P. R. Whitfeld, W. Bottomley, and A. M. Wheeler. 1971. The nature of the proteins and nucleic acids synthesized by isolated chloroplasts. in Autonomy and Biogenesis of Mitochondria and Chloroplasts (N. K. Boardman, A. W. Linnane, and R. M. Smillie, eds.). Amsterdam: North-Holland. pp. 372-382.

Stegeman, W. J., and J. K. Hoober. 1974. Mitochondrial protein synthesis in Chlamydomonas reinhardtii y-1. Polypeptide products of mitochondrial transcription and translation in vivo as revealed by selective labeling with ^{3}H-leucine. J. Biol. Chem. 249:6866-6873.

Stephenson, M. L., K. V. Thimann, and P. C. Zamecnik. 1956. Incorporation of ^{14}C-amino acids into proteins of leaf disks and cell-free fractions of tobacco leaves. Arch. Biochem. Biophys. 65: 194-209.

Stepka, W., A. A. Benson, and M. Calvin. 1948. The path of carbon in photosynthesis. II. Amino acids. Science 108:304.

Stolarsky, L., N. Slavik, and C. L. Hershberger. 1973. Purification of chloroplast DNA from Euglena gracilis. Biochim. Biophys. Acta 335:30-34.

Stutz, E. 1970. The kinetic complexity of Euglena gracilis chloroplast DNA. FEBS Lett. 8:25-28.

Stutz, E. 1971. Characterization of Euglena gracilis chloroplast single strand DNA. in Autonomy and Biogenesis of Mitochondria and Chloroplasts (N. K. Boardman, A. W. Linnane, and R. M. Smillie, eds.). Amsterdam: North-Holland. pp. 277-281.

Stutz, E., and H. Noll. 1967. Characterization of cytoplasmic and chloroplast polysomes in plants: evidence for three classes of ribosomal RNA in nature. Proc. Nat. Acad. Sci. US 57:774-781.

Sueoka, N., K.-S. Chiang, and J. R. Kates. 1967. Deoxyribonucleic acid replication in meiosis of Chlamydomonas reinhardi. I. Isotopic transfer experiments with a strain producing eight zoospores. J. Mol. Biol. 25:47-66.

Sugiyama, T., and T. Akazawa. 1970. Subunit structure of spinach leaf ribulose 1,5-diphosphate carboxylase. Biochemistry 9:4499-4504.

Surzycki, S. J. 1969. Genetic function of the chloroplast of *Chlamydomonas reinhardi*: effect of rifampin on chloroplast DNA-dependent RNA polymerase. *Proc. Nat. Acad. Sci. US 63*:1327-1334.

Surzycki, S. J., and N. W. Gillham. 1971. Organelle mutations and their expression in *Chlamydomonas reinhardi*. *Proc. Nat. Acad. Sci. US 68*:1301-1306.

Surzycki, S. J., and J. D. Rochaix. 1971. Transcriptional mapping of ribosomal RNA genes of the chloroplast and nucleus of *Chlamydomonas reinhardi*. *J. Mol. Biol. 62*:89-109.

Svetailo, E. N., I. I. Philippovich, and N. M. Sissakian. 1967. Differences in sedimentation properties of chloroplast and cytoplasmic ribosomes from pea seedlings. *J. Mol. Biol. 24*:405-415.

Sy, J., and K. S. McCarty. 1970. Characterization of 5.8 S RNA from a complex with 26 S ribosomal RNA from *Arbacia punctulata*. *Biochim. Biophys. Acta 199*:86-94.

Taylor, M. M., and R. Storck. 1964. Uniqueness of bacterial ribosomes. *Proc. Nat. Acad. Sci. US 52*:958-965.

Tewari, K. K., and S. G. Wildman. 1968. Function of chloroplast DNA. I. Hybridization studies involving nuclear and chloroplast DNA with RNA from cytoplasmic (80 S) and chloroplast (70 S) ribosomes. *Proc. Nat. Acad. Sci. US 59*:569-576.

Tewari, K. K., and S. G. Wildman. 1970. Information content in the chloroplast DNA. *Symp. Soc. Exptl. Biol. 24*:147-179.

Thomas, J. R. 1972. Distinct cistrons for the 23 S and 16 S RNA in the chloroplast DNA of pea leaves. *Fed. Proc. 31*:914 (abstract).

Tiboni, O., B. Parisi, A. Perani, and O. Ciferri. 1970. Presence of two sets of ribosome-specific transfer factors in the cell-free extracts from the non-photosynthetic alga *Prototheca zopfii*. *J. Mol. Biol. 47*:467-476.

Togasaki, R. K., and R. P. Levine. 1970. Chloroplast structure and function in *ac-20*, a mutant strain of *Chlamydomonas reinhardi*. I. CO_2 fixation and ribulose 1,5-diphosphate carboxylase synthesis. *J. Cell Biol. 44*:531-539.

Udem, S. A., K. Kaufman, and J. R. Warner. 1971. Small ribosomal ribonucleic acid species of *Saccharomyces cerevisiae*. *J. Bacteriol. 105*:101-106.

Uzzell, T., and C. Spolsky. 1974. Mitochondria and plastids: a revival of special creation? *Am. Scient. 62*:334-343.

Vanyushin, B. F., and D. B. Dunn. 1967. The dissociation of 80-S ribosomes with polyvinylsulfate. *Biochim. Biophys. Acta 134*:91-105.

Vasconcelos, A. C. L., and L. Bogorad. 1971. Proteins of cytoplasmic, chloroplast, and mitochondrial ribosomes of some plants. *Biochim. Biophys. Acta 228*:492-502.

Vasconcelos, A., M. Pollack, L. R. Mendiola, H.-P. Hoffmann, D. H. Brown, and C. A. Price. 1971. Isolation of intact chloroplasts from *Euglena gracilis* by zonal centrifugation. *Plant Physiol. 47:* 217-221.

Wallach, D., S. Bar-Nun, and I. Ohad. 1972. Biogenesis of chloroplast membranes. IX. Development of photophosphorylation and proton pump activities in greening *Chlamydomonas reinhardi* y-1 as measured with an open-cell preparation. *Biochim. Biophys. Acta 267:*125-137.

Weisblum, B., C. Siddhikol, C. J. Lai, and V. Demohn. 1971. Erythromycin-inducible resistance in *Staphylococcus aureus:* requirements for induction. *J. Bacteriol. 106:*835-847.

Wells, R., and M. Birnstiel. 1969. Kinetic complexity of chloroplastal deoxyribonucleic acid and mitochondrial deoxyribonucleic acid from higher plants. *Biochem. J. 112:*777-786.

Wells, R., and R. Sager. 1971. Denaturation and the renaturation kinetics of chloroplast DNA from *Chlamydomonas reinhardi. J. Mol. Biol. 58:*611-622.

Wetmur, J. G., and N. Davidson. 1968. Kinetics of renaturation of DNA. *J. Mol. Biol. 31:*349-370.

von Wettstein, D., K. W. Henningsen, J. E. Boynton, G. C. Kannangara, and O. F. Nielsen. 1971. The genic control of chloroplast development in barley. in *Autonomy and Biogenesis of Mitochondria and Chloroplasts* (N. K. Boardman, A. W. Linnane, and R. M. Smillie, eds.). Amsterdam: North-Holland. pp. 205-223.

Wildman, S. G. 1971. An approach towards ascertaining the function of chloroplast DNA in tobacco plants. in *Autonomy and Biogenesis of Mitochondria and Chloroplasts* (N. K. Boardman, A. W. Linnane, and R. M. Smillie, eds.). Amsterdam: North-Holland. pp. 402-412.

Williams, G. R., and A. S. Williams. 1970. Hybridization of bean leaf leucyl-tRNA with nuclear DNA and with chloroplast DNA. *Biochem. Biophys. Res. Commun. 39:*858-863.

Yoshikami, D., and E. B. Keller. 1971. Chemical modification of the fluorescent base in phenylalanine transfer ribonucleic acid. *Biochemistry 10:*2969-2976.

Zalik, S., and B. L. Jones. 1973. Protein biosynthesis. *Ann. Rev. Plant Physiol. 24:*47-68.

Zimmerman, R. A., Y. Ikeya, and P. F. Sparling. 1973. Alteration of ribosomal protein S4 by mutation linked to kasugamycin-resistance in *Escherichia coli. Proc. Nat. Acad. Sci. US 70:*71-75.

Chapter 5

MITOCHONDRIAL PROTEIN SYNTHESIS

Thomas W. O'Brien

Department of Biochemistry
University of Florida
Gainesville, Florida

I. INTRODUCTION

Mitochondria are prominent membranous organelles occurring in virtually all cells of eukaryotic organisms. Most eukaryotes, and certainly all multicellular organisms, as obligate aerobes, are vitally
dependent on these ubiquitous organelles to meet their energy needs.
To fulfill their primary role in cellular energy metabolism, mitochondria contain a characteristic system of internal membranes that
is highly specialized for the aerobic production of ATP by oxidative
phosphorylation. Whereas most of the many mitochondrial proteins are
made outside the mitochondria, on cytoplasmic ribosomes, certain of
the proteins in this mitochondrial inner membrane are made inside mitochondria, on mitoribosomes. This phenomenon is probably the *rai-
son-d'etre* for the semiautonomous mitochondrial biogenetic system.
We now appreciate that the biogenesis and the proper functioning of
mitochondria ultimately depend on mitochondrial protein synthesis, a
process occurring within mitochondria on specific ribosomes, using
mitochondrial-specific factors and macromolecules as well.

More than fifteen years have elapsed since the first demonstration of mitochondrial protein synthesis (McLean et al., 1958). During most of this period, our understanding of mitochondrial protein
synthesis has progressed painstakingly slowly, despite considerable
effort by many investigators and often amidst an air of controversy
and skepticism. As those who witnessed the growth of the area are
well aware, this field has suffered more than its share of conflicting or artifactual results that often have led to major conceptual
reversals.

The confusion unfortunately started promptly, after mitochondria
were shown capable of protein synthesis in vitro (McLean et al., 1958),
with the reports that cytochrome c is among the proteins synthesized
by mitochondria in vitro. This announcement, retracted later, was
followed by the news that among the products of mitochondrial protein
synthesis were the mitochondrial structural protein and the mitochondrial contractile protein. In fact, at one point, it was even proposed that mitochondria synthesize and export structural proteins for
assimilation into general cellular membranes. These reports contri-

buted greatly to the controversial status of mitochondrial protein
synthesis for, as future work showed, such proteins as mitochondrial
contractile protein and mitochondrial structural protein do not even
exist.

In view of recurring, grave implications that amino acid incorporation by mitochondria is due solely to the bacteria present in the
preparations used, there has been a healthy preoccupation with mitochondrial purification conditions and the adoption of measures to
circumvent the problems of bacterial contamination. As if bacterial
contamination were not enough to worry about, the problem of contamination of mitochondrial preparations by microsomal ribosomes has
dominated many studies. Indeed, whatever protein synthetic activity
was not ascribed by critics to bacteria, or to microsomal contaminants present in the mitochondrial preparations, was most recently at-
tributed to the artifactual labeling of preexisting proteins by various nonsynthetic side reactions.

If the demonstration of protein synthesis by mitochondria were
suspect and controversial, how much more so was the demonstration of
specific intramitochondrial ribosomes. The first reports identified
55S ribosomes in rat liver mitochondria and 70S ribosomes in *Neurospora* mitochondria. From these seemingly disparate findings has
sprung the surprising result that mitochondria, depending on the organism, contain either 55S, 70S, or 80S monoribosomes. It is noteworthy that no other system, prokaryotic or eukaryotic, shows such
diversity in the size, composition, and physical-chemical properties
of ribosomes as do mitochondria.

A more optimistic picture of mitochondrial protein synthesis is
now emerging. Indeed, impressive advances have been made in many
aspects of this field, as attested by the proliferation of reviews
on this subject in recent years (Schatz and Mason, 1974; Mahler, 1973;
Borst and Grivell, 1973; Wheeldon, 1973; Tzagoloff et al., 1973;
Borst, 1972; Dawid, 1972; Kroon et al., 1972; Linnane et al., 1972;
Borst and Grivell, 1971; Beattie, 1971; Kuntzel, 1971; Ashwell and
Work, 1970a; Linnane and Haslam, 1970). In the present review, we
will focus on mitochondrial protein synthesis in vitro and the nature,

distribution, and distinctive structural and functional properties
of mitochondrial ribosomes.

II. GENERAL ASPECTS OF PROTEIN SYNTHESIS BY
MITOCHONDRIA IN VITRO

One of the more interesting general features of protein synthesis in
mitochondria is that these organelles, as isolated from active cells,
contain all the components and factors necessary to support protein
synthesis for limited periods in vitro. The initial description of
this phenomenon (McLean et al., 1958) formed the basis for many of
the early studies of protein synthesis by mitochondria, and variations
of this in vitro system have been most useful in deciphering many aspects of the process.

A. Requirements

The minimum requirements of this process are few. All that mitochondria require to synthesize proteins in vitro is essentially a buf-
fered isotonic medium containing salts (magnesium and potassium, chloride and phosphate) and oxygen (Roodyn et al., 1961). Several variations of this reaction medium have been described, which include addi-
tions to boost the rate or lengthen the duration of amino acid incorporation in vitro. Beattie (1971) has presented a critique of several
of the systems that have been used to study mitochondrial protein synthesis in vitro. To someone in search of such a system it may be disconcerting to learn that different laboratories disagree, not only on
the optimum incubation conditions, but also on the effects of specific
additives. The in vitro mitochondrial system is particularly labile,
also depending somewhat on the conditions under which mitochondria are
prepared. Their activity depends, for example, not only on the ionic
and osmotic properties of their isolation media (Haldar and Freeman,
1969) but also on methods used to release them from cells (Roodyn,
1965). Most recently, Mockle (1972) has conducted a systematic comparison of three of the more popular in vitro systems (Kroon and De-
Vries, 1971; Wheeldon and Lehninger, 1966; and Beattie et al., 1967a),

considering the effects of different conditions for the preparation of mitochondria as well. He concluded that maximal rates of amino acid incorporation are achieved using the general incubation conditions of Kroon and DeVries (1971). It is noteworthy that in this system, the energy for protein synthesis is provided by endogenous ATP, produced by mitochondrial oxidative phosphorylation.

From careful studies of in vitro systems, it is apparent that certain factors, such as temperature (Mockle, 1972) and amino acid concentration, are more important in establishing maximal rates of mitochondrial protein synthesis. Whereas mitochondria contain a significant pool of amino acids (Roodyn et al., 1961; Truman and Korner, 1962) that can be drawn upon to support protein synthesis for short periods, the overall incorporation is increased during extended incubations when the reaction mixture is supplemented with amino acids (Roodyn et al., 1961), especially proline, serine, and methionine (Beattie et al., 1967b). Most importantly, optimum incorporation rates are not achieved unless the radioactively labeled amino acid is also present at relatively high concentrations (Kroon and DeVries, 1971). The reaction also proceeds better in the presence of added oxidizable substrates, such as succinate, a-keto-glutarate, or pyruvate and malate (Mockle, 1972; Beattie and Ibrahim, 1973), especially when the energy is provided by oxidative phosphorylation. Finally, it should be noted that certain buffering agents, such as bicine, appear more suitable (Beattie et al., 1967b).

To illustrate some vagaries of the in vitro systems used to study mitochondrial protein synthesis, and also to acknowledge that divergent results have been obtained in different laboratories, attention should be called to reports of the artifactual labeling of mitochondrial proteins during in vitro incubations (Hochberg et al., 1972). Depending somewhat on the incubation conditions, but more so on the particular radioactively labeled amino acid used to follow the incorporation, variable amounts of artifactual labeling were observed. With some amino acids, methionine, for example, the incorporation into denatured (heat or TCA) mitochondria was high enough to mask any real protein synthesis occurring in control mitochondria. Artifactual labeling probably occurs in variable amounts during most in vitro studies but it appears to be less of a problem with leucine (Ibrahim et al., 1973). This disconcerting phenomenon, which occurs at 0°C as well, and can mimic normal reaction kinetics, is though to represent a covalent attachment of amino acids to preexisting proteins by some poorly understood, nonbiosynthetic side reactions (Suttie, 1962; Hochberg et al., 1972). Workers are admonished that these side reactions are apt to predominate in incubations carried out under suboptimal conditions, especially if mitochondria are used that have been "damaged" or compromised somehow during their preparation.

Under appropriate conditions, high protein synthetic rates can be achieved (Kroon and DeVries, 1971; Mockle, 1972; Ibrahim et al., 1973) and specific controls can be employed to verify the integrity of the biosynthetic reaction. Systems relying on endogenous ATP, for example, are depressed by inhibitors of oxidative phosphorylation (Mockle, 1972; Beattie and Ibrahim, 1973). Even more consoling, puromycin or chloramphenicol and many other antibiotics that act on prokaryotic ribosomes will inhibit amino acid incorporation by mitochondria (Kroon and DeVries, 1971; Lederman and Attardi, 1970; Mockle, 1972; Ibrahim et al., 1973). Additional support for the contention that protein synthesis continues as a real biosynthetic process in mitochondria for brief periods after their isolation from other cellular components, follows from observations that the rate of protein synthesis in isolated mitochondria reflects their prior physiological state. Thus, mitochondria from rapidly growing fetal and neonatal tissues exhibit higher protein synthetic rates than those from the same tissues of adult animals (Roodyn, 1965; Graffi et al., 1965; Malkin, 1970). Similarly, some hormones modulate mitochondrial protein synthetic rates (Roodyn et al., 1965; Delaney et al., 1967; Pegg and Williams-Ashman, 1968). The specific effects of thyroid hormone on mitochondria of responsive tissues is especially significant since this hormone is involved in the regulation of total mitochondrial mass and content of functional respiratory assemblies (Gustafsson et al., 1965). Coincident with the stimulation of oxidative metabolism, administered thyroid hormones also increase the RNA content and protein synthetic rate of mitochondria (Roodyn et al., 1965).

B. Properties of the Products of Protein Synthesis in vitro

While none of the in vitro products have been identified, and there is still disagreement on the fundamental questions of number and size of these products, there is unanimous accord on their relative insolubility in aqueous media.

1. Identity

One of the first questions raised by the discovery that mitochondria could incorporate amino acids in vitro concerned the identity of the products. Despite some early unfounded claims that cytochrome c was made by mitochondria (Bates and Simpson, 1959; Bates et al., 1960; Simpson et al., 1961), it appears that the obvious candidates, the cytochromes, are not among the products of mitochondrial protein synthesis (Sect. VII). Furthermore, neither are the well known, relatively soluble mitochondrial proteins, such as malate dehydrogenase (Roodyn et al., 1962; Bingham and Campbell, 1972). The major incorporation of radioactive amino acids occurred into a poorly characterized, relatively insoluble subfraction of mitochondria and for some time it was thought that the major product of mitochondrial protein synthesis was the so-called "structural protein" (Criddle et al., 1962). However, this position was abandoned later after a critical reexamination of the material showed it to be a heterogeneous mixture of predominantly denatured mitochondrial proteins (Senior and MacLennan, 1970). Indeed, the apparent in vitro labeling of "structural protein" was due to the presence in this fraction of denatured components of the mitochondrial ATPase, itself shown to contain products of mitochondrial protein synthesis (Schatz and Saltzgaber, 1969).

After "structural protein," the next candidate for a product of mitochondrial protein synthesis was mitochondrial "contractile protein" (Kalf and Grece, 1964). This protein, having an ATPase acti-

vity and some physical-chemical properties in common with contractile
proteins of muscle origin (Ohnishi and Ohnishi, 1962; Blair et al.,
1964) had been implicated in conformational transformation of the mitochondrial "mechano-enzyme system" (Vignais et al., 1963). However,
other laboratories were unable to substantiate the existence of a
"contractile protein" in mitochondria (Conover and Barany, 1966; Bemis et al., 1968) and further scrutiny of this extract ascribed the
ATPase activity to a mitochondrial nucleoside triphosphatase (Kalf
and Grece, 1970b) that was not labeled significantly during in vitro
incubations (Kalf and Grece, 1970a).

There are two major pitfalls inherent in the use of the in vitro
system, as applied above, to identify products of mitochondrial protein synthesis. The first of these, which was not generally appreciated until relatively recently, is the concerted nature of the biosynthetic and assembly processes, whereby products of extramitochondrial origin are also required for assembly of the nascent mitochondrial products into recognizable structural or functional entities
(Vary et al., 1970; Mahler et al., 1968; Work, 1968; Henson et al.,
1968; Yu et al., 1968; Chen and Charalampous, 1969). Thus, bona
fide mitochondrial products destined for one enzyme complex may, under conditions (in vitro) of impaired assembly, appear to accompany
a different complex or protein fraction during the subfractionation
of mitochondria. This phenomenon alone invalidates any conclusions
based on negative findings with this system (Roodyn et al., 1962).
That is to say, the failure to obtain labeling of a multisubunit enzyme during in vitro incubations need not imply an extramitochondrial
origin for all subunits of the enzyme because of the likely possibility that labeled nascent mitochondrial products neither exchange with
unlabeled counterparts in preexisting enzymes, nor integrate into new
enzymes in the absence of a sufficient pool of some complementary
subunit made on extramitochondrial ribosomes. The second and more
obvious problem in the above studies concerns the need to work with
homogeneous mitochondrial components. Unless the purity of a putative product is beyond doubt, the mere association of radioactivity

with the material after in vitro incubation cannot be taken as evidence that the component is a product of mitochondrial protein synthesis.

2. *Properties*

Although the identity of few of the products of mitochondrial protein synthesis in vitro has been established, it does appear that all are relatively insoluble proteins that are localized mainly in the inner mitochondrial membrane (Neupert et al., 1967; Beattie et al., 1967a; Ashwell and Work, 1970a; Neupert and Ludwig, 1971; Werner and Neupert, 1972). The affinity of the mitochondrial products for this membrane reflects their solubility properties and their intramitochondrial site of synthesis. These proteins are synthesized mainly by ribosomes bound to the inner mitochondrial membrane, probably via the nascent protein itself. As would be expected for intrinsic membrane proteins, the mitochondrial products show a marked tendency to aggregate, and they may be dispersed in aqueous media only in the presence of detergents (Ashwell and Work, 1970a). Most of these products are readily soluble in chloroform:methanol (Tzagoloff and Akai, 1972; Kadenbach and Hadvary, 1973; Hadvary and Kadenbach, 1973; Murray and Linnane, 1972; Burke and Beattie, 1973) evincing their strongly hydrophobic nature (Michel and Neupert, 1973). This property apparently follows from their high content of hydrophobic residues (Costantino and Attardi, 1973). Indeed, one of the more hydrophobic products of yeast mitochondrial protein synthesis contains an unusually high amount of hydrophobic amino acids (Tzagoloff et al., 1973). In addition, mitochondrial products may contain, like proteolipids (Folch and Lees, 1951), significant amounts of covalently attached lipoid residues. Some of these may be esterified fatty acids, incorporated directly in the course of regular posttranslational modification of precursor proteins, while others may be coupled after integration of the products into functional complexes in the mitochondrial inner membrane.

The products of mitochondrial protein synthesis have been examined by electrophoresis in polyacrylamide gels containing SDS in order to determine the number and size of these proteins. The products
are detected in gels by their content of radioactively labeled amino
acids that have been incorporated in vitro or in vivo in the presence
of inhibitors of extramitochondrial protein synthesis. Estimates for
the number of different (size) proteins in various mitochondrial systems range from as few as 1 or 2 to as many as 10 to 15 (Sebald et
al., 1968; Yang and Criddle, 1969; Coote and Work, 1971; Premkumar
and Bhargava, 1972; Burke and Beattie, 1973; Lederman and Attardi,
1973; England and Attardi, 1974; Jeffreys and Craig, 1974).

Electrophoresis in SDS-containing gels has also been employed to
compare the size distribution of the products of protein synthesis
in vitro with that of the products in vivo. It is encouraging to find
that both sorts of products give similar profiles in SDS gels, and
these observations are advanced in support of the notion that legiti-
mate mitochondrial protein synthesis continues in vitro, at least for
limited periods (Sebald et al., 1968; Coote and Work, 1971; Lederman
and Attardi, 1973; Ibrahim et al., 1973). The significance of these
observations, however, remains to be established in view of the growing evidence that the products of mitochondrial protein synthesis are
relatively low molecular-weight precursor polypeptides (Michel and
Neupert, 1973), that show a marked tendency to aggregate, even in the
presence of SDS (Kuntzel and Blossey, 1974). Also, it is not unreasonable that the posttranslational modification of the products will
occur to a different extent in vitro than in vivo. Furthermore, the
"nonspecific" aggregation of products with each other or additional
polypeptides will have the artifactual effect of increasing the size
and apparent number of products of mitochondrial protein synthesis.

It should also be appreciated that the nascent products of mitochondrial protein synthesis may exhibit a marked tendency toward
crosslinking with other proteins, especially during attempts to characterize these components. Reactions of this sort, mediated by autooxidation products of lipids that may be intimately associated with
the hydrophobic mitochondrial products, have been implicated in the

inactivation of mitochondrial enzymes and in the concomitant cross-linking of mitochondrial membrane proteins during aerobic incubations in vitro (Minssen and Munkres, 1973). Accordingly, there is a high probability that mitochondrial products will react with mono- and bifunctional lipid autooxidation products, especially when they are manipulated in the absence of antioxidants, resulting in marked changes in their physical and chemical properties. Moreover, if abnormal crosslinking of nascent proteins occurs in detectable amounts, it will also have the effect of increasing the apparent size and number of products of mitochondrial protein synthesis.

If the nascent products of mitochondrial protein synthesis are an undetermined number of relatively low molecular-weight precursor polypeptides (Michel and Neupert, 1973; Werner, 1974b), crosslinking of some of these proteins must occur normally during their processing and integration into the inner mitochondrial membrane, because some products that have been identified as constituents of known mitochondrial functional complexes are significantly larger molecules (Mahler, 1973; Schatz and Mason, 1974). While this crosslinking may be of the sort alluded to above, it is more likely to occur in a fashion analogous to that described for the serum lipoprotein apoproteins that appear to serve as precursors for membrane proteins of the erythrocyte (Langdon, 1974).

Another property of the products of mitochondrial protein synthesis and one deriving from their synthesis as precursor polypeptides, is their relative instability. Under the usual conditions of in vitro protein synthesis, it is known that nascent mitochondrial products show an increased lability relative to bulk mitochondrial proteins (Wheeldon and Lehninger, 1966; Wheeldon, 1973; Wheeldon et al., 1974). This high turnover of nascent mitochondrial proteins probably results from their abortive processing and assimilation into stable complexes in vitro. It would appear that the stability of a mitochondrial product depends on its being integrated into a membrane protein complex, a process that may occur inefficiently, if at all, in the absence of complementary subunits from the extramitochondrial pool.

The very properties of the products of mitochondrial protein synthesis have complicated what at the outset appeared to be a straight-
forward task: to determine the number and size of these products.
Some of the divergent results reported by different laboratories, no
doubt, arise from the instability, differential processing, and crosslinking of precursor polypeptides, and especially from the marked tendency of these hydrophobic products to aggregate under analytical conditions.

Optimistically, we may expect that the in vitro system will be
more fully exploited in the near future, both to identify additional
products of mitochondrial protein synthesis and to study the posttranslational processing of some of the nascent mitochondrial products. These studies are feasible now, in view of emerging methods
for the resolution and analysis of insoluble membrane proteins (Chignell and Wingfield, 1974; Capaldi, 1974) coupled with the use of antibodies to the proteins of well-defined mitochondrial components (Tza-
galoff and Meagher, 1971; Shakespeare and Mahler, 1971; Mason et al.,
1972; Werner, 1974a).

C. Discrimination from Bacterial and Microsomal Protein Synthesis

Many of the earlier studies of mitochondrial protein synthesis
in vitro were directed toward authenticating the process and defining
its requirements. Other applications of the in vitro system have included: (a) product identification; (b) studies of the antibiotic
susceptibility of mitoribosomes; and (c) its use as a criterion for
identifying intramitochondrial ribosomes. In all of these applications, it is important to be able to exclude possible contributions
of contaminating bacteria or of extramitochondrial ribosomes to the
observed activity.

Despite some allegations (von der Decken et al., 1966; Sandell
et al., 1967) that most or all of the protein synthetic activity of
mitochondrial preparations derives from the bacteria they contain,
it appears that bacteria are not usually a serious problem in studies

using mitochondria prepared under semisterile conditions by careful
workers (Roodyn, 1968; Kroon et al., 1967; Beattie et al., 1967a).

Indeed, most of the bacteria and bacterial spores normally present in such mitochondrial preparations are relatively inactive and
contribute little in the way of protein synthetic activity during
brief assay periods (Kroon et al., 1968). This is an especially fortunate situation since, in contrast to the microsomal system, no general inhibitors of protein synthesis exist that will allow the ready
discrimination of the bacterial and mitochondrial activities (see
Sect. IV). While it is possible to inhibit mitochondrial and not
bacterial protein synthesis with some agents (Kroon et al., 1968),
there are no selective inhibitors available to suppress protein synthesis in bacteria without also affecting that in mitochondria. In
critical applications, however, advantage can be taken of the resistance of bacteria to lysis by detergents. Pulse-labeled mitoribosomes
and radioactively labeled products of mitochondrial protein synthesis
can be extracted by treating mitochondria with detergents under conditions that do not extract significant amounts of ribosomes or labeled proteins from bacteria (Wheeldon, 1966; O'Brien, 1971).

The presence of variable amounts of cytoribosomes in mitochondrial preparations may pose more serious problems in the identification of mitoribosomes, and also in the study of products of mitochondrial protein synthesis in vitro. This contamination of mitochondria
by extramitochondrial ribosomes is a particular nuisance for tissues
containing significant amounts of membrane-bound cytoribosomes. It
can be reduced by use of media and procedures that serve to minimize
both the adsorption of ribosomes and microsomes to mitochondria, and
their entrapment within clumps of mitochondria (O'Brien and Kalf,
1967a). Relatively small amounts of cytoribosomes can interfere with
the detection of mitoribosomes, particularly in mitochondria from
adult mammals, because these contain so few ribosomes (O'Brien, 1971).
Still, mitoribosomes may be discriminated from present cytoribosomes
by application of a useful functional criterion. Only intraorganellar
ribosomes should become labeled when intact mitochondria are incubated

with radioactive amino acids using conditions adequate for mitochondrial, but not microsomal protein synthesis.   There are two notable
exceptions, however, where this functional criterion fails.   Active,
extramitochondrial ribosomes are obtained in preparations of mitochondria from brain (Morgan, 1970; Haldar, 1971; Mahler et al., 1971;
Gambetti et al., 1972) and some fetal tissues (O'Brien, 1972a), where
they apparently occur within structures furnishing the support factors required for cell-free protein synthesis.   In these cases, and
perhaps as a routine precautionary measure, it is advisable to use
one of the selective inhibitors of cytoribosomal function (Sect. IV)
to restrict the labeling to mitoribosomes.

Finally, it should be appreciated that it is often possible to
remove even residual cytoribosomal contaminants from mitochondrial
preparations by selective disruption of the microsomal and outer mitochondrial membranes by digitonin (Malkin, 1971; DeVries and van der
Koogh-Schurring, 1973).   Presumably because of its lower cholesterol
content, the mitochondrial inner membrane resists dissolution by
amounts of digitonin adequate to disperse the other membranes.   This
treatment also offers a bonus of lysosome-free mitochondria, because
lysosomal membranes are particularly labile to digitonin (Loewenstein
et al., 1970).   This measure carries a special benefit in cases where
exposure to lysosomal hydrolases during isolation might alter the properties of mitoribosomes.

III. MITOCHONDRIAL RIBOSOMES

Ribosomes were implicated in mitochondrial protein synthesis relatively early, by the observation that ribosomal inhibitors of protein
synthesis also inhibit the incorporation of amino acids into mitochondrial proteins in vitro (Rendi, 1959; Mager, 1960; Kalf, 1963; Kroon,
1963).   Yet, several years lapsed before mitoribosomes were isolated
and identified.   A major reason for this delay was the tendency of
most investigators to use rat liver mitochondria in their studies.
As already mentioned, these contain very few ribosomes, and further-

more, they are heavily contaminated with extramitochondrial ribosomes as they had usually been prepared (O'Brien and Kalf, 1967a). However, by using procedures adequate for the purification of mitochondria, and by application of the functional criterion discussed above, mito-ribosomes were eventually isolated and identified from rat liver mitochondria (O'Brien and Kalf, 1967b).

Quite unexpectedly, these mitoribosomes were found to have a sedimentation coefficient of only 55S. Also surprising were the findings that they contained less than half the RNA of other kinds of ribosomes (O'Brien and Kalf, 1967b) and that their buoyant density (in CsCl) was unusually low (Perlman and Penman, 1970b). Because all other ribosomes known fall into two main sedimentation classes of 70S (mainly bacterial ribosomes) or 80S (mainly cytoplasmic ribosomes), many investigators initially considered the 55S mitoribosome to be either a degraded ribosome, or a subunit of a larger ribosome.

These suspicions were, of course, strengthened by the report of more plausible, 73S ribosomes in *Neurospora* mitochondria (Kuntzel and Noll, 1967). Indeed, the difficulty with which mitoribosomes were detected in some systems led ultimately to proposals that mammalian mitochondria do not even contain ribosomes (Attardi and Attardi, 1969) and that the functional protein synthetic structure in yeast mitochondria is actually a specialized region of the inner membrane (Bunn et al., 1970; Dixon et al., 1971; Forrester et al., 1971).

The situation became even more unsettled when it was reported that mitoribosomes from *Tetrahymena*, like the cytoribosomes in this organism, have a sedimentation coefficient of 80S (Chi and Suyama, 1970). However, from continuing studies in these and other systems, it is now apparent that mitoribosomes occur in widely diverse forms (see Fig. 1 for a comparison of ribosome types). Moreover, it is also clear that what at once appeared to be a curious anomaly, the 55S ribosome, is apparently the predominant form of mitoribosome occurring in all multicellular animals thus far examined. Appropriately then, we will begin our discussion of mitoribosomes with this most unusual kind of ribosome.

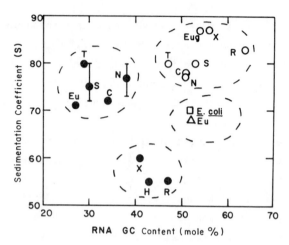

FIG. 1. Comparison of ribosome types on the basis of sedimentation coefficient and RNA base composition. Ribosomes of *E. coli* and
chloroplasts (open triangles) cluster separately from cytoribosomes
(open circles) and mitoribosomes (closed circles). Mitoribosomes of
animals and primitive eukaryotes segregate not only from bacterial
ribosomes and cytoribosomes, but also from each other. C, *Candida;*
Eu, *Euglena;* H, human; N, *Neurospora;* R, rat: S, *Saccharomyces;* T,
Tetrahymena; *Xenopus.*

A. Mitoribosomes of Animals

Following their discovery in rat liver mitochondria, 55-60S mitoribosomes have been studied in several animal systems. Among mammals, this kind of ribosome has been demonstrated in the rat (O'Brien
and Kalf, 1967b; Ashwell and Work, 1970b; O'Brien, 1971; Hernandez
et al., 1971; Aaij et al., 1972; Sacchi et al., 1971; Sacchi et al.,
1973; DeVries and van der Koogh-Schurring, 1973; Greco et al., 1973),
the human (Perlman and Penman, 1970b; Attardi and Ojala, 1971; Brega
and Vesco, 1971; Wengler et al., 1972), the rabbit, the pig, the cow
(O'Brien, 1971), and in the hamster (Coote et al., 1971). The 55S
form of mitoribosome also occurs in the chicken (Rabbitts and Work,
1971), the frog (Swanson and Dawid, 1970; Leister and Dawid, 1974),
the shark (O'Brien, 1972b), and even in an insect, the locust (Kleinow et al., 1971).

1. Functional Characterization

The 55S particle has been established as the monoribosome form on the basis of functional as well as physical-chemical criteria. It is virtually the only particle that rapidly attains high specific radioactivities when mitochondria are incubated with radioactive amino acids in vitro (O'Brien and Kalf, 1967b; Sacchi et al., 1973), or in vivo (O'Brien, 1967a; Ashwell and Work, 1970b; Kleinow et al., 1971). This labeling of the 55S particles is prevented or markedly diminished by many ribosomal inhibitors of protien synthesis. Furthermore, the radioactive polypeptide products are only transiently associated with the mitoribosome; treatment of active mitoribosomes with puromycin will strip them of their nascent, radioactively labeled proteins (Ashwell and Work, 1970b; Kleinow et al al., 1971; O'Brien, 1971; Sacchi et al., 1973). While isolated 55S ribosomes are competent to translate poly(U), compelling evidence that 55S particles represent mitochondrial monoribosomes came from studies of this kind of ribosome in the frog. Both subribosomal particles (below), neither of which is active alone, are required to translate poly(U) (Swanson and Dawid, 1970; Leister and Dawid, 1974).

Under some conditions the mitoribosomes are obtained in aggregates that have the properties of polyribosomes containing 2-7 monoribosomes (Perlman and Penman, 1970b; Brega and Vesco, 1971; Ojala and Attardi, 1972; Kleinow et al., 1971; Avadhani et al., 1974; Michel and Neupert, 1973).

Probably because of the hydrophobic properties of the protein products, and their strong tendency to associate with each other and membranous material, mitochondrial polyribosomes often show some resistance to RNase (Ojala and Attardi, 1972; Michel and Neupert, 1973). Complete disaggregation of such mitochondrial polyribosomes to 55S monoribosomes is achieved by RNase treatment, however, if the polyribosomes are also treated briefly with pronase to remove the nascent polypeptides (Ojala and Attardi, 1972). This problem can be overcome by concentrating the polyribosome fraction before it is dispersed with detergents (Avadhani et al., 1974).

2. Physical and Chemical Properties

a. *Sedimentation coefficients of the monoribosome and its sub-ribosomal particles.* The sedimentation coefficient of mitoribosomes is usually estimated from their sedimentation rate in sucrose gradients relative to a standard ribosome. Reported values for the sedimentation coefficient of vertebrate mitoribosomes range from 54 to 61S. While there seem to be slight differences in the sedimentation rate of mitoribosomes, depending on their functional state, the average sedimentation coefficient of mammalian mitoribosomes is closer to 55S than 60S (O'Brien, 1971; Hamilton and O'Brien, 1974). We therefore choose to refer to them categorically as 55S ribosomes, to discriminate them from ribosomes of the 70S or 80S classes.

Like bacterial ribosomes, the mitoribosomes require relatively high concentrations of divalent cation (magnesium) for stabilization of the monoribosome form. The 55S monoribosomes dissociate into dissimilar subribosomal particles when the magnesium concentration is lowered or the ionic strength of the medium is increased (O'Brien, 1971; Sacchi et al., 1971; Kleinow and Neupert, 1971; Greco et al., 1973; Leister and Dawid, 1974; Kleinow, 1974). In sucrose gradients, the large and small subribosomal particles have apparent sedimentation coefficients of about 40S and 30S, respectively. The derived subribosomal particles are obtained in roughly stoichiometric amounts and they will reassociate, but only in the presence of factors released during their dissociation in high salt buffers (O'Brien et al., 1974).

b. *Buoyant density.* In addition to their unusually low sedimentation coefficient, these mitoribosomes are also distinguished by their very low buoyant density. The buoyant density in cesium chloride of formalin-fixed mitoribosomes and subribosomal particles ranges from 1.40 to 1.45 g/cm^3 (Wengler et al., 1972; Sacchi et al., 1973; DeVries and Van der Koogh-Schurring, 1973; O'Brien et al., 1974; Leister and Dawid, 1974; Hamilton and O'Brien, 1974). While there are questions regarding the relation of RNA and protein content to particle buoyant density (McConkey, 1974), there is ample reason to believe that the low buoyant density of these mitoribosomes reflects an unusually high protein content, in the range of 65-70% (Leister and Dawid, 1974; Hamilton and O'Brien, 1974). Alternative possibilities to account for the low buoyant density, such as adherent membrane fragments, and atypical, low density constituents, are less attractive, especially since the subribosomal particles appear to contain an appropriately large number of different proteins (O'Brien et al., 1974; Leister and Dawid, 1974).

c. *Size.* The first indications that mitoribosomes were significantly smaller than cytoribosomes came from an electron microscopic study of rat liver mitochondria by Andre and Marinozzi (1965). These size differences were confirmed by electron microscopic examination of the isolated ribosomes (O'Brien and Kalf, 1967b). After negative staining of the formalin-fixed particles with phosphotungstate, the average diameter of 55S mitoribosomes was found to be only 145 Å, while that for 80S cytoribosomes was 190 Å. Somewhat larger dimensions for rat mitoribosomes (169 x 199 Å) and cytoribosomes (194 x 247 Å) were reported by Aaij et al. (1972), who examined positively stained, unfixed particles. Although the mitoribosomes had the same general appearance as cytoribosomes in this study, they were somewhat smaller and less asymmetric. Similar conclusions regarding the relative size and appearance of insect mitoribosomes and cytoribosomes were reached in an electron microscopic study of isolated ribosomes from the locust (Kleinow et al., 1974a,b), although these mitoribosomes were reported to have significantly larger dimensions (210 x 215 x 270 Å).

As the scientific community grew less uncomfortable with the notion of 55S monoribosomes, the term "miniribosome" was coined to denote this "smallest of ribosomes" (Borst and Grivell, 1971). Because these ribosomes were known to contain small RNA molecules (below), and also because of their small size in the electron microscope, and especially because of their low sedimentation coefficient, it was generally assumed that 55S mitoribosomes would be appreciably smaller than 70S bacterial ribosomes, as well. However, the surprises were

not over for this ribosome. The 55S ribosomes are actually somewhat
larger than bacterial (*E. coli*) ribosomes, both on the basis of particle molecular weight (Sacchi et al., 1973; Leister and Dawid, 1974;
Hamilton and O'Brien, 1974), and physical dimensions (DeVries and Van
der Koogh-Schurring, 1973; DeVries and Kroon, 1974).

    d. RNA content. Probably the main reason for thinking of the
55S ribosome as a miniribosome, in addition to its low sedimentation
coefficient, was its content of RNA species much smaller than those
found in other kinds of ribosomes. The mitochondrial large and small
subribosomal particles contain RNA molecules that sediment in sucrose
gradients as a 16-17S species and a 12-13S species, respectively
(Borst and Grivell, 1971). Some confusion accompanied early efforts
to correlate the results of studies using sedimentation analysis or
electrophoresis in polyacrylamide gels to characterize the RNA molecules. In the first place, the sedimentation coefficient (S) of mitoribosomal RNA determined relative to a standard rRNA in sucrose gradients may differ considerably from the sedimentation coefficient
(S_E) of the same RNA determined relative to the electrophoretic mobility of the standard RNA in polyacrylamide gels. Secondly, the sedimentation coefficients determined by either of these methods vary depending on the experimental conditions (mainly temperature, ionic
strength, and magnesium concentration). This phenomenon is generally
ascribed to a less stable secondary structure of mitoribosomal RNA,
apparently a consequence of the atypical base composition of these
ribosomal RNA's (Groot et al., 1970; Edelman et al., 1970, 1971; Forrester et al., 1970; Mitra et al., 1972; Dawid and Chase, 1972). Unlike most ribosomal RNA's, the G + C content of mitoribosomal RNA's
is relatively low, ranging from only 30 to 45% (Freeman et al., 1973).
Because of their lower melting temperature, these molecules generally
exhibit less secondary structure under usual analytical conditions,
than do the rRNA standards. Besides introducing additional variability into the apparent sedimentation coefficients estimated for these
molecules under different conditions, failure to appreciate the effect often resulted in overestimates for the molecular weight of mi-

toribosomal RNA's. Clearly, to compare different mitoribosomal RNAs, and to obtain realistic estimates of their molecular weight by comparison to RNA standards, it is essential that the molecules have the same overall conformation. Consequently, to achieve these conditions, investigators have resorted to carrying out their molecular weight measurements of mitoribosomal RNAs under denaturing conditions, either by sedimentation analysis (Dawid and Chase, 1972), by electrophoresis (Reijnders et al., 1973), or by electron microscopic length measurements (Wellauer and Dawid, 1974; Robberson et al., 1972). Under these conditions, the molecular weight determinations made by different methods agree closely, and the values so obtained allow comparisons of different mitoribosomal RNA's. Values for the molecular weight of the RNA in mitochondrial small subribosomal particles from man and the frog range from 0.3 to 0.36 x 10^6, and the molecular weight of RNA in the large subribosomal particles is in the range of 0.53 to 0.58 x 10^6 (Robberson et al., 1971; Dawid and Chase, 1972; Leister and Dawid, 1974). Thus, analogous subribosomal particles of diverse animal mitoribosomes appear to contain RNA molecules of roughly equivalent mass, remarkably smaller than the corresponding cytoribosomal RNA's, and only about half the size of bacterial ribosomal RNA's (Borst, 1972).

We have seen that the RNA molecules of animal mitoribosomes differ from bacterial and cytoribosomal RNA's on the basis of size as well as base composition. Animal mitoribosomal RNA's differ also in their very low content of methylated bases, relative to the other ribosomal RNA's (Vesco and Penman, 1969; Dubin, 1974).

Another intriguing aspect of mitoribosomes is the question of whether they contain 5S RNA. Bacterial ribosomes, as well as cytoplasmic ribosomes of eukaryotes contain a single 5S RNA molecule in the large subribosomal particle. Indispensable for ribosome function (Erdmann et al., 1971), the 5S RNA appears to play an important role in the binding of aminoacyl-tRNA to the ribosome (Richter et al., 1973). By analogy, it might be anticipated that mitoribosomes should also possess a 5S RNA, or its functional equivalent. However, mitoribosomal 5S RNA escaped detection in *Neurospora* mitoribosomes in a

study by Lizardi and Luck (1971), and until recently, it was largely
held that mitoribosomes in general do not contain 5S RNA. Yet, some
studies in animal systems suggest that these mitoribosomes, at least,
contain a separate RNA molecule analogous to the 5S RNA of other ribosomes. An unusual, low molecular weight RNA ($3S_E$), distinct from
mitochondrial transfer RNA, has been isolated from hamster mitochondria (Dubin et al., 1974). It is unmethylated, has a base composition resembling that of mitoribosomal RNA, and appears to be transcribed from mitochondrial DNA. Although it has not yet been demonstrated to be present in mitoribosomes, per se, it does appear to occur in the expected 1:1 stoichiometry, relative to mitoribosomal RNA
(Dubin et al., 1974). A similar low molecular weight RNA molecule
has been detected in HeLa mitochondria (Gray and Attardi, 1973). It,
too, appears to be a mitochondrial transcription product, and on the
basis of its electrophoretic mobility in polyacrylamide gels, it appears distinct from, and slightly smaller than mitochondrial transfer
RNA. Significantly, it has been extracted from HeLa mitoribosomes
and large subribosomal particles, where it occurs in unit stoichiometry (Gray and Attardi, 1973). In view of these reports, then, it
appears likely that mitoribosomes do contain a separate RNA molecule
serving as a functional analog of the 5S RNA found in the large subribosomal particle of all other ribosomes.

     e. *Protein content.* One of the more unusual distinguishing pro-
perties of animal mitoribosomes is their high protein content. Early
indications of this feature came from chemical analyses (O'Brien and
Kalf, 1967b) and buoyant density measurements on the isolated mitoribosomes (Perlman and Penman, 1970b), but only relatively recently was
it generally appreciated that animal mitoribosomes are truly protein
rich (Table 1).

     Compared to bacterial ribosomes, animal mitoribosomes appear to
contain a surprisingly large number of different proteins. Leister
and Dawid (1974) analyzed the proteins of *Xenopus* mitoribosomes by
two-dimensional electrophoresis in polyacrylamide gels, and they were
able to resolve as many as 84 different proteins in these ribosomes.

TABLE 1 Major Subclasses of Mitoribosomes[a]

Kingdom	Varieties	Occurrence	Sedimentation coefficient	Buoyant density[b]	Subunits	RNA (%)	GC (%)	Total RNA (mol wt)[b]
Animals	1[c]	Vertebrates and insects	55–60S	1.42–1.45	40S, 30S	30–35	40–45	0.90
Protists	2	Tetrahymena	80S	1.46	55S, 55S		28	1.38
		Euglena	71S		50S, 32S	57	27	1.52
Fungi	3	Saccharomyces	72–80S	1.64	50S, 38S		30	2.0
		Candida	72S	1.48	50S, 36S		34	1.9
		Neurospora	73–80S	1.52	50S, 38S		38	2.0
Plants	1	Higher plants	78S	1.56	60S, 40S			2.0

[a]Data taken from references in text.
[b]The units for buoyant density in CsCl are grams per cubic centimeter; for molecular weight, x 10^6.
[c]The presence of 70S mitoribosomes in Drosophila (Chooi and Laird, 1976) suggests there may be more than one subclass of animal mitoribosomes.

Forty of these appeared to be unique to the large subribosomal particle, while the remaining 44 were contained in the small subribosomal particle. In a similar electrophoretic system, as many as 90
different proteins are resolved from bovine mitoribosomes; 52 of them
appear to be localized to the large subribosomal particle, while the
other 38 are in the small one (O'Brien et al., 1974; Matthews and
O'Brien, unpublished). As was also inferred for proteins of the locust mitoribosome (Kleinow and Neupert, 1971), virtually all of the
mitoribosomal proteins in the Xenopus and bovine systems can be discriminated from the corresponding cytoribosomal proteins on the basis
of their different electrophoretic mobility.

In the two cases just cited where animal mitoribosomal proteins
have been examined by two-dimensional electrophoresis, a surprisingly large number of proteins have been resolved. The number of discrete subunit-specific proteins detected in these animal mitoribosomes
(84-90) exceeds even those figures reported for the larger, 80S cytoribosomes (Sherton and Wool, 1972). In this regard, it is reasur-
ing that estimates of total protein content for these subribosomal
particles based on the number and size of the proteins disclosed by
two-dimensional electrophoresis are in general agreement with estimates made on the basis of buoyant density or chemical measurements
(O'Brien et al., 1974; Leister and Dawid, 1974).

Obviously, one of the first questions raised by these findings
is whether all of the apparent mitoribosomal proteins visualized are
indeed structural or functional constituents of the mitoribosome.
Although the mitoribosomal subunits are prepared under conditions to
minimize the adsorption of nonribosomal proteins, they probably do
contain a few. For the present, however, and lacking any more definitive information, it should be emphasized that these proteins
are characteristically and reproducibly associated with one or the
other subribosomal particle. In attempts to discern which of these
proteins are characteristically and tightly associated with the subribosomal particles, we have washed them in various high-salt buffers.
Relatively few proteins are removed by this procedure, even using
conditions that "strip" bona fide ribosomal proteins from bacterial

and cytoplasmic ribosomes, and also result in the loss of some mitoribosomal functions (Matthews and O'Brien, 1974). It thus appears
that these ribosomes, which are somewhat larger than bacterial ribosomes but contain only half the RNA of bacterial ribosomes, make up
the difference in mass by containing more proteins.

B. Mitoribosomes of Fungi

The two Ascomycete systems in which mitoribosomes have been studied most extensively are Neurospora and yeast. Like animal mitoribosomes, they resemble bacterial ribosomes more than cytoribosomes.
However, fungal mitoribosomes do not appear to form a homogeneous
class.

1. Functional Characterization

Incubation of intact mitochondria from Neurospora (Neupert et al.,
1969b) and yeast (Schmitt, 1970; Yu et al., 1972b; Vignais et al.,
1972) with radioactive amino acids for brief periods has been used
to identify active intramitochondrial ribosomes in these systems.
This activity is suppressed by the common antibiotic inhibitors of
bacterial, but not cytoplasmic protein synthesis (Vignais et al.,
1972; Datema et al., 1974; Yu et al., 1972b), suggesting that these
ribosomes are of the prokaryotic type with respect to antibiotic susceptibility. The above approach has also been used to identify active polysomes isolated from Neurospora mitochondria (Agsteribbe et
al., 1974). Arrays of mitoribosomes characteristic of polysomes have
also been observed electron microscopically in yeast mitochondria,
in situ (Vignais et al., 1969) and in isolated preparations of yeast
mitoribosomes (Cooper and Avers, 1974).

2. Physical and Chemical Properties

a. Sedimentation coefficient of the monoribosome and its subribosomal particles. From the many studies of mitoribosomes in diverse
fungi, it appears that these mitoribosomes generally sediment at
rates intermediate between 70S E. coli ribosomes and 80S cytoribo-

somes (Table 1). However, because of conflicting reports it is not
always possible to assign fungal mitoribosomes unequivocally to either the 70S or 80S class. Indeed, there are reasons to believe that
different fungal mitoribosomes may belong to different classes, on
the basis of significant differences in their physical and chemical
properties.

Until recently, it appeared that *Neurospora* mitoribosomes belonged
to the 70S class, having monoribosomes of 73S (Kuntzel and Noll, 1967).
However, under certain conditions, *Neurospora* mitoribosomes are obtained as 80S particles (Datema et al., 1974), substantiating an earlier claim to this effect (Rifkin et al., 1967). Because these 80S
mitoribosomes can be obtained in significant amounts only under favorable conditions, in the presence of RNAse inhibitors (Agsteribbe
et al., 1974), and because they are converted into 73S ribosomes by
aging or by incubation with the supernatant fraction of mitochondrial
lysates, Datema et al. (1974) concluded that they represent the monoribosome form of *Neurospora* mitoribosomes. Under conditions favoring the dissociation of bacterial ribosomes, but not cytoribosomes,
both the 80S and the 73S forms of the *Neurospora* mitoribosome give
rise to the same subribosomal particles of 50-52S and 37-39S (Kuntzel, 1969a; Datema et al., 1974). *Neurospora* mitoribosomes resemble
bacterial ribosomes, not only with respect to the magnesium levels
required to stabilize the monoribosome form, but also in their susceptibility to dissociation by the bacterial protein factor that causes dissociation of bacterial ribosomes (Agsteribbe and Kroon, 1973;
Datema et al., 1974).

An element of uncertainty also exists concerning the assignment
of yeast mitoribosomes to a single sedimentation class. Not only
are different yeasts reported to have different kinds of ribosomes,
based on their sedimentation coefficients, but divergent values are
often reported for mitoribosomes of given strains (Table 1). The
sedimentation coefficients reported for mitoribosomes of the yeast
Saccharomyces cerevisiae, for example, range from 80S (Schmitt, 1969;
Morimoto and Halvorson, 1971; Yu et al., 1972b) to 72S (Schmitt,
1970). Although it has been proposed that the 80S ribosomes are cy-

toribosomal contaminants that can be eliminated by washing mitochondria with EDTA-containing buffers (Schmitt, 1970), the 80S form has
been identified as a mitoribosome on the basis of functional as well
as physical-chemical criteria in other laboratories (Yu et al., 1972b;
Morimoto and Halvorson, 19721; Morimoto et al., 1971). Yet, it appears that the mitoribosomes may be distinguished from cytoribosomes
under certain conditions by their somewhat slower sedimentation rate
(75S) (Stegeman et al., 1970). In view of these observations, a
sedimentation coefficient of 75S seems to be a reasonable compromise,
and it is a value that, furthermore, corresponds to that (74S) reported for mitoribosomes of *Saccharomyces carlsbergensis* (Grivell et al.,
1971).

A similar dilemma exists for the mitoribosomes in another yeast,
Candida. The reported sedimentation coefficients for these mitoribosomes, too, range from 80S (Vignais et al., 1969) for *C. utilis,* to
70S for *C. parapsilosis* (Yu et al., 1972a). In the case of *Candida*
mitoribosomes, however, it appears more likely that the mitochondrial
monoribosome is of the 70S type since, in an extensive follow-up study of mitoribosomes from *C. utilis* (Vignais et al., 1972) they were
found to have a relative sedimentation coefficient of 72S. The 80S
particles previously identified as mitoribosomes are considered to
be dimers of the 50S large subribosomal particle of *C. utilis* mitoribosomes.

Except for those from *C. parapsilosis* (Yu et al., 1972a), yeast
mitoribosomes appear to dissociate more readily than cytoribosomes,
giving rise to two dissimilar particles of 50S and 36-38S (Vignais
et al., 1972; Schmitt, 1970).

Admittedly, it is unwise to pay undue attention to small differences in relative sedimentation coefficients, since these depend on
the functional state and prior treatment of ribosomes as well as upon many other parameters. Accordingly, it is not unreasonable that
the different sedimentation behavior reported for individual yeast
mitoribosomes varies as a function of experimental conditions. What
is important, however, is the ability to discriminate ribosome types
on the basis of several characteristic properties. Yet if it is ac-

cepted that the monoribosome form of Neurospora mitoribosomes may be
an 80S particle, there is some basis for considering that multiple
kinds of mitoribosomes exist in fungal systems. The various kinds of
mitoribosomes, only marginally discriminated on the basis of relative
sedimentation coefficients, are easily distinguished by other physical-chemical criteria (see the following).

b. Buoyant density. In view of the range of sedimentation coefficients reported for fungal mitoribosomes, it is also interesting
to note the wide range of buoyant densities reported for these ribosomes. The 74S mitoribosome from S. carlsbergensis has a buoyant
density in CsCl of 1.64 g/cm^3 (Grivell et al., 1971), a value characteristic of bacterial ribosomes, while that of C. utilis has a
buoyant density of only 1.48 g/cm^3 (Vignais et al., 1972) a value
more in line with animal mitoribosomes. Furthermore, it also comes
as some surprise that the Neurospora mitoribosome has a buoyant density of 1.52 g/cm^3 (Agsteribbe and Kroon, personal communication), a
value characteristic of most cytoribosomes. Whether these buoyant
densities are accurate predictors of the chemical constitution of
mitoribosomes is open to question (McConkey, 1974). However, these
divergent values must underlie basic structural, if not composition-
al, differences among the various fungal mitoribosomes.

c. Size. As might be anticipated from the variable sedimentation coefficients reported for different types of fungal mitoribosomes, they appear to come in different sizes as well. Because their
sedimentation coefficients approximate values characteristic of cytoribosomes, it should come as no great surprise that some fungal
mitoribosomes are very large, about the size of many cytoribosomes.

Only the mitoribosomes from Candida have been examined critically
by electron microscopic methods (Vignais et al., 1972). The dimensions of the 72S monoribosome visualized by negative staining methods
(265 x 210 x 200 Å) are surprisingly close to those reported under
similar conditions for the 60S monoribosomes of locust mitochondria
(270 x 210 x 215 Å) (Kleinow et al., 1974). On the basis of its
buoyant density and RNA mass, the molecular weight of the Candida

mitoribosome is estimated to be about 4.2 million (Vignais et al., 1972), a figure more in line with particle weights of cytoribosomes, and certainly in excess of those values determined for animal mitoribosomes. The molecular weight of *Candida* mitoribosomes compares with that of *Neurospora* mitoribosomes, 4.4 million, estimated on a similar basis. Both of these mitoribosomes appear significantly larger than those of *Saccharomyces* which, by similar reasoning, have a molecular weight of about 3 million.

On the basis of the available evidence, it appears that mitoribosomes of fungi are also heterogeneous with respect to size and, while they do show some homologies with bacterial ribosomes, they are considerably larger.

d. RNA content. Although the mitoribosomes of different fungi exhibit diverse physical properties, they appear more uniform with respect to their RNA content. The mitoribosomal RNA's, in those fungi for which we have information, appear to be uniformly large. Indeed, these molecules correspond in size to the RNA's normally found in many cytoribosomes, and this feature alone distinguishes them from mitoribosomal RNA's of animals and protists. Molecular-weight estimates for the small mitoribosomal RNA range from 0.65 to 0.72 x 10^6 and that of the large RNA from 1.25 to 1.3 x 10^6 in *Neurospora* (Neupert et al., 1969b), *Candida* (Vignais et al., 1972), *Saccharomyces* (Reijnders et al., 1973), and *Aspergillus* (Verma et al., 1970). Like animal mitoribosomal RNA's, these molecules manifest anomalous sedimentation and electrophoretic migration rates, relative to bacterial (*E. coli*) and cytoribosomal RNA standards (Edelman et al., 1970, 1971). Consequently, meaningful molecular weight measurements are obtained for these molecules only under denaturing conditions, or conditions under which the RNA samples have qualitatively equivalent conformations. The lower degree and relative lability of the secondary structure of these RNA's reflect their unusual base composition. Their G + C content is exceedingly low, for ribosomal RNA's, ranging from only 30 to 37% (Edelman et al., 1970; Morimoto and Halvorson, 1971; Vignais et al., 1972; Kuntzel and Noll, 1967; Rifkin et al., 1967).

On the basis of reported evidence, it would appear that fungal mitoribosomes contain no 5S RNA (Lizardi and Luck, 1971). However, in view of the possible existence of a 5S RNA analog in animal mitoribosomes, and of its presence in mitoribosomes of protists and plants (see the following), the possibility of 5S RNA in fungal mitoribosomes bears reexamination.

Overall, then, mitoribosomes of fungi, as a group, contain characteristically large RNA molecules of very low G + C content.

    e. *Protein content*. The proteins of two kinds of fungal mitoribosomes have been studied. In *Saccharomyces* and *Neurospora*, the mitoribosomal proteins appear distinct from those of the corresponding cytoribosomes on the basis of several criteria.

Morimoto and Halvorson (1971) used a split gel technique to contrast the mitoribosomal and cytoribosomal proteins in *Saccharomyces* by electrophoresis. Several of the approximately 18 bands resolved in each set of proteins appeared unique to one or the other ribosome. The same general conclusion was reached by Schmitt (1971), who extended the comparison to include bacterial ribosomal proteins. The *Saccharomyces* mitoribosomal and cytoribosomal proteins have also been discriminated on the basis of cochromatography of differentially labeled samples on carboxymethyl cellulose columns (Schmitt, 1972).

From the failure of mitoribosomal protein to become labeled when yeast were grown in the presence of cycloheximide, Schmitt (1972) inferred an extramitochondrial origin for the mitoribosomal proteins. However, the particular experimental design allowed the possibility that some mitoribosomal proteins are made within mitochondria. That some mitoribosomal proteins are actually coded for by mitochondrial DNA was suggested by the isolation of several yeast mutants that are resistant to antibiotic inhibitors acting on mitoribosomes (Grivell et al., 1973). Because separate mutations resulting in chloramphenicol, erythromycin, or spiramycin resistance map in different regions of the mitochondrial DNA, hopes were raised that the resistant mitoribosomes in these mutants would contain altered ribosomal proteins. However, no such differences were detected upon analysis of the mitoribosomal proteins by electrophoresis in polyacrylamide gels (Grivell et al., 1973; Grivell, 1974).

A subsequent search for mitoribosomal proteins that are products of mitochondrial protein synthesis in the yeast *S. cerevisiae* yielded a protein of molecular weight 35,000, which appears to reside on the small subunit of yeast mitoribosomes (Groot, 1974). Whether this protein is an actual ribosomal protein, or simply a mitochondrial product that shows a particular affinity for the small, but not large, subribosomal particle is unknown. The protein does not appear to be required for normal functioning of the mitoribosome (Groot, 1974).

Neurospora mitoribosomes and cytoribosomes are also known to contain qualitatively different proteins (Gualerzi, 1969). Kuntzel (1969b) pointed out their very different chromatographic behavior on carboxymethyl cellulose columns by cochromatography of the differentially labeled proteins. It has also been suggested that the majority of *Neurospora* mitoribosomal proteins are made on cytoribosomes, because mitoribosomes are not labeled when *Neurospora* is grown with radioactive amino acids in the presence of cycloheximide (Neupert et al., 1969a; Kuntzel, 1969b). This notion was corroborated in a subsequent, more critical study by Lizardi and Luck (1972). Using a judicious combination of isoelectric focusing and electrophoresis to achieve a two-dimensional separation, they were able to resolve 53 different mitoribosomal proteins. Because the in vivo labeling of all 53 proteins was sensitive to cycloheximide or anisomycin and not to chloramphenicol, it was concluded that all were products of extramitochondrial protein synthesis.

We are left then with the very intriguing situation that the proteins of these mitoribosomes are not made within mitochondria, as are the RNA components of the mitoribosome (Borst, 1972). The protein constituents of mitoribosomes and cytoribosomes certainly appear different on the basis of their electrophoretic and chromatographic properties, and they can also be discriminated immunochemically (Hallermayer and Neupert, 1974). It is indeed intriguing to find that cells make two mutually exclusive sets of ribosomal proteins, one of which appears destined for use only within mitochondria.

C. Mitoribosomes of Protists

The two protists from which mitoribosomes have been most extensively characterized are *Tetrahymena* and *Euglena*. From studies in these and other organisms, it is apparent that protist mitoribosomes are also heterogeneous, comprising at least two separate categories of ribosomes.

1. Sedimentation coefficient of the Monoribosome and its Subribosomal Particles

In *Tetrahymena*, both the mitoribosome and cytoribosome have a sedimentation coefficient of 80S. The 80S mitoribosome contains two 55S subribosomal particles (Chi and Suyama, 1970; Curgy et al., 1974). It also appears that the mitoribosomes of *Paramecium* are 80S, with 55S subunits (Tait, 1972).

In contrast to the above examples, the mitoribosome of *Euglena* has a sedimentation coefficient of 71S, relative to 70S *E. coli* ribosomes and 87S *Euglena* cytoribosomes (Avadhani and Buetow, 1972a). Upon dissociation, the 71S monoribosomes give rise to subunits of 50S and 32S (Avadhani and Buetow, 1972a).

There is some evidence suggesting that still another kind of protist mitoribosome may exist. The mitochondrial monoribosome of the trypanosome *Crithidia* has a sedimentation coefficient of 60S, relative to 83S *Crithidia* cytoribosomes and 50S and 30S subribosomal particles from *E. coli* (Laub-Kupersztejn and Thirion, 1974). In the presence of low magnesium concentrations, the 60S mitoribosomes dissociate into 45S and 32S subribosomal particles.

The ionic conditions required to maintain protist mitoribosomes in the monoribosome form are those usually required by bacterial ribosomes. However, *Tetrahymena* monoribosomes dissociate less readily upon lowering the magnesium concentration, and complete dissociation requires treatment of the particles with EDTA (Chi and Suyama, 1970; Curgy et al., 1974; Stevens et al., 1974).

2. Buoyant Density

The *Tetrahymena* mitoribosome, indistinguishable from 80S cytoribosomes on the basis of sedimentation rate, can be distinguished readily on the basis of its buoyant density in CsCl. The mitoribosomes have a buoyant density of only 1.46 g/cm^3, as compared to 1.56 g/cm^3 for *Tetrahymena* cytoribosomes (Chi and Suyama, 1970). Fortuitously, the two 55S subribosomal particles of *Tetrahymena* mitoribosomes can be separated on the basis of buoyant density; one has a density of 1.46 g/cm^3, and the other 1.52 g/cm^3 (Chi and Suyama, 1970).

Unfortunately, buoyant densities have not been reported for other protist ribosomes. On the basis of RNA and protein content, however, we might infer a rather high buoyant density for *Euglena* mitoribosomes (Avadhani and Buetow, 1974), and an intermediate buoyant density for *Crithidia* mitoribosomes (Laub-Kupersztejn and Thirion, 1974). Also by this criterion, then, protist mitoribosomes appear heterogeneous.

3. Size

Tetrahymena mitoribosomes are physically very large. Their dimensions (370 x 240 Å) are significantly greater than those even of *Tetrahymena* cytoribosomes (275 x 230 Å), as visualized by negative staining in the electron microscope (Stevens et al., 1974; Curgy et al., 1974). From its buoyant density and RNA mass, a particle mass of about 3.2 million daltons has been calculated for the mitoribosome (Chi and Suyama, 1970). Probably as a result of both a larger volume and lower RNA content, the mitoribosomes show a much lower electrophoretic mobility in polyacrylamide gels than do *Tetrahymena* cytoribosomes (Curgy et al., 1974).

A particle mass of only 2.7 million daltons can be estimated for *Euglena* mitoribosomes on the basis of their RNA mass and protein content (Avadhani and Buetow, 1974). Not only are these significantly smaller than *Tetrahymena* mitoribosomes, but they actually appear to resemble bacterial ribosomes with respect to molecular weight as well as sedimentation coefficient. Therefore, protist mitoribosomes also appear heterogeneous on the basis of size.

4. RNA Content

Tetrahymena and Euglena mitoribosomes contain RNA molecules of
the same general size and base composition, despite other differences
noted above for these ribosomes. The 21S and 14S RNA's of Tetrahymena mitoribosomes have molecular weights of 0.91 and 0.47 x 10^6,
respectively (Chi and Suyama, 1970; Schutgens et al., 1973). Euglena
mitoribosomes contain similar RNA species of 21S and 16S (Avadhani
and Buetow, 1972a).

Despite an earlier indication that Euglena mitoribosomes may contain a 5S RNA (Avadhani and Buetow, 1972a), it appears to be lacking
(Avadhani and Buetow, 1974). In this respect, Euglena mitoribosomes
may differ from those of Tetrahymena, which are reported to contain
5S RNA (Chi and Suyama, 1970).

Protist mitoribosomes appear to be singularly distinguished in
having RNA's whose G + C content is the lowest known for any ribosome. The G + C content of Euglena and Tetrahymena ribosomal RNA's
ranges from 27-30% (Chi and Suyama, 1970; Avadhani and Buetow, 1972a).

Thus, in spite of their apparent heterogeneity (noted previously), protist mitoribosomes as a group are clearly delineated from
other kinds of ribosomes, including animal and fungal mitoribosomes,
by the criteria of RNA base composition and total RNA mass.

5. Protein Content

Only limited studies have been made of the proteins of protist
mitoribosomes. Chi and Suyama (1970) compared the proteins extracted
from Tetrahymena mitoribosomes and cytoribosomes by electrophoresis
in polyacrylamide gels. About 27 components were resolved in each
set of proteins and it was apparent that the proteins in each kind
of ribosome were different. It was suggested (Millis and Suyama,
1972) that some mitoribosomal proteins are made within mitochondria.
The proteins of Paramecium mitoribosomes have been analyzed with a
view toward detecting possible differences in erythromycin-resistant
cytoplasmic mutants. Although a mitoribosomal protein was implicated
in the mutation to drug resistance (Beale et al., 1972), the question
remains open (Tait, 1972).

D. Mitoribosomes of Plants

In contrast to the above systems, little work has been done on plant mitoribosomes. These mitoribosomes have not yet been characterized functionally, and advantage has not been taken of many of the criteria normally used to identify intramitochondrial ribosomes.

1. Sedimentation Coefficient of the Monoribosome and Its Subribosomal Particles

Although there are some reports to the contrary (Wilson et al., 1968; Vasconcelos and Bogorad, 1971), most of the evidence indicates that the mitoribosomes of higher plants have a sedimentation coefficient of 77-78S (Leaver and Harmey, 1972, 1973; Pring, 1974). It is unfortunate that functional criteria have not been applied to these ribosomes since they are indistinguishable from cytoribosomes on the basis of sedimentation rate. They can be discriminated from cytoribosomes, however, by their RNA content (see the following) and their dissociability. Like most mitochondrial ribosomes, relatively high levels of magnesium are required to stabilize the monoribosome form. Cytoplasmic monoribosomes are relatively stable under low magnesium conditions that allow the mitoribosome to dissociate into 60S and 44S subribosomal particles (Pring, 1974).

2. RNA Content

The mitochondrial ribosomes of higher plants appear to contain relatively large RNA molecules that are usually distinguishable from cytoribosomal RNA on the basis of their differential electrophoretic mobility. The mitoribosomal RNA's of six different Angiosperms (mung bean, pea, cauliflower, maize, turnip, and potato) appear to range in mass from 1.12 to 1.26 million daltons for the large component, and 0.69 to 0.75 million daltons for the smaller component (Leaver and Harmey, 1973; Pring, 1974). Under the same conditions, the two cytoribosomal RNA's have apparent molecular weights of 1.3 to 1.5 x 10^6 and 0.68 to 0.70 x 10^6.

In contrast, another higher plant, *Parthenocissus* (Virginia creeper), is reported to contain mitoribosomal RNA molecules of molecular weight of only 0.84 and 0.42 x 10^6 (Quetier and Vedel, 1974).

Finally, it should be remarked that plant mitoribosomal RNA also appears to have low G + C content relative to cytoribosomal RNA (Baxter and Bishop, 1968), and plant mitoribosomes also seem to contain 5S RNA (Leaver and Harmey, 1973).

3. Protein Content and Buoyant Density

In the only instance where proteins of plant mitoribosomes have been analyzed, they appear qualitatively different from the proteins of cytoribosomes, chloroplast ribosomes, and *E. coli* ribosomes (Vasconcelos and Bogorad, 1971). As judged from electrophoretic patterns, plant mitoribosomes and cytoribosomes probably contain a similar number of proteins.

Maize mitoribosomes have essentially the same buoyant density, 1.56 g/cm^3, as do maize cytoribosomes (Pring et al., 1974). In this case where mitoribosomes resemble cytoribosomes so closely with respect to sedimentation coefficient and buoyant density, the use of additional criteria to identify and characterize the mitoribosome is certainly warranted. The differential antibiotic susceptibility of pulse-labeled ribosomes and base composition analyses on defined ribosomal RNA's should be of special help in this regard.

IV. HOMOLOGY BETWEEN MITOCHONDRIAL AND BACTERIAL RIBOSOMES

Apparent homologies between bacterial and mitochondrial protein synthetic systems are often cited to support the notion of an endosymbiotic origin of mitochondria. In view of the tremendous diversity of mitoribosomes it is no longer appropriate to consider them homologous to bacterial ribosomes on the basis of size or sedimentation coefficient. Several homologies do remain, however.

A. Antibiotic Susceptibility

One of the first homologies described between bacterial and mitochondrial ribosomes was their susceptibility to chloramphenicol (Rendi, 1959; Mager, 1960). Since then, the effects of many antibiotics
on mitochondrial protein synthesis have been studied at several levels
of organization, and these initial observations have been extended
to include several antibiotics (Lamb et al., 1968; Kroon et al.,
1972; Kroon and Arendzen, 1972; Mahler, 1973; Borst and Grivell,
1971). In general, the antibiotic inhibitors of bacterial protein
synthesis seem capable of binding to mitochondrial ribosomes or affecting a mitoribosomal function (Table 2). On the other hand, mitoribosomes are largely unaffected by inhibitors that do not affect the
function of bacterial ribosomes. So, by this criterion of antibiotic

TABLE 2   *Some Ribosomal Inhibitors Used in the Study of Mitochon-*
drial Protein Synthesis

Specificity		
Bacterial and mitoribosomes only[a]	Cytoribosomes only[a]	All ribosomes[a]
Chloramphenicol (1, 2)	Cycloheximide (1, 9)	Puromycin (9)
Lincomycin (1, 4)	Anisomysin (9, 10)	
Erythromycin (1, 4)	Pederine (11)	
Carbomycin (1, 4)	(Emetine) (5)	
Tylosin (6, 7)		
Vernamycin (3, 7)		
Streptogramins (7)		
Aminoglycosides (9, 12)		
Thiostrepton (8)		
Siomycin (8)		

[a]Numbers in parentheses are references: 1. Lamb et al. (1968);
2. Freeman (1970); 3. Towers et al. (1972); 4. Kroon and DeVries
(1971); 5. Lietman (1971); 6. DeVries et al. (1973); 7. Denslow
and O'Brien (1974); 8. Richter et al. (1971); 9. Pestka (1971);
10. DeVries et al. (1971); 11. Brega and Vesco (1971); 12. Davey
et al. (1970).

susceptibility, mitochondrial ribosomes of all organisms show homologies with bacterial ribosomes.

Some additional comments are warranted. Emetine, which has been
used as a specific inhibitor of cytoribosomal protein synthesis in
studies of mitochondrial protein synthesis (Perlman and Penman, 1970a;
Ojala and Attardi, 1972) is not the best choice for this purpose,
since it also inhibits mitoribosomes (Chakrabarti et al., 1972; Lietman, 1971; Kroon and Arendzen, 1972). Emetine is the only known example of an inhibitor affecting cytoribosomes and mitoribosomes, but
not bacterial ribosomes.

Certain of the antibiotics have quantitatively different effects
on intramitochondrial ribosomes and isolated mitoribosomes. In particular, the diminished sensitivity of protein synthesis in mammalian
mitochondria, relative to yeast mitochondria, was cited as a basis
for phylogenetic differences in mitoribosomes (Firkin and Linnane,
1969; Towers et al., 1972). Although most of this effect can be ascribed to a permeability barrier at the level of the mitochondrial
membrane (Kroon and DeVries, 1971; DeVries et al., 1973; Ibrahim et
al., 1973), some differences in antibiotic susceptibility among mitochondrial ribosomes remain when the antibiotics are tested with isolated mitoribosomes (DeVries et al., 1973; Denslow and O'Brien, 1974).

In view of the multiplicity of mitoribosome types and the limited
studies of antibiotic susceptibility that have been conducted, it is
premature to speak of phylogenetic differences among mitoribosomes
with respect to antibiotic susceptibility.

B. Physical-chemical Properties

Although some mitoribosomes may resemble bacterial ribosomes with
respect to sedimentation coefficient, molecular weight, or buoyant
density, the real diversity of mitoribosomal types and physical-chemical properties renders such comparisons meaningless.

C. Ribosome Dissociability and Subribosomal Particles

1. *Divalent Cation Requirement*

Most mitoribosomes resemble bacterial ribosomes, but not cytoribosomes, in their requirement of relatively high levels of magnesium for stabilization of the monoribosome form. As noted above, *Tetrahymena* mitoribosomes are an exception.

2. *Interaction with Bacterial IF-3*

One of the prokaryotic initiation factors, IF-3, promotes dissociation of bacterial ribosomes. This protein factor from *E. coli* has been shown capable of specific interaction with *Neurospora* mitoribosomes but not cytoribosomes (Agsteribbe and Kroon, 1973; Datema et al., 1974). It would be interesting to learn whether this homology extends to other mitoribosomes.

3. *Interchangeability of Subribosomal Particles*

Attempts to form hybrid ribosomes using subribosomal particles from yeast mitochondria (*Saccharomyces*) and *E. coli* were unsuccessful (Grivell and Walg, 1972). At first sight it seems most unlikely that sufficient homology would be conserved over the multiple binding regions on the subunit interfaces (Garrett and Wittmann, 1973). However, chloroplast ribosomes do exhibit this high level homology with *E. coli* ribosomes (Grivell and Walg, 1972), and many eukaryotic ribosomes can form hybrid couples with each other, but not with subunits of prokaryotic ribosomes (Martin and Wool, 1969). The one instance in which functional hybrid ribosomes have been formed between a eukaryote (*Artemia*) ribosome subunit and the *E. coli* large subribosomal particle should encourage the search for this kind of homology among other mitoribosomes.

D. Partial Reactions of Protein Synthesis

1. *Initiation*

a. *Involvement of fMet-tRNA.* Like bacterial ribosomes, mitoribosomes from yeast, *Neurospora,* and mammals initiate protein synthesis with fMet-tRNA (Galper and Darnell, 1969; Sala and Kuntzel, 1970;
Feldman and Mahler, 1974; Mahler et al., 1974; Galper, 1974).

b. *Interaction with prokaryotic initiation factors.* In addition
to the example with IF-3, mitoribosomes from *Neurospora* have also been
shown to be capable of specific interaction with *E. coli* initiation
factors IF-1 and IF-2 (Sala and Kuntzel, 1970). Factors required for
the AUG-directed binding of fMet-tRNA can be stripped from *Neurospora*
subribosomal particles, and they can be replaced by bacterial initiation factors.

2. *Elongation*

The general interchangeability of the elongation factors EF-G
and EF-T between mitochondrial and bacterial, but not cytoplasmic,
systems manifests additional homologies between mitochondrial and
bacterial ribosomes. Such homologies have been demonstrated for
yeast mitoribosomes (Richter and Lipmann, 1970; Scragg, 1971; Morimoto et al., 1971), *Neurospora* mitoribosomes (Grandi and Kuntzel,
1970; Kuntzel, 1969c) and *Xenopus* mitoribosomes (Swanson, 1973). In
addition, mitoribosomes from *Euglena* (Avadhani and Buetow, 1972b)
and rat liver (Greco et al., 1973; Ibrahim et al., 1974) function
with *E. coli* elongation factors. It has also been shown that cytoplasmic factors cannot support phenylalanine polymerization on *Tetrahymena* mitoribosomes (Allen and Suyama, 1972). *Euglena* mitoribosomes,
on the other hand, can utilize cytoplasmic supernatant factors, although not as well as homologous factors, or those from *Euglena* chloroplasts (Avadhani and Buetow, 1974).

E. Ribosome Proteins

1. Functional Substitution

Bovine liver mitoribosomal proteins appear unable to substitute functionally for several individual *E. coli* 30S ribosomal proteins when tested (Held et al., 1974) by the ribosome reconstitution assay (Higo et al., 1973).

2. Immunochemical Homologies

So far, attempts to demonstrate immunochemical homologies between ribosomal proteins of bacteria and mitochondria have yielded negative results. *Neurospora* mitoribosomes showed no cross reaction with antisera prepared against *E. coli* ribosomes and subribosomal particles (Gualerzi et al., 1974). Similarly, bovine mitoribosomes do not show a cross reaction with antisera prepared against several individual proteins of the *E. coli* 30S subribosomal particle (Kahan et al., 1974). The possibility that mitoribosomes contain proteins homologous to bacterial ribosomal proteins remains open. Further studies of this possibility should be spurred by the demonstration of immunochemical homology between the *E. coli* ribosomal proteins L7 and L12, and their analogs L40 and L41 in rat cytoribosomes (Stoffler et al., 1974).

V. CONCLUDING REMARKS

We have seen that the mitochondria in all eukaryotic systems studied have their own exclusive protein synthetic system. The presence of unique ribosomes within all mitochondria attests to the vital role of mitochondrial protein synthesis in eukaryotes, if only to make approximately 10 polypeptides. The semiautonomous character of mitochondrial biogenesis extends even to mitochondrial ribosomes. They represent a grand example of concerted biosynthesis, in which RNA transcripts of the mitochondrial genome assemble with protein products of the nuclear genome to produce a functional organellar ribosome.

In spite of the wide-ranging differences among mitoribosomes with respect to their physical and chemical properties, they do share certain distinguishing features: ribosomal RNA's of uncommonly low G + C content, and multiple homologies with bacterial ribosomes. Their surprising diversity suggests that many of the different mitoribosomes may have arisen from separate endosymbioses. Alternatively, we are led to consider that the heterogeneity of mitoribosomes arose in response to very different selective pressures than those resulting in much less apparent diversity among the corresponding cytoribosomes. Indeed, in view of their heterogeneity, it is noteworthy that mitoribosomes exhibit so many homologies with bacterial, rather than cytoplasmic ribosomes.

REFERENCES

Aaij, C., N. Nanninga, and P. Borst. 1972. The structure of ribosome-like particles from rat liver mitochondria. *Biochim. Biophys. Acta 277*:140-148.

Agsteribbe, E., and A. M. Kroon. 1973. Dissociation of mitochondrial ribosomes of *Neurospora crassa* by a bacterial dissociation factor. *Biochem. Biophys. Res. Commun. 51*:8-13.

Agsteribbe, E., R. Datema, and A. Kroon. 1974. Mitochondrial polysomes from *Neurospora crassa*. in: *The Biogenesis of Mitochondria* (A. Kroon and C. Saccone, eds). New York: Academic Press. pp. 305-314.

Allen, N. E., and Y. Suyama. 1972. Protein synthesis in vitro with *Tetrahymena* mitochondrial ribosomes. *Biochim. Biophys. Acta 259*: 369-377.

Andre, J., and V. Marinozzi. 1965. Presence, dans les mitochondries, de particules ressemblant aux ribosomes. *J. de Microscopie 4*: 615-626.

Ashwell, M., and T. S. Work. 1970a. The biogenesis of mitochondria. *Ann. Rev. Biochem. 39*:251-290.

Ashwell, M., and T. S. Work. 1970b. The functional characterization of ribosomes from rat liver mitochondria. *Biochem. Biophys. Res. Commun. 39*:204-211.

Attardi, B., and G. Attardi. 1969. Sedimentation properties of RNA species homologous to mitochondrial DNA in HeLa cells. *Nature 224*:1079-1083.

Attardi, G., and D. Ojala. 1971. Mitochondrial ribosomes in HeLa cells. *Nature New Biol. 229*:133-136.

Avadhani, N. G., and D. E. Buetow. 1972a. Isolation of active poly-ribosomes from the cytoplasm, mitochondria and chloroplasts of *Euglena gracilis*. *Biochem. J. 128*:353-365.

Avadhani, N. G., and D. E. Buetow. 1972b. Protein synthesis with isolated mitochondrial polysomes. *Biochem. Biophys. Res. Commun. 46*:773-778.

Avadhani, N. G., and D. E. Buetow. 1974. Mitochondrial and cytoplasmic ribosomes: distinguishing characteristics and a requirement for the homologous ribosomal salt-extractable fraction for protein synthesis. *Biochem. J. 140*:73-78.

Avadhani, N. G., F. S. Lewis, and R. J. Rutman. 1974. Messenger ribonucleic acid metabolism in mammalian mitochondria. Quantitative aspects of structural information coded by the mitochondrial genome. *Biochemistry 13*:4638-4645.

Bates, H. M., and M. V. Simpson. 1959. The net synthesis of cytochrome c in calf heart mitochondria. *Biochim. Biophys. Acta 32*:597-599.

Bates, H. M., V. M. Craddock, and M. V. Simpson. 1960. The biosynthesis of cytochrome c in cell-free systems. *J. Biol. Chem. 235*:140-148.

Baxter, R., and D. H. L. Bishop. 1968. The high molecular weight ribonucleic acid species of soya-bean mitochondria. *Biochem. J. 109*:13p-14p.

Beale, G. H., J. K. C. Knowles, and A. Tait. 1972. Mitochondrial genetics in *paramecium*. *Nature 235*:396-397.

Beattie, D. S. 1971. The synthesis of mitochondrial proteins. *Sub-Cell. Biochem. 1*:1-23.

Beattie, D. S., and N. G. Ibrahim. 1973. Optimal conditions for amino acid incorporation by isolated rat liver mitochondria. Stimulation by valinomycin and other agents. *Biochemistry 12*:176-180.

Beattie, D. S., R. E. Basford, and S. B. Koritz. 1967a. Bacterial contamination and amino acid incorporation by isolated mitochondria. *J. Biol. Chem. 242*:3366-3368.

Beattie, D., R. Basford, and S. Koritz. 1967b. The inner membrane as the site of the in vitro incorporation of L-(14C)-leucine into mitochondrial protein. *Biochemistry 6*:3099-3106.

Bemis, J. A., G. M. Bryant, J. C. Arcos, and M. F. Argus. 1968. Swelling and contraction of mitochondrial particles: a re-examination of the existence of a contractile protein extractable with 0.6M-potassium chloride. *J. Mol. Biol. 33*:299-307.

Bingham, R. W., and P. N. Campbell. 1972. Studies on the biosynthesis of mitochondrial malate dehydrogenase and the location of its synthesis in the liver cell of the rat. *Biochem. J. 126*:211-215.

Blair, P. V., J. F. Perdue, and D. E. Green. 1964. Properties of mitochondrial contractile protein and actomyosin of heart muscle. *Science 144*:560.

Borst, P. 1972. Mitochondrial nucleic acids. *Ann. Rev. Biochem.* 41:333-376.

Borst, P., and L. Grivell. 1971. Mitochondrial ribosomes. *FEBS Lett.* 13:73-88.

Borst, P., and L. A. Grivell. 1973. Mitochondrial nucleic acids. *Biochimie* 55:801-804.

Brega, A., and C. Vesco. 1971. Ribonucleoprotein particles involved in HeLa mitochondrial protein synthesis. *Nature New Biol.* 229: 136-139.

Bunn, C. L., C. H. Mitchell, H. B. Lukins, and A. W. Linnane. 1970. Biogenesis of mitochondria. XVIII. A new class of cytoplasmically determined antibiotic resistant mutants in *Saccharomyces cerevisiae*. *Proc. Nat. Acad. Sci. US* 67:1233-1240.

Burke, J. P., and D. S. Beattie. 1973. The synthesis of proteolipid protein by isolated rat liver mitochondria. *Biochem. Biophys. Res. Commun.* 51:349-356.

Capaldi, R. A. 1974. Identification of the major enzymic activities of the mitochondrial inner membrane in terms of their migration in sodium dodecyl sulfate polyacrylamide gel electrophoresis. *Arch. Biochem. Biophys.* 163:99-105.

Chakrabarti, S., D. K. Dube, and S. C. Roy. 1972. Effects of emetine and cycloheximide on mitochondrial protein synthesis in different systems. *Biochem. J.* 128:461-462.

Chen, W. L., and F. C. Charalampous. 1969. Mechanism of induction of cytochrome oxidase in yeast. I. Kinetics of induction and evidence for accumulation of cytoplasmic and mitochondrial precursors. *J. Biol. Chem.* 244:2767-2776.

Chi, J. C. H., and Y. Suyama. 1970. Comparative studies on mitochondrial and cytoplasmic ribosomes of *Tetrahymena pyriformis*. *J. Mol. Biol.* 53:531-556.

Chignell, D. A., and P. T. Wingfield. 1974. Two dimensional separation of the proteins of the inner mitochondrial membrane on polyacrylamide gel. *Fed. Proc.* 33:1283.

Chooi, W. Y., and C. D. Laird. 1976. Association of transcription and translation in mitochondria of *Drosophila melanogaster*. In press.

Conover, T. E., and M. Barany. 1966. The absence of a myosin-like protein in liver mitochondria. *Biochim. Biophys. Acta* 127:235-238.

Cooper, C., and C. Avers. 1974. Evidence of involvement of mitochondrial polysomes and messenger RNA in synthesis of organelle proteins. in: *The Biogenesis of Mitochondria* (A. Kroon and C. Saccone, eds.). New York: Academic Press. pp. 289-303.

Coote, J. L., and T. S. Work. 1971. Proteins coded by mitochondrial DNA of mammalian cells. *Eur. J. Biochem.* 23:564-574.

Coote, J. L., T. H. Rabbitts, and T. S. Work. 1971. The mitochondrial ribosomes of baby hamster kidney cells. *Biochem. J. 123:*
279-281.

Costantino, P., and G. Attardi. 1973. Atypical pattern of utilization of amino acids for mitochondrial protein synthesis in HeLa
cells. *Proc. Nat. Acad. Sci. US 70:*1490-1494.

Criddle, R., R. Bock, D. Green, and H. Tisdale. 1962. Physical characteristics of proteins of the electron transfer system and interpretation of the structure of the mitochondrion. *Biochemistry 1:*
827-842.

Curgy, J. J., G. Ledoigt, B. J. Stevens, and J. Andre. 1974. Mitochondrial and cytoplasmic ribosomes from *Tetrahymena pyriformis:*
correlative analysis by gel electrophoresis and electron microscopy. *J. Cell Biol. 60:*628-640.

Datema, R., E. Agsteribbe, and A. M. Kroon. 1974. The mitochondrial
ribosomes of *Neurospora crassa*. I. On the occurrence of 80S ribosomes. *Biochim. Biophys. Acta 335:*386-395.

Davey, P. J., J. M. Haslam, and A. W. Linnane. 1970. Biogenesis of
mitochondria 12. *Arch. Biochem. Biophys. 136:*54-64.

Dawid, I. B. 1972. Mitochondrial protein synthesis. *Eighth FEBS
Meeting.* (S. G. Van den Bergh, P. Borst, L. L. M. Van Deenen, J.
C. Riemersma, E. C. Slater, and J. M. Tager, eds.) Vol. *28:*53-69.

Dawid, I. B., and J. W. Chase. 1972. Mitochondrial RNA in *Xenopus
laevis*. II. Molecular weights and other physical properties of
mitochondrial ribosomal and 4S RNA. *J. Mol. Biol. 63:*217-231.

Delaney, P., M. Keyes, and T. Soule. 1967. Growth hormone stimulation of protein synthesis in isolated mouse liver mitochondria.
*Endocrin. 80:*772-774.

Denslow, N. D., and T. W. O'Brien. 1974. Susceptibility of 55S mitochondrial ribosomes to antibiotics inhibitory to prokaryotic
ribosomes, lincomycin, chloramphenicol and PA114A. *Biochem. Biophys. Res. Commun. 57:*9-16.

DeVries, H., and A. Kroon. 1974. Physicochemical and functional
characterization of the 55S ribosomes from rat liver mitochondria.
in: *The Biogenesis of Mitochondria* (A. Kroon and C. Saccone, eds.).
New York: Academic Press. pp. 357-365.

DeVries, H., and R. van der Koogh-Schurring. 1973. Physicochemical
characteristics of isolated 55S mitochondrial ribosomes from rat
liver. *Biochem. Biophys. Res. Commun. 54:*308-314.

DeVries, H., E. Agsteribbe, and A. M. Kroon. 1971. The "fragment
reaction": a tool for the discrimination between cytoplasmic and
mitochondrial ribosomes. *Biochim. Biophys. Acta 246:*111-122.

DeVries, H., A. J. Arendzen, and A. M. Kroon. 1973. The interference of the macrolide antibiotics with mitochondrial protein synthesis. *Biochim. Biophys. Acta 331:*264-275.

Dixon, H., G. M. Kellerman, C. H. Mitchell, N. H. Towers, and A. W. Linnane. 1971. Mikamycin, an inhibitor of both mitochondrial protein synthesis and respiration. *Biochem. Biophys. Res. Commun.* *43*:780-786.

Dubin, D. T. 1974. Methylated nucleotide content of mitochondrial ribosomal RNA from hamster cells. *J. Mol. Biol.* *84*:257-273.

Dubin, D. T., T. H. Jones, and G. R. Cleaves. 1974. An unmethylated "3S" RNA in hamster mitochondria: a 5S RNA equivalent? *Biochem. Biophys. Res. Commun.* *56*:401-406.

Edelman, M., I. M. Verma, and U. Z. Littauer. 1970. Mitochondrial ribosomal RNA from *Aspergillus nidulans*: characterization of a novel molecular species. *J. Mol. Biol.* *49*:67-83.

Edelman, M., I. M. Verma, R. Herzog, E. Galun, and U. Z. Littauer. 1971. Physico-chemical properties of mitochondrial ribosomal RNA from fungi. *Eur. J. Biochem.* *19*:372-378.

England, J. M., and G. Attardi. 1974. Expression of the mitochondrial genome in HeLa cells. XXI. Mitochondrial protein synthesis during the cell cycle. *J. Mol. Biol.* *85*:433-444.

Erdmann, V. A., S. Fahnestock, K. Higo, and M. Nomura. 1971. Role of 5S RNA in the functions of 50S ribosomal subunits. *Proc. Nat. Acad. Sci. US* *68*:2932-2936.

Feldman, F., and H. R. Mahler. 1974. Mitochondrial biogenesis: retention of terminal formylmethionine in membrane proteins and regulation of their synthesis. *J. Biol. Chem.* *249*:3702-3709.

Firkin, F. C., and A. W. Linnane. 1969. Phylogenetic differences in the sensitivity of mitochondrial protein synthesizing systems to antibiotics. *FEBS Lett.* *2*:330-332.

Folch, J., and N. Lees. 1951. Proteolipids, a new type of tissue lipoprotein--their isolation from brain. *J. Biol. Chem.* *191*: 807-817.

Forrester, I. T., P. Nagley, and A. W. Linnane. 1970. Yeast mitochondrial ribosomal RNA: a new extraction procedure and unusual physical properties. *FEBS Lett.* *11*:59-61.

Forrester, I. T., K. Watson, and A. W. Linnane. 1971. Mitochondrial membrane organization, a determinant of mitochondrial ribosomal RNA synthesis. *Biochem. Biophys. Res. Commun.* *43*:409-415.

Freeman, K. B. 1970. Inhibition of mitochondrial and bacterial protein synthesis by chloramphenicol. *Can. J. Biochem.* *48*:479-485.

Freeman, K. B., R. S. Mitra, and B. Bartoov. 1973. Characteristics of the base composition of mitochondrial ribosomal RNA. *Sub-Cell. Biochem.* *2*:183-192.

Galper, J. B. 1974. Mitochondrial protein synthesis in HeLa cells. *J. Cell Biol.* *60*:755-763.

Galper, J. B., and J. E. Darnell. 1969. The presence of N-formyl-methionyl-tRNA in HeLa cells. *Biochem. Biophys. Res. Commun.* *34*: 205-214.

Gambetti, P., L. A. Autilio-Gambetti, N. Gonatas, and B. Shafer. 1972. Protein synthesis in synaptosomal fractions. Ultrastructural radioautographic study. *J. Cell Biol.* *52*:526-535.

Garrett, R. A., and H. G. Wittmann. 1973. Structure of bacterial ribosomes. *Adv. Prot. Syn.* *27*:277-347.

Graffi, A., G. Butschak, and E. J. Schneider. 1965. Differences of mitochondrial protein synthesis in vitro between tumour and normal tissues. *Biochem. Biophys. Res. Commun.* *21*:418-423.

Grandi, M., and H. Kuntzel. 1970. Mitochondrial peptide chain elongation factors from *Neurospora crassa*. *FEBS Lett.* *10*:25-28.

Gray, P. H., and G. Attardi. 1973. An attempt to identify a presumptive 5S RNA species in mitochondrial ribosomes. *J. Cell Biol.* *59*:120a.

Greco, M., P. Cantatore, G. Pepe, and C. Saccone. 1973. Isolation and characterization of rat liver mitochondrial ribosomes highly active in poly(U)-directed polyphenylalanine synthesis. *Eur. J. Biochem.* *37*:171-177.

Grivell, L. A. 1974. Mutations affecting mitochondrial ribosomes. in: *The Biogenesis of Mitochondria* (A. Kroon and C. Saccone, eds.). New York: Academic Press. pp. 275-288.

Grivell, L. A., and H. L. Walg. 1972. Subunit homology between *Escherichia coli*, mitochondrial and chloroplast ribosomes. *Biochem. Biophys. Res. Commun.* *49*:1452-1458.

Grivell, L. A., L. Reijnders, and P. Borst. 1971. Isolation of yeast mitochondrial ribosomes highly active in protein synthesis. *Biochim. Biophys. Acta* *247*:91-103.

Grivell, L. A., P. Netter, P. Borst, and P. P. Slonimski. 1973. Mitochondrial antibiotic resistance in yeast ribosomal mutants resistant to chloramphenicol erythromycin and spiramycin. *Biochim. Biophys. Acta* *312*:358-367.

Groot, G. S. 1974. The biosynthesis of mitochondrial ribosomes in *Saccharomyces cerevisiae*. in: *The Biogenesis of Mitochondria* (A. Kroon and C. Saccone, eds.). New York: Academic Press. pp. 443-452.

Groot, P. H. E., C. Aaij, and P. Borst. 1970. Variation with temperature of the apparent molecular weight of rat liver mitochondrial RNA, determined by gel electrophoresis. *Biochem. Biophys. Res. Commun.* *41*:1321-1327.

Gualerzi, C. 1969. Electrophoretic comparison of cytoplasmic and mitochondrial ribosomal proteins from *Neurospora crassa*. *Italian J. Biochem.* *18*:418.

Gualerzi, C., H. G. Janda, H. Passow, and G. Stoffler. 1974. Studies on the protein moiety of plant ribosomes: enumeration of the proteins of the ribosomal subunits and determination of the

degree of evolutionary conservation by electrophoretic and immunochemical methods. *J. Biol. Chem. 249:*3347-3355.

Gustafsson, R., J. Tata, O. Lindberg, and L. Ernster. 1965. The
relationship between the structure and activity of rat skeletal
muscle mitochondria after thyroidectomy and thyroid hormone treatment. *J. Cell Biol. 26:*555-578.

Hadvary, P., and B. Kadenbach. 1973. Isolation and characterization
of chloroform-soluble proteins from rat liver mitochondria and
other fractions. *Eur. J. Biochem. 39:*11-20.

Haldar, D. 1971. Protein synthesis in isolated rat brain mitochondria. *Biochem. Biophys. Res. Commun. 42:*899-904.

Haldar, D., and K. B. Freeman. 1969. Importance of the osmolarity
of the incubation medium on amino acid incorporation into protein
by isolated rat liver mitochondria. *Biochem. J. 111:*653-663.

Hallermayer, G., and W. Neupert. 1974. Immunological difference of
mitochondrial and cytoplasmic ribosomes of *Neurospora crassa.*
*FEBS Lett. 41:*264-268.

Hamilton, M. G., and T. W. O'Brien. 1974. Ultracentrifugal characterization of the mitochondrial ribosome and subribosomal particles of bovine liver: molecular size and composition. *Biochemistry 13:*5400-5403.

Henson, C. P., P. Perlman, C. N. Weber, and H. R. Mahler. 1968.
Formation of yeast mitochondria. II. Effects of antibiotics on
enzyme activity during derepression. *Biochemistry 7:*4445-4454.

Hernandez, A., I. Burdett, and T. S. Work. 1971. Protein synthesis
by brain-cortex mitochondria characterization of a 55S mitochondrial ribosome as the functional unit in protein synthesis by cortex mitochondria and its distinction from a contaminant cytoplasmic protein-synthesizing system. *Biochem. J. 124:*327-336.

Higo, K., W. Held, L. Kahan, and M. Nomura. 1973. Functional corres-
pondence between 30S ribosomal proteins of *Escherichia coli* and
Bacillus stearothermophilus. *Proc. Nat. Acad. Sci. US 70:*944-948.

Hochberg, A. A., F. W. Stratman, R. N. Zahlten, and H. A. Lardy.
1972. Artifacts in protein synthesis by mitochondria in vitro.
*FEBS Lett. 25:*1-7.

Ibrahim, N. G., J. P. Burke, and D. S. Beattie. 1973. Mitochondrial
protein synthesis in vitro is not an artifact. *FEBS Lett. 29:*73-
76.

Ibrahim, N. G., J. P. Burke, and D. S. Beattie. 1974. The sensitivity of rat liver and yeast mitochondrial ribosomes to inhibitors
of protein synthesis. *J. Biol. Chem. 249:*6806-6811.

Jeffreys, A., and I. Craig. 1974. Differences in the products of
mitochondrial protein synthesis in vivo in human and mouse cells

and their potential use as markers for the mitochondrial genome in human-mouse somatic cell hybrids. *Biochem. J. 144*:161-164.

Kadenbach, H., and P. Hadvary. 1973. Demonstration of two types of proteins synthesized in isolated rat liver mitochondria. *Eur. J. Biochem. 32*:343-349.

Kalf, G. 1963. The incorporation of leucine-1-^{14}C into the protein of rat heart sarcosomes: an investigation of optimal conditions. *Archiv. Biochem. Biophys. 101*:350-359.

Kalf, G. F., and M. A. Grece. 1964. The in vitro incorporation of ^{14}C-amino acids into the contractile protein of intact lamb heart mitochondria. *Biochem. Biophys. Res. Commun. 17*:674-679.

Kalf, G. F., and M. A. Grece. 1970a. The incorporation of ^{14}C-amino acids into the contractile protein of intact mitochondria: a correction. *Biochem. Biophys. Res. Commun. 40*:1488-1491.

Kalf, G. F., and M. A. Grece. 1970b. Purification and properties of an acid nucleoside triphosphatase from rat liver mitochondria. *Biochemistry 9*:4049-4056.

Kleinow, W. 1974. Influence of cations on the dissociation of mitochondrial and cytoplasmic ribosomes from locusta migratoria. *Hoppe-Seyler's Z. Physiol. Chem. 355*:1027-1034.

Kleinow, W., and W. Neupert. 1971. The mitochondrial ribosome from locusta migratoria: dissociation into subunits. *FEBS Lett. 15*: 359-364.

Kleinow, W., W. Neupert, and T. Bucher. 1971. Small sized ribosomes from mitochondria of locusta migratoria. *FEBS Lett. 12*:129-133.

Kleinow, W., W. Neupert, and F. Miller. 1974a. Electron microscope study of mitochondrial 60S and cytoplasmic 80S ribosomes from locusta migratoria. *J. Cell Biol. 62*:860-875.

Kleinow, W., W. Neupert, and F. Miller. 1974b. Fine structure of mitochondrial ribosomes of locust flight muscle. in: *The Biogenesis of Mitochondria* (A. Kroon and C. Saccone, eds.). New York: Academic Press. pp. 337-346.

Kroon, A. M. 1963. Inhibitors of mitochondrial protein synthesis. *Biochim. Biophys. Acta 76*:165-167.

Kroon, A. M., and A. J. Arendzen. 1972. The inhibition of mitochondrial biogenesis by antibiotics. *Eighth FEBS Meeting* (S. G. Van den Bergh, P. Borst, L. L. M. van Deenen, J. C. Riemersma, E. C. Slater, and J. M. Tager, eds.). Vol. *28*, pp. 53-69.

Kroon, A. M., and H. DeVries. 1971. Mitochondriogenesis in animal cells: studies with different inhibitors. in: *Autonomy and Biogenesis of Mitochondria and Chloroplasts* (N. K. Boardman, A. W. Linnane, R. M. Smillie, eds). New York: American Elsevier. pp. 218-327.

Kroon, A., C. Saccone, and M. Botman. 1967. RNA and protein synthesis by sterile rat liver mitochondria. *Biochim. Biophys. Acta*
142:552-554.

Kroon, A. M., M. J. Botman, and C. Saccone. 1968. Practical procedures for the isolation of mitochondrial preparations suitable for
the study of mitochondrial macromolecules and with minimal contamination by other cell fractions or bacteria. in: *Biochemical Aspects of the Biogenesis of Mitochondria* (E. C. Slater, J. M. Tager,
S. Papa, and E. Quagliariello, eds.). Bari: Adriatica Editrice.
pp. 439-455.

Kroon, A. M., E. Agsteribbe, and H. DeVries. 1972. Protein synthesis in mitochondria and chloroplasts. in: *The Mechanism of Protein Synthesis and its Regulation* (L. Bosch, ed.). New York:
American Elsevier. pp. 539-582.

Kuntzel, H. 1969a. Mitochondrial and cytoplasmic ribosomes from
Neurospora crassa: characterization of their subunits. *J. Mol.
Biol. 40*:315-320.

Kuntzel, H. 1969b. Proteins of mitochondrial and cytoplasmic ribosomes from *Neurospora crassa*. *Nature 222*:142-146.

Kuntzel, H. 1969c. Specificity of mitochondrial and cytoplasmic
ribosomes from *Neurospora crassa* in poly-U dependent cell-free
systems. *FEBS Lett. 4*:140-142.

Kuntzel, H. 1971. The genetic appearance of mitochondria from *Neurospora* and yeast. *Curr. Topics Microb. Immun. 54*:94-118.

Kuntzel, H., and H. C. Blossey. 1974. Translation products in vitro
of mitochondrial messenger RNA from *Neurospora crassa*. *Eur. J.
Biochem. 47*:165-171.

Kuntzel, H., and H. Noll. 1967. Mitochondrial and cytoplasmic polysomes from *Neurospora crassa*. *Nature 215*:1340-1345.

Lamb, A. J., G. D. Clark-Walker, and A. W. Linnane. 1968. The biogenesis of mitochondria. 4. The differentiation of mitochondrial
and cytoplasmic protein synthesizing systems in vitro by antibiotics. *Biochim. Biophys. Acta 161*:415-427.

Langdon, R. G. 1974. Serum lipoprotein apoproteins as major protein
constituents of the human erythrocyte membrane. *Biochim. Biophys.
Acta 342*:213-228.

Laub-Kupersztejn, R., and J. Thirion. 1974. Existence of two distinct protein synthesis systems in the trypanosomatid *Crithidia
luciliae*. *Biochim. Biophys. Acta 340*:314-322.

Leaver, C. J., and M. A. Harmey. 1972. Isolation and characterization of mitochondrial ribosomes from higher plants. *Biochem. J.
129*:37-38.

Leaver, C. J., and M. A. Harmey. 1973. Ribosomal RNAs of higher
plant mitochondria. in: *Ribosomes and RNA Metabolism* (J. Zelin-

ka and J. Balan, eds). Bratislava: Publishing House of the Slovak Academy of Sciences. pp. 407-417.

Lederman, M., and G. Attardi. 1970. In vitro protein synthesis in a mitochondrial fraction from HeLa cells: sensitivity to antibiotics and ethidium bromide. *Biochem. Biophys. Res. Commun. 40:* 1492-1500.

Lederman, M., and G. Attardi. 1973. Expression of the mitochondrial genome in HeLa cells. XVL. Electrophoretic properties of the products of in vivo and in vitro mitochondrial protein synthesis. *J. Mol. Biol. 78:*275-283.

Leister, D. E., and I. B. Dawid. 1974. Physical properties and protein constituents of cytoplasmic and mitochondrial ribosomes of *Xenopus laevis. J. Biol. Chem. 249:*5108-5118.

Lietman, P. S. 1971. Mitochondrial protein synthesis: inhibition by emetine hydrochloride. *Mol. Pharm. 7:*122-128.

Linnane, A. W., and J. M. Haslam. 1970. The biogenesis of yeast mitochondria. *Curr. Topics Cell. Reg. 2:*101-172.

Linnane, A. W., J. M. Haslam, H. B. Lukins, and P. Nagley. 1972. The biogenesis of mitochondria in microorganisms. *Ann. Rev. Micro. 26:*163-198.

Lizardi, P. M., and D. J. L. Luck. 1971. Absence of a 5S RNA component in the mitochondrial ribosomes of *N. crassa. Nature New Biol. 229:*140-142.

Lizardi, P. M., and D. J. L. Luck. 1972. The intracellular site of synthesis of mitochondrial ribosomal proteins in *Neurospora crassa. J. Cell Biol. 54:*56-74.

Loewenstein, J., H. R. Scholte, and E. M. Wit-Peeters. 1970. A rapid and simple procedure to deplete rat liver mitochondria of lysosomal activity. *Biochim. Biophys. Acta 223:*432-436.

Mager, J. 1960. Chloramphenicol and chlortetracycline inhibition of amino acid incorporation into proteins in a cell-free system from *Tetrahymena pyriformis. Biochim. Biophys. Acta 38:*150-152.

Mahler, H. R. 1973. Biogenetic autonomy of mitochondria. *CRC Critical Reviews in Biochemistry 1:*381-460.

Mahler, H. R., P. Perlman, C. Henson, and C. Weber. 1968. Selective effects of chloramphenicol, cycloheximide, and nalidixic acid on the biosynthesis of respiratory enzymes in yeast. *Biochem. Biophys. Res. Commun. 31:*474-480.

Mahler, H. R., L. R. Jones, and W. J. Moore. 1971. Mitochondrial contribution to protein synthesis in cerebral cortex. *Biochem. Biophys. Res. Commun. 42:*384-389.

Mahler, H., F. Feldman, S. Phan, P. Hamill, and K. Dawidowicz. 1974. Initiation, identification, and integration of mitochondrial proteins. in: The Biogenesis of Mitochondria (A. Kroon and C. Saccone, eds.). New York: Academic Press. pp. 423-441.

Malkin, L. I. 1970. Amino acid incorporation by isolated rat liver mitochondria during liver regeneration. Proc. Nat. Acad. Sci. US 67:1695-1702.

Malkin, L. I. 1971. Minor species of ribonucleic acid associated with rat liver mitochondria. Biochemistry 10:4752-4756.

Martin, T. E., and I. G. Wool. 1969. Active hybrid 80S particles formed from subunits of rat, rabbit and protozoan (Tetrahymena pyriformis) ribosomes. J. Mol. Biol. 43:151-161.

Mason, T., E. Ebner, R. O. Poyton, J. Saltzgaber, D. C. Wharton, L. Mennucci, and G. Schatz. 1972. The participation of mitochondrial and cytoplasmic protein synthesis in mitochondrial formation. in: Mitochondria: Biomembranes, Eighth FEBS Meeting (S. G. Van Den Bergh, P. Borst, L. L. M. Van Deenen, J. C. Riemersma, E. C. Slater, and J. M. Tager, eds.). Vol. 28:53-69.

Matthews, D. E., and T. W. O'Brien. 1974. Personal observation.

McConkey, E. H. 1974. Composition of mammalian ribosomal subunits: a re-evaluation. Proc. Nat. Acad. Sci. US 71:1379-1383.

McLean, J. C., G. L. Cohn, I. K. Brandt, and M. V. Simpson. 1958. Incorporation of labeled amino acids into the protein of muscle and liver mitochondria. J. Biol. Chem. 233:657-663.

Michel, R., and W. Neupert. 1973. Mitochondrial translation products before and after integration into the mitochondrial membrane in Neurospora crassa. Eur. J. Biochem. 36:53-67.

Michel, R., and W. Neupert. 1974. Nascent polypeptide chains on mitochondrial ribosomes and their integration into the inner mitochondrial membrane. in: The Biogenesis of Mitochondria (A. Kroon and C. Saccone, eds.). New York: Academic Press. pp. 315-326.

Millis, A. J. T., and Y. Suyama. 1972. Effects of chloramphenicol and cycloheximide on the biosynthesis of mitochondrial ribosomes in Tetrahymena. J. Biol. Chem. 247:4063-4073.

Minssen, M., and K. D. Munkres. 1973. Preparation of mitochondrial membrane proteins from Neurospora crassa: prevention of lipid autoxidation damage by an antioxidant. Biochim. Biophys. Acta 291: 398-410.

Mitra, R. S., B. Bartoov, J. Monahan, and K. B. Freeman. 1972. Comparison of mammalian mitochondrial ribosomal ribonucleic acid from different species. Biochem. J. 128:1033-1041.

Mockel, J. 1972. Amino acid incorporation into rat liver mitochondria. Biochim. Biophys. Acta 277:628-638.

Morgan, I. G. 1970. Protein synthesis in brain mitochondrial and synaptosomal preparations. FEBS Lett. 10:273.

Morimoto, H., and H. O. Halvorson. 1971. Characterization of mito-chondrial ribosomes from yeast. *Proc. Nat. Acad. Sci. US 68*:324-328.

Morimoto, H., A. H. Scragg, J. Nekhorocheff, V. Villa, and H. O. Hal-vorson. 1971. Comparison of the protein synthesizing systems from mitochondria and cytoplasm of yeast. in: *Autonomy and Bio-genesis of Mitochondria and Chloroplasts* (N. K. Boardman, A. W. Linnane, and R. M. Smillie, eds.). New York: American Elsevier. pp. 282-292.

Murray, D. R., and A. W. Linnane. 1972. Synthesis of proteolipid protein by yeast mitochondria. *Biochem. Biophys. Res. Commun. 49:* 855-862.

Neupert, W., and G. D. Ludwig. 1971. Sites of biosynthesis of outer and inner membrane proteins of *Neurospora crassa* mitochondria. *Eur. J. Biochem. 19*:523-532.

Neupert, W., D. Brdiczka, and T. Bucher. 1967. Incorporation of amino acids into the outer and inner membrane of isolated rat liver mitochondria. *Biochem. Biophys. Res. Commun. 27*:488-493.

Neupert, W., W. Sebald, A. J. Schwab, P. Massinger, and T. Bucher. 1969a. Incorporation in vivo of ^{14}C-labeled amino acids into the proteins of mitochondrial ribosomes from *Neurospora crassa* sensi-tive to cycloheximide and insensitive to chloramphenicol. *Eur. J. Biochem. 10*:589-591.

Neupert, W., W. Sebald, A. J. Schwab, A. Pfaller, and T. Bucher. 1969b. Puromycin sensitivity of ribosomal label after incorpora-tion of ^{14}C-labeled amino acids into isolated mitochondria from *Neurospora crassa*. *Eur. J. Biochem. 10*:585-588.

O'Brien, T. W. 1971. The general occurrence of 55S ribosomes in mammalian liver mitochondria. *J. Biol. Chem. 246*:3409-3417.

O'Brien, T. W. 1972a. Occurrence of active 80S ribosomes in subcel-lular particles in the mitochondrial fraction of fetal bovine li-ver. *J. Cell Biol. 53*:590-594.

O'Brien, T. W. 1972b. Occurrence of 55S miniribosomes in mitochon-dria of the shark. *J. Cell Biol. 55*:191a.

O'Brien, T. W., and G. F. Kalf. 1967a. Ribosomes from rat liver mitochondria. I. Isolation procedure and contamination studies. *J. Biol. Chem. 242*:2172-2179.

O'Brien, T. W., and G. F. Kalf. 1967b. Ribosomes from rat liver mitochondria. II. Partial characterization. *J. Biol. Chem. 242:* 2180-2185.

O'Brien, T. W., N. D. Denslow, and G. R. Martin. 1974. The struc-ture, composition and function of 55S mitochondrial ribosomes. in: *The Biogenesis of Mitochondria* (A. Kroon and C. Saccone, eds.). New York: Academic Press. pp. 347-356.

Ohnishi, T., and T. Ohnishi. 1962. Extraction of contractile protein from liver mitochondria. *J. Biochem. 51:*379-381.

Ojala, D., and G. Attardi. 1972. Expression of the mitochondrial genome in HeLa cells. X. Properties of mitochondrial polysomes. *J. Mol. Biol. 65:*273-289.

Pegg, A., and H. Williams-Ashland. 1968. Effects of androgens on incorporation of labeled amino acids into proteins by prostate mitochondria. *Endocrin. 82:*603-610.

Perlman, S., and S. Penman. 1970a. Mitochondrial protein synthesis: resistance to emetine and response to RNA synthesis inhibitors. *Biochem. Biophys. Res. Commun. 40:*941-948.

Perlman, S., and S. Penman. 1970b. Protein-synthesizing structures associated with mitochondria. *Nature 227:*133-137.

Pestka, S. 1971. Inhibitors of ribosome functions. *Ann. Rev. Microbiol. 25:*487-562.

Premkumar, E., and P. M. Bhargava. 1972. Transcription and translation in bovine spermatozoa. *Nature New Biol. 240:*139-143.

Pring, D. R. 1974. Maize mitochondria: purification and characterization of ribosomes and ribosomal ribonucleic acid. *Plant Physiol. 53:*677-683.

Pring, D., W. Clark, M. Critoph, and T. W. O'Brien. 1974. Unpublished observation.

Quetier, F., and F. Vedel. 1974. Identification of mitochondrial rRNA from plant cells. *FEBS Lett. 42:*305-308.

Rabbitts, T. H., and T. S. Work. 1971. The mitochondrial ribosome and ribosomal RNA of the chick. *FEBS Lett. 14:*214-218.

Reijnders, L., P. Sloof, J. Sival, and P. Borst. 1973. Gel electrophoresis of RNA under denaturing conditions. *Biochim. Biophys. Acta 324:*320-333.

Rendi, R. 1959. The effect of chloramphenicol on the incorporation of labelled amino acids into proteins by isolated subcellular fractions from rat liver. *Exp. Cell Res. 18:*187-189.

Richter, D., and F. Lipmann. 1970. Separation of mitochondrial and cytoplasmic peptide chain elongation factors from yeast. *Biochemistry 9:*5065-5070.

Richter, D., L. Lin, and J. W. Bodley. 1971. Studies on translocation. IX. The pattern of action of antibiotic translocation inhibitors in eukaryotic and prokaryotic systems. *Arch. Biochem. Biophys. 147:*186-191.

Richter, D., V. A. Erdmann, and M. Sprinzl. 1973. Specific recognition of GTψC loop (loop IV) of tRNA by 50S ribosomal subunits from *E. coli. Nature New Biol. 246:*132-135.

Rifkin, M., D. Wood, and D. Luck. 1967. Ribosomal RNA and ribosomes from mitochondria of *Neurospora crassa*. *Proc. Nat. Acad. Sci. US* *58*:1025-1032.

Robberson, D., Y. Aloni, and G. Attardi. 1971. Expression of the mitochondrial genome in HeLa cells. VL. Size determination of mitochondrial ribosomal RNA by electron microscopy. *J. Mol. Biol.* *60*:473-484.

Robberson, D., Y. Aloni, G. Attardi, and N. Davidson. 1972. Expression of the mitochondrial genome in HeLa cells. VIII. The relative position of ribosomal genes in mitochondrial DNA. *J. Mol. Biol.* *64*:313-317.

Roodyn, D. 1965. Further study of factors affecting amino acid incorporation into protein by isolated mitochondria. *Biochem. J.* *97*:782-793.

Roodyn, D. B. 1968. Mitochondrial biogenesis: germ-free mitochondria. *FEBS Lett.* *1*:203-205.

Roodyn, D., P. Reis, and T. Work. 1961. Protein synthesis in mitochondria. Requirements for the incorporation of radioactive amino acids into mitochondrial protein. *Biochem. J. 80*:9-21.

Roodyn, D., J. Suttie, and T. Work. 1962. Protein synthesis in mitochondria. Rate of incorporation in vitro of radioactive amino acids into soluble proteins in the mitochondrial fraction, including catalase, malic dehydrogenase and cytochrome. *Biochem. J.* *83*:29-39.

Roodyn, D., K. Freeman, and J. Tata. 1965. The stimulation by treatment in vivo with tri-iodothyronine of amino acid incorporation ing catalase, malic dehydrogenase and cytochrome C. *Biochem. J.* *83*:29-39.

Sacchi, A., U. Ferrini, and P. Cammarano. 1971. Isolation and characterization of mitochondrial ribosomes in rat liver. *Ital. J. Biochem. 20*:191-203.

Sacchi, A., F. Cerbone, P. Cammarano, and U. Ferrini. 1973. Physico-chemical characterization of ribosome-like (55S) particles from rat liver mitochondria. *Biochim. Biophys. Acta 308*:390-403.

Sala, F., and H. Kuntzel. 1970. Peptide chain initiation in homologous and heterologous systems from mitochondria and bacteria. *Eur. J. Biochem. 15*:280-286.

Sandell, S., H. Low, and A. von der Decken. 1967. A critical study of amino acid incorporation into protein by isolated liver mitochondria from adult rats. *Biochem. J. 104*:575-584.

Schatz, G., and T. L. Mason. 1974. The biosynthesis of mitochondrial proteins. *Ann. Rev. Biochem. 43*:51-87.

Schatz, G., and J. Saltzgaber. 1969. Identification of denatured mitochondrial ATPase in "structural protein" from beef heart mitochondria. *Biochim. Biophys. Acta 180*:186-189.

Schmitt, H. 1969. Characterization of mitochondrial ribosomes from *Saccharomyces cerevisiae*. *FEBS Lett.* *4:*234-238.

Schmitt, H. 1970. Characterization of a 72S mitochondrial ribosome from *Saccharomyces cerevisiae*. *Eur. J. Biochem.* *17:*278-283.

Schmitt, H. 1971. Core particles and proteins from mitochondrial ribosome of yeast. *FEBS Lett.* *15:*186-190.

Schmitt, H. 1972. Analysis and site of synthesis of ribosomal proteins from yeast mitochondria. *FEBS Lett.* *26:*215-220.

Schutgens, R. B. H., L. Reijnders, S. P. Hoekstra, and P. Borst. 1973. Transcription of *Tetrahymena* mitochondrial DNA in vivo. *Biochim. Biophys. Acta* *308:*372-380.

Scragg, A. H. 1971. Chain elongation factors of yeast mitochondria. *FEBS Lett.* *17:*111-114.

Sebald, W., T. Bucher, B. Olbrich, and F. Kaudewitz. 1968. Electrophoretic pattern of and amino acid incorporation in vitro into the insoluble mitochondrial protein of *Neurospora crassa* wild type and MI-1 mutants. *FEBS Lett.* *1:*235-240.

Senior, A. E., and D. H. MacLennan. 1970. Mitochondrial "structural protein"; a reassessment. *J. Biol. Chem.* *245:*5086-5095.

Shakespeare, P. G., and H. R. Mahler. 1971. Purification and some properties of cytochrome c oxidase from the yeast *Saccharomyces cerevisiae*. *J. Biol. Chem.* *246:*7649-7655.

Sherton, C. C., and I. G. Wool. 1972. Determination of the number of proteins in liver ribosomes and ribosomal subunits by two-dimensional polyacrylamide gel electrophoresis. *J. Biol. Chem.* *247:*4460-4467.

Simpson, M. V., D. M. Skinner, and J. M. Lucas. 1961. On the biosynthesis of cytochrome c. *J. Biol. Chem.* *236:*PC81-PC82.

Stegeman, W. J., C. S. Cooper, and C. J. Avers. 1970. Physical characterization of ribosomes from purified mitochondria of yeast. *Biochem. Biophys. Res. Commun.* *39:*69-76.

Stevens, B., J. Curgy, G. Ledoigt, and J. Andre. 1974. Analysis of mitoribosomes from *Tetrahymena* by polyacrylamide gel electrophoresis and electron microscopy. in: *The Biogenesis of Mitochondria* (A. Kroon and C. Saccone, eds.). New York: Academic Press. pp. 327-335.

Stoffler, G., I. G. Wool, A. Lin, and K. H. Rak. 1974. The identification of the eukaryotic ribosomal proteins homologous with *Escherichia coli* proteins L7 and L12. *Proc. Nat. Acad. Sci. US* *71:*4723-4726.

Suttie, J. W. 1962. The existence of two routes for incorporation of amino acids into protein of isolated rat liver mitochondria. *Biochem. J.* *84:*382-386.

Swanson, R. F. 1973. Specificity of mitochondrial and cytoplasmic ribosomes and elongation factors from *Xenopus laevis*. *Biochemistry 12*:2142-2146.

Swanson, R. F., and I. B. Dawid. 1970. The mitochondrial ribosome of *Xenopus laevis*. *Proc. Nat. Acad. Sci. US 66*:117-124.

Tait, A. 1972. Altered mitochondrial ribosomes in an erythromycin resistant mutant of paramecium. *FEBS Lett. 24*:117-120.

Towers, N. R., H. Dixon, G. M. Kellerman, and A. W. Linnane. 1972. Biogenesis of mitochondria. XXII. The sensitivity of rat liver mitochondria to antibiotics; a phylogenetic difference between a mammalian system and yeast. *Arch. Biochem. Biophys. 151*:361-369.

Truman, D., and A. Korner. 1962. Incorporation of amino acids into the protein of isolated mitochondria. *Biochem. J. 83*:588-596.

Tzagoloff, A., and A. Akai. 1972. Assembly of the mitochondrial membrane system. VIII. Properties of the products of mitochondrial protein synthesis in yeast. *J. Biol. Chem. 247*:6517-6523.

Tzagoloff, A., and P. Meagher. 1971. Assembly of the mitochondrial membrane system. V. Properties of a dispersed preparation of the rutamycin-sensitive adenosine triphosphatase of yeast mitochondria. *J. Biol. Chem. 246*:7328-7336.

Tzagoloff, A., M. Rubin, and M. F. Sierra. 1973. Biosynthesis of mitochondrial enzymes. *Biochim. Biophys. Acta 301*:71-104.

Vary, M. J., P. R. Stewart, and A. W. Linnane. 1970. Biogenesis of mitochondria. XVII. The role of mitochondrial and cytoplasmic ribosomal protein synthesis in the oxygen-induced formation of yeast mitochondrial enzymes. *Arch. Biochem. Biophys. 141*:430-439.

Vasconcelos, A. C. L., and L. Bogorad. 1971. Proteins of cytoplasmic, chloroplast, and mitochondrial ribosomes of some plants. *Biochim. Biophys. Acta 228*:492-502.

Verma, I. M., M. Edelman, M. Herzhog, and U. Z. Littauer. 1970. Size determination of mitochondrial ribosomal RNA from *Aspergillus nidulans* by electron microscopy. *J. Mol. Biol. 52*:137-140.

Vesco, C., and S. Penman. 1969. The cytoplasmic RNA of HeLa cells: new discrete species associated with mitochondria. *Proc. Nat. Acad. Sci. US 62*:218-225.

Vignais, P., P. Vignais, C. Rossi, and A. Lehninger. 1963. Restoration of ATP-induced contraction of pre-treated mitochondria by "contractile protein." *Biochem. Biophys. Res. Commun. 11*:307-312.

Vignais, P. V., J. Huet, and J. Andre. 1969. Isolation and characterization of ribosomes from yeast mitochondria. *FEBS Lett. 3*: 177-181.

Vignais, P. V., B. J. Stevens, J. Huet, and J. Andre. 1972. Mito-ribosomes from *Candida utilis*. Morphological, physical, and chemical characterization of the monomer form and of its subunits. *J. Cell Biol.* *54*:468-492.

von der Decken, A., H. Low, and S. Sandell. 1966. On the biological significance of the incorporation of ^{14}C-amino acids into protein in mitochondrial preparations. in: *Regulation of Metabolic Processes in Mitochondria,* Vol. 7 (J. Tager, S. Papa, E. Quagliariello, and E. Slater, eds.). New York: American Elsevier. pp. 415-425.

Wellauer, P. W., and I. B. Dawid. Unpublished data cited in: D. E. Leister and I. B. Dawid. 1974. *J. Biol. Chem. 249*:5108-5118.

Wengler, G., G. Wengler, and K. Scherrer. 1972. Ribonucleoprotein particles in HeLa cells: the contamination of the postmitochondrial-cytoplasmic fraction from HeLa cells with ribonucleoprotein particles of mitochondrial origin. *Eur. J. Biochem. 24*:477-484.

Werner, S. 1974a. Antibodies to subunits of cytochrome c oxidase and their relation to precursor proteins of this enzyme. in: *The Biogenesis of Mitochondria* (A. Kroon and C. Saccone, eds.). New York: Academic Press. pp. 505-509.

Werner, S. 1974b. Isolation and characterization of a mitochondrially synthesized precursor protein of cytochrome oxidase. *Eur. J. Biochem. 43*:39-48.

Werner, S., and W. Neupert. 1972. Functional and biogenetical heterogeneity of the inner membrane of rat liver mitochondria. *Eur. J. Biochem. 25*:379-396.

Wheeldon, L. 1966. The problem of bacterial contamination in studies of protein synthesis by isolated mitochondria. *Biochem. Biophys. Res. Commun. 24*:407-411.

Wheeldon, L. W. 1973. Products of mitochondrial protein synthesis. *Biochimie 55*:805-814.

Wheeldon, L., and A. Lehninger. 1966. Energy-linked synthesis and decay of membrane proteins in isolated rat liver mitochondria. *Biochemistry 5*:3533-3545.

Wheeldon, L. W., A. C. Dianoux, M. Bof, and P. V. Vignais. 1974. Stable and labile products of mitochondrial protein synthesis in vitro. *Eur. J. Biochem. 46*:189-199.

Wilson, R. H., J. B. Hanson, and H. H. Mollenhauer. 1968. Ribosome particles in corn mitochondria. *Plant Physiol. 43*:1874-1877.

Work, T. S. 1968. The biogenesis of mitochondria: evidence for a dual origin of mitochondrial proteins and its bearing on the theories regarding the function of mitochondrial DNA. in: *Biochemical Aspects of the Biogenesis of Mitochondria* (E. C. Slater, S. Papa, J. M. Tager, and E. Quagliariello, eds.). Bari: Adriatica Editrice. pp. 367-391.

Yang, S., and R. Criddle. 1969. Identification of a major membrane fraction as a product of synthesis by isolated yeast mitochondria. *Biochem. Biophys. Res. Commun.* *35*:429-436.

Yu, R., H. B. Lukins, and A. W. Linnane. 1968. Selective in vivo action of cycloheximide on the synthesis of soluble mitochondrial proteins. in: *Biochemical Aspects of the Biogenesis of Mitochondria* (E. C. Slater, J. M. Tager, S. Papa, and E. Quagliariello, eds.). Bari: Adriatica Editrice. pp. 359-366.

Yu, R., R. Poulson, and P. R. Stewart. 1972a. Comparative studies on mitochondrial development in yeasts. I. Mitochondrial ribosomes from *Candida parapsilosis*. *Molec. Gen. Genet.* *114*:325-338.

Yu, R., R. Poulson, and P. R. Stewart. 1972b. Comparative studies on mitochondrial development in yeasts. II. Mitochondrial ribosomes from *Saccharomyces cerevisiae*. *Mol. Gen. Genet.* *114*:339-349.

Chapter 6

THE REGULATION OF RIBOSOMAL RNA SYNTHESIS AND DEGRADATION IN BACTERIA

Jonathan Gallant

Department of Genetics
University of Washington
Seattle, Washington

Robert A. Lazzarini

Laboratory of Molecular Biology
NINCDS, National Institutes of Health
Bethesda, Maryland

I. INTRODUCTION

In bacterial cells the rate of protein synthesis varies with the
growth rate. On the other hand, the rate of amino acid incorporation
into protein per ribosome is constant over a wide range of growth
rates (Maaloe and Kjeldgaard, 1966). Thus bacterial cells adjust the
overall rate of protein synthesis by controlling the number of ribosomes per cell, which is to say the rate at which ribosomes are produced. Although the pathway of ribosome assembly has been the subject
of vigorous investigation (see Nomura, 1970, for review), relatively
little is known about the mechanisms that regulate the synthesis of
the various ribosome proteins, or their assembly together with ribosomal RNA into complete subunits. On the other hand, much has been
learned in the last few years about the regulation of ribosomal RNA
accumulation.

The studies that we will review here focus on the events that occur during an adjustment in the rate of ribosomal RNA accumulation.
A rapid change in the rate of ribosomal RNA accumulation occurs when
conditions of steady-state growth are disturbed in either of the following ways.

1. The supply of any species of aminoacyl-tRNA is reduced, either
   through starving an amino acid auxotroph, inactivating a temperature-sensitive aminoacyl-tRNA synthetase, or blocking the
   production of formylmethionyl-tRNA through selective inhibition of formyl transfer reactions. In each case, the accumulation of ribosomal RNA is quickly reduced in rate by a factor of 10-20 (see Fig. 1) (reviewed most recently by Ryan and
   Borek, 1971). We will refer to this control mechanism as
   amino acid control, since the trigger is a reduction in the
   supply of that form of the amino acid which is involved in
   protein synthesis, namely the aminoacyl -tRNA derivative.
   Amino acid control is often referred to as "stringent" control and the overall pattern of which it is a part as the
   "stringent" response. As we will see, this control mechanism
   appears to operate primarily on the synthesis of ribosomal
   RNA.

2. Bacteria are shifted from one growth medium to another. When
   the shift is to a medium where the overall growth rate is reduced (a "downshift") then the rate of ribosomal RNA accumulation is quickly reduced (see Fig. 1). Conversely, when the
   shift is to a medium supporting an increase in growth rate
   (an "upshift") then the rate of ribosomal RNA accumulation

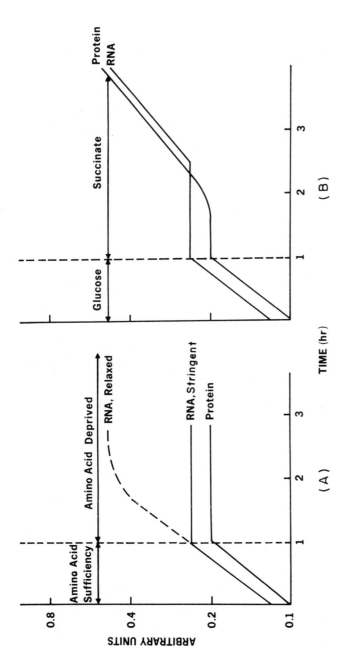

FIG. 1. A schematic representation of the response of bacterial cells to amino acid deprivation (left panel) and to metabolic downshift (right panel). "Stringent" refers to bacteria carrying the wild type allele of the *rel* gene; "relaxed" to any *rel⁻* mutant.

is increased. We will refer to this kind of response as
"growth-rate" control.

Growth-rate control of rRNA accumulation is a complex of at least
two different mechanisms. One apparently affects the synthesis of
rRNA by a process that overlaps partly with that of amino acid control.
A second mechanism affects the stability of newly made rRNA. The net
accumulation of rRNA is, of course, a reflection of the balance between rates of synthesis and rates of degradation. Control over rRNA
stability is most clearly demonstrated during an extreme form of downshift: shift to a medium where, for lack of a required purine or pyrimidine, the growth rate is zero. Under these conditions, the total
rate of RNA synthesis is reduced by a moderate amount, because of the
diminished supply of a ribonucleoside triphosphate. However the complete failure of RNA to accumulate is, as we will see, due principally
to the turnover of newly produced rRNA.

In practice, it is often difficult to disentangle these two modes
of control. Shifts from one medium to another undoubtedly affect the
generation of both amino acids and nucleotides and may thus involve
both amino acid and growth-rate control. Nevertheless, a judicious
choice of experimental manipulations permits at least a partial isolation of two kinds of triggering events and suggests that these two
control mechanisms, although often compounded, correspond to distinct
responses.

II. AMINO ACID CONTROL

The early work on amino acid control, some of which goes back more
than 20 years, has been carefully reviewed by Ryan and Borek (1971).
Before going on to present the current status of the field, as we see
it, we wish to warn the reader about certain complexities that have
received little attention.

The literature on amino acid control is rather rich in discrepancies and contradictory reports. One reason, we suspect, is that the
conditions that have been used for imposing amino acid starvation
vary in at least three different ways. The first and most obvious

source of variation is simply the severity of the nutritional block.
Amino acid starvation of a leaky auxotroph will probably elicit a less
severe form of the stringent response than starvation of a tightly
blocked mutant. Second, even with tight mutants, the physiological
disturbance produced by amino acid starvation can vary depending on
the amino acid involved and even on the position of the defective
mutant enzyme within its metabolic pathway. This is because some
amino acid pathways consume high-energy phosphate and some do not.
Removal of a required amino acid often derepresses that pathway and
abolishes feedback control by the end product. The wide-open pathway can produce a serious metabolic drain of high-energy phosphate
if the pathway does consume ATP at one point or another, and if the
mutationally defective step is located after that point.

Finally, the details of the stringent response are undoubtedly
subject to modifying influence by more than one polymorphic locus,
i.e., strain differences undoubtedly exist. Although the major genetic determinant of the stringent response is the *rel* locus, described
below, the existence of strain-specific modifiers is clear. Early
genetic crosses between *rel*⁺ and *rel*⁻ strains that were not isogenic
showed an extraordinarily wide range of stringency among the recom-
binants (Alföldi et al., 1963), which strongly suggests the segrega-
tion of polymorphic loci in addition to *rel*. Strain differences
among *rel*⁺ strains have been reported in connection with stringent
control of T4 messenger RNA synthesis (Olsnes and Hauge, 1968). Finally, one such modifying locus has recently been identified and
mapped (see Sect. II-G).

For all these reasons, clear characterization of the stringent
response demands the utilization of isogenic strains. We will con-
centrate on experiments with such pairs of strains in the following
discussion.

A. The *rel* Gene

Shortly after the discovery of amino acid control, Borek et al.
(1955) encountered an anomolous strain in which methionine starvation
failed to restrict RNA accumulation at all. Stent and his associates

later showed that this "relaxation" of amino acid control, which applied to starvation for any amino acid, was due to a spontaneous mutation in a new locus that they named RC for RNA control (Stent and
Brenner, 1961; Alföldi et al., 1962, 1963). The normal wild-type behavior was termed "stringent" control. Subsequently, the RC gene has
been renamed *rel* to fit the conventions of bacterial genetics; stringent bacteria bear the *rel$^+$* allele and relaxed mutants the *rel$^-$* allele.

Fiil and Friesen (1968) have isolated many more such relaxed mutants and have shown that they map in the *rel* locus. The existence
of these near isogenic *rel$^+$/rel$^-$* pairs has proven of inestimable value in further analysis of the *rel* function. Fiil (1969) showed that
rel$^-$ is recessive to *rel$^+$* in merodiploids, proving that the *rel$^+$* gene
product is a cytoplasmic component that acts, directly or indirectly,
to inhibit RNA accumulation.

B. Control of RNA Synthesis

In principle, a decrease in the rate of RNA accumulation could
mean either a decrease in the rate at which stable (ribosomal and
transfer) RNA molecules are made, or the onset of degradation of these
molecules, or both. Moreover, since the messenger RNA pool represents
only 3% of total RNA in *E. coli,* decreases in the rate of overall RNA
accumulation tell nothing about the response of messenger synthesis
and turnover.

It was originally suggested that the *rel* control system mediated
a general repression of the synthesis of all types of RNA during
amino acid starvation (Stent and Brenner, 1961), a notion that came
to be termed "coordinate control" (Friesen, 1966). Subsequently, experimental results that seemed to support this proposal were reported
(Gros et al., 1963; Friesen, 1966).

The proposal of simple coordinate control, however, failed to an-
ticipate that amino acid starvation actually derepresses the synthesis of some messenger RNA's, i.e., those coding for enzymes concerned
with the biosynthesis of the missing amino acid. This phenomenon was
first demonstrated through DNA:RNA hybridization studies of a *rel$^+$*
strain of *E. coli* (Imamoto et al., 1965). Subsequently, Venetianer

(1968, 1969) made similar measurements with a stringent *Salmonella* strain and acutely pointed out the discrepancy between this observation and the notion of coordinate control.

Lavalle and DeHauwer (1968) and Edlin et al. (1968) made use of DNA from φ80 transducing phage carrying the *trp* operon to study the labeling of specifically hybridizable *trp* mRNA during tryptophan starvation in isogenic *rel*⁺ and *rel*⁻ strains. Both groups found little effect of the *rel* function on the derepressed synthesis of this messenger. (One minor and still puzzling discrepancy was that Edlin et al. reported that their *rel*⁻ strain accumulated about one-half as much *trp* mRNA as its *rel*⁺ partner, whereas Lavalle and DeHauwer reported identical quantities of *trp* mRNA.)

One possible source of ambiguity in these studies is that tryptophan starvation elicits a rather weak stringent response, as judged by the rate of net RNA synthesis. Accordingly, Stubbs and Hall (1968) constructed isogenic *rel*⁺ and *rel*⁻ strains that were constitutive, i.e., genetically derepressed for the *trp* operon, and examined the labeling of *trp* mRNA during arginine starvation, which elicits a severe reduction in net RNA synthesis in *rel*⁺ cells. They also found little effect of the *rel* system, although, surprisingly, they found that arginine starvation greatly reduced the synthesis of *trp* mRNA in both *rel*⁺ and *rel*⁻ cells. This effect probably reflects a species of "metabolic" regulation of the *trp* operon (Stubbs and Stubbs, 1970; Rose and Yanofsky, 1972) governed by metabolites unrelated to tryptophan.

These three studies thus showed that stringent control did not apply to the synthesis of, at least, the messenger RNA species corresponding to the *trp* operon. More recently, Primakoff and Berg (1970) have used hybridization techniques to show that the synthesis of phage φ80 messenger RNA is also insensitive to stringent control. These results, taken together with those summarized below, have quite buried the notion of coordinate stringent control of the synthesis of all types of RNA. However, we would caution the reader against jumping to the diametrically opposite conclusion that the synthesis of all

messengers is entirely independent of the *rel* function. The true
situation, we believe, is rather more complicated.

Measurements of enzyme-forming capacity (Kepes, 1963; Hartwell
and Magasanik, 1963) have also been used to estimate messenger levels.
Morris and Kjeldgaard (1968) reported that the level of β-galactosidase messenger was independent of the *rel* function in cells starved
for both an amino acid and a carbon source. However, as we will see,
carbon-source starvation obliterates a crucial difference between the
behavior of *rel*⁺ and *rel*⁻ strains. Gallant et al. (1970) studied the
capacity for formation of ornithine transcarbamylase during arginine
starvation of two isogenic pairs; they reported a moderate (two- to
fourfold) reduction in the synthesis of this messenger in *rel*⁺ as
compared to *rel*⁻ bacteria.

These results all concur in showing that the synthesis of a variety of messengers responds to stringent control much less dramatically
(if indeed at all) than the accumulation of stable RNA. The results
on ornithine transcarbamylase, however, raise the possibility that
the *rel* function may have a moderate effect on the synthesis of at
least some messengers.

A similar conclusion emerges from quite a different kind of experiment. In this experiment the instantaneous rate of RNA synthesis
is estimated from the rates of incorporation of a labeled precursor
into RNA and the specific activity of the corresponding intracellular
ribonucleoside triphosphate (Nierlich, 1968; Winslow and Lazzarini,
1969). When the period of labeling is small compared to the rate of
decay of mRNA, the value obtained approximates the global rate of RNA
synthesis, i.e., the sum of the rates of synthesis of stable and unstable RNA. Measurements of this kind have shown that the global
rate of RNA synthesis is reduced to 30% of normal during threonine
starvation (Lazzarini and Dahlberg, 1971) and 30-45% of normal after
inactivation of a thermolabile valyl-tRNA synthetase (Nierlich, 1968).

Since the rate of RNA accumulation is reduced to 5-10% of normal
under these conditions, it is clear that the synthesis of unstable
RNA species is much less affected than is the accumulation of stable
RNA. However, the data do suggest a modest reduction in the rate of

synthesis of unstable RNA as well. The synthesis of stable RNA species comprises about 40% of the global rate of RNA synthesis (Kennell, 1968; Winslow and Lazzarini, 1969). If the synthesis of these species were abolished completely by stringent control, then the global rate should decrease to no lower than 60% of its normal value. Since the global rate falls to a value closer to 30%, it appears that there is a small, perhaps twofold, effect on the synthesis of unstable messengers.

This interpretation is strengthened by the observation that *Bacillus subtilis* cells, in which the formation of unstable RNA represents more than 80% of the global rate (Salser et al., 1968), also show a reduction to about 30% in global rate during amino acid starvation (Gallant and Margason, 1972). In this organism, moreover, there is one report suggesting an effect of this magnitude on the formation of a specific messenger. The flagellar protein of *B. subtilis* does not contain tryptophan, yet tryptophan starvation reduces the rate of flagellar protein synthesis by a factor of four (Dimmitt et al., 1968).

Taken all together, these results suggest that the average rate of messenger RNA synthesis is somewhat affected by stringent control. It is impossible to decide, of course, whether this effect applies to all messengers or only to some, producing a twofold effect on average. All the evidence, however, points to a preferential and much more severe reduction in the rate of stable RNA accumulation.

This effect of stringent control could, as we pointed out earlier, reflect either breakdown of normally stable RNA molecules or a preferential inhibition of their synthesis. In order to distinguish between these alternatives, Lazzarini and Dahlberg (1971) examined the labeling of ribosomal RNA, the major stable species, during short pulses. Labeled ribosomal RNA was identified by the fact that its hybridization to DNA would suffer a competitive disadvantage in the presence of an excess of unlabeled, purified ribosomal RNA. Approximately 30% of the RNA labeled during a short pulse in growing cells was ribosomal by this criterion, while the proportion was reduced to 10% in amino acid starved *rel*+ cells. Stamatos and Pettijohn (1971),

utilizing similar techniques, found values of 36 and 9%, respectively.
Donini (1972) reported a decrease from 36 to 19% under similar conditions; the smaller decrease he reports may be due to the fact that
mispaired hybrids between nonribosomal RNA and ribosomal DNA were not
eliminated by ribonuclease treatment, as was done in the other two
studies.

These results suggest that a decrease in the synthesis of ribosomal RNA molecules is largely responsible for the decreased rate of
rRNA accumulation during the stringent response. If the global rate
of RNA synthesis is reduced by a factor of three, and the proportion
of ribosomal RNA molecules under construction is further reduced by
a factor of three or four, then the rate of rRNA accumulation should
be reduced by a factor of nine to twelve. This is close to the observed ten- to twentyfold reduction in the rate of rRNA accumulation.
The possibility remains that degradation of rRNA may play an additional, minor role in the stringent response (Donini, 1972), but a degradative mode of control is more evident under other circumstances, as
we will see.

These observations appear to rule out degradation of completed
ribosomal RNA molecules as the major factor in amino acid control,
since the experiments employ periods of pulse labeling so short that
most of the ribosomal chains labeled are nascent; were degradation of
completed ribosomal chains the mechanism, then there would be no decrease in the proportion of nascent ribosomal chains labeled. On the
other hand, these experiments do not rule out the possibility that
rapid degradation of nascent ribosomal chains occurs simultaneously
with synthesis, since this situation would also eventuate in a decrease in the fraction of pulse-labeled nascent RNA that is ribosomal.
The experiments of Kossman et al. (1971), however, address this question, making use of the fact that the 16 and 23S ribosomal RNA chains
are evidently produced by the sequential transcription of the two adjacent genes for these RNA molecules, starting at the beginning of
the gene for 16S ribosomal RNA and "reading through" to the gene for
23S ribosomal RNA. If amino acid control operates on initiation of
this transcriptional unit, then a strong prediction can be made: The

relief of amino acid control, through restoration of the missing amino acid, should lead to a burst of *synchronous* initiation of the ribosomal RNA transcriptional unit, with the consequent production of a burst of 16S RNA followed by 23S RNA. This has been the observed result.

In summary, the present data indicate that the synthesis of ribosomal RNA is preferentially inhibited by the amino acid control mechanism, probably at the level of initiation. Amino acid control also appears to reduce the overall rate of messenger RNA synthesis by a modest and somewhat variable amount, although some messengers may be exempt from this effect. It is possible that this latter effect reflects an overall reduction in RNA chain growth rate observed by Winslow and Lazzarini (1969).

C. Formation of the MS Nucleotides

In the last several years, it has become apparent that the function of the *rel* gene product pertains to many other cellular processes besides the synthesis of ribosomal RNA, or even RNA synthesis in general. Comparison of *rel*$^+$ and *rel*$^-$ strains has revealed that the response of the *rel* gene product to amino acid starvation also affects the following unrelated aspects of cellular physiology: the replication of T4 DNA (Kellenberg et al., 1962; Donini, 1970); the uptake and anabolism of exogenous purines and pyrimidines (Edlin and Neuhard, 1967; Nierlich, 1968; Winslow and Lazzarini, 1969), phosphate (Cashel and Gallant, 1968; Irr and Gallant, 1969), and glucose (Sokawa and Kaziro, 1969); endogenous pathways of nucleotide (Gallant et al., 1971) and lipid (Sokawa et al., 1968, 1970; Golden and Powell, 1972; Merlie and Pizer, 1973; Polakis et al., 1973) biosynthesis; and the turnover of protein (Sussman and Gilvarg, 1969).

Several of these metabolic consequences of *rel* function cannot be dismissed as side effects of blocked RNA accumulation, inasmuch as they did not occur when RNA accumulation was blocked by such trivial means as the addition of RNA polymerase inhibitors or starvation for RNA precursors (Cashel and Gallant, 1968; Irr and Gallant, 1969; Lazzarini et al., 1969; Gallant and Harada, 1969). Thus, these experiments argued for a direct intervention of the *rel* function in metabolism and lent plausibility to the suggestion (Cashel and Gallant, 1968) that the actual function of the *rel* gene product was the generation of a metabolic inhibitor.

A search for such an inhibitor led to the discovery of two unusual nucleotides, originally designated MSI and MSII, which accumulated rapidly in response to amino acid starvation in *rel*$^+$ but not in *rel*$^-$ cells (Cashel and Gallant, 1969; Cashel, 1969). Figure 2 shows the

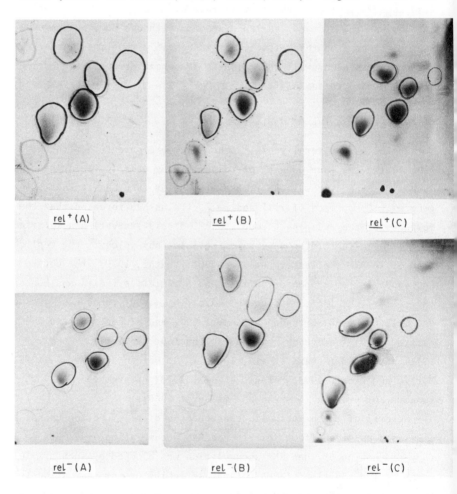

FIG. 2. The MS nucleotides under various conditions in *rel*$^+$ and *rel*$^-$ cells (see text).

autoradiograms of these nucleotides under various conditions in *rel*⁺
and *rel*⁻ cells. The nucleotide pools were labeled with ^{32}P, extracted, resolved by two-dimensional thin layer chromatography, and visualized by autoradiography, as described by Cashel and Gallant (1969).
The top row, labeled *rel*⁺ (A), (B), and (C), shows autoradiograms of
the nucleotide pools of strain CP78 (*rel*)⁺ growing under three conditions: exponential growth on glucose medium (A); shortly after the
imposition of isoleucine starvation (B); and shortly after downshift
from glucose to succinate (C). Note the accumulation of both pppGpp
and ppGpp in (B), and of ppGpp only in (C).

The bottom row, labeled *rel*⁻ (A), (B), and (C), shows autoradiograms of the nucleotide pools of an isogenic *rel*⁻ strain growing under
the same three conditions. Note the absence of either MS nucleotide
in (B), and the accumulation of ppGpp in (C). The nucleotides are
as follows:

UTP

GDP

CTP

ATP

GTP

ppGpp

pppGpp

ORIGIN⟶X

The positions of marker compounds are indicated, as above, by circles
drawn on the film. Radioactivity present in each nucleotide produces
film darkening in its spot. The (A) and (B) autoradiograms of each
set are from Cashel and Gallant (1969). The (C) autoradiograms were
made separately (from J. Gallant).

The accumulation of MSI proved to be an invariant manifestation
of amino acid control in various *rel*⁺ strains, although some failed
to accumulate MSII, while several different *rel*⁻ mutants failed to
accumulate either nucleotide (Cashel, 1969). Cashel and Kalbacher
(1970) have isolated MSI and shown that it is a tetraphosphate of
guanosine with an esterified 2' or 3' position. More recent studies

have shown that it is 5'ppG3'pp, conventionally referred to as ppGpp
(Sy and Lipmann, 1973; Que et al., 1973). MSII is a pentaphosphate
of guanosine (Cashel and Kalbacher, 1970) and probably pppGpp (Haseltine et al., 1972), although definitive evidence for this chemical
structure remains to be shown.

A large amount of circumstantial evidence links the accumulation
of ppGpp to the function of the *rel* gene product and the control of
RNA accumulation. In addition to Cashel's initial survey (1969) of
rel[+] and *rel*[−] strains, many other workers have confirmed this correlation in strains of *E. coli* (Lazzarini et al., 1971; Harshman and
Yamazaki, 1971, 1972; Khan and Yamazaki, 1972; Lund and Kjeldgaard,
1972; Fiil et al., 1972; Kaplan et al., 1973; DeBoer et al., 1971)
and other bacteria (Gallant and Margason, 1972; Swanton and Edlin,
1972). The accumulation of ppGpp, like the stringent response, occurs
when amino acid activation is blocked even in the presence of all required amino acids (Cashel, 1969; Fiil et al., 1972; Kaplan et al.,
1973), and there is a simple, monotonic relationship between the rate
of RNA accumulation and the level of ppGpp in cells where valine activation is restricted to varying extents (Fiil et al., 1972).

When the stringent response is reversed, either through restoration of the missing amino acid or the addition of certain antibiotics,
then the rate of RNA accumulation returns to its normal value in step
with the declining level of ppGpp (Cashel, 1969; Stamminger and Lazzarini, 1974). In certain strains, the genetics of which is discussed
in Sect. II-G, the level of ppGpp falls very slowly, and normal RNA
accumulation resumes correspondingly slowly (Stamminger and Lazzarini,
1974). These observations argue against the possibility that high
levels of ppGpp and restricted RNA accumulation are independent consequences of amino acid starvation, for their correlation persists
even in the presence of required amino acids.

The circumstantial link between high levels of ppGpp and restricted RNA accumulation does not, in itself, tell which is cause and
which is effect. However, when RNA accumulation is blocked by direct,
trivial means, ppGpp does not accumulate (Cashel and Gallant, 1969;
Gallant et al., 1970; Lazzarini et al., 1971). More significantly,

the accumulation of ppGpp during amino acid starvation commences
slightly earlier than the fall in rates of RNA accumulation (Cashel,
1969; Gallant et al., 1970). These observations eliminate the possibility that accumulation of ppGpp is a side effect of blocked RNA accumulation.

Indeed, when RNA accumulation is blocked by the RNA polymerase
inhibitor rifampicin, then the accumulation of the MS nucleotides
during amino acid starvation is itself prevented (Wong and Nazar,
1970). Kinetic (Erlich et al., 1971) and physiological (Lund and
Kjeldgaard, 1972) studies of this effect have suggested that it is
the disappearance of messenger RNA, rather than any direct effect of
rifampicin, that renders the cells unable to produce the MS nucleotides.

Virtually any treatment that inhibits residual protein synthesis
during amino acid starvation also prevents the accumulation of the
MS nucleotides and antagonizes the stringent response. In most cases,
this result is simply due to the fact that inhibition of residual protein synthesis allows the trickle of amino acids generated by protein
turnover to charge up the tRNA cognate to the missing amino acid (Kurland and Maaloe, 1962). In at least one case, however, this indirect
mechanism is not the explanation: The antibiotic tetracycline, which
binds to the A site of the ribosome (Zagorska et al., 1971), inhibits
MS nucleotide generation even when one species of tRNA is completely
deacylated (Kaplan et al., 1973).

Thus, even in rel^+ cells, the presence of deacylated tRNA is a
necessary but not sufficient condition for the generation of the MS
nucleotides and the occurrence of the stringent response: An intact
protein synthetic apparatus is also required. This latter condition
is not met when mRNA decays, leaving the ribosomes unprogrammed for
protein synthesis, or when tetracycline ties up the ribosomal A site.
The dual requirement for deacylated tRNA and an intact protein synthetic apparatus leads to the notion that the MS nucleotides are produced in an "idling reaction" of protein synthesis (Cashel and Gallant, 1969).

Inhibition of formyl transfer reactions also gives rise to the
stringent response (Shih et al., 1966) and accumulation of the MS nucleotides (Khan and Yamazaki, 1972; Lund and Kjeldgaard, 1972). Since
this condition blocks the formation of N-formylmethionyl-tRNA, Lund
and Kjeldgaard (1972) have suggested that the "idling reaction" pertains both to peptide chain initiation and to peptide chain growth,
thus implicating both the P and A sites of the ribosome. However,
Then and Angehrn (1972) and Smith and Midgley (1973a,b) have shown
that inhibition of formyl transfer gives rise to the stringent response only via starvation for glycine and methionine. It follows,
therefore, that the postulated idling reaction pertains to chain elongation and not to chain initiation.

This hypothesis predicts that the *rel* gene product is in fact a
ribosomal or ribosome-associated factor. Indirect support for this
hypothesis, through the demonstration of abnormal translation in *rel*[−]
cells, is discussed in Sect. II-E. First, we will describe recent
experiments on the in vitro synthesis of the MS nucleotides, which
have provided direct confirmation for the general terms of the hypothesis and have opened the way for study of the *rel* gene product at
the molecular level.

D. In Vitro Synthesis of the MS Nucleotides

Haseltine et al. (1972) first described a system for the in vitro
synthesis of the MS nucleotides. Radioactive GDP or GTP was employed
as a substrate, and ATP was a pyrophosphorylating agent; conversion
of labeled substrate to ppGpp and pppGpp occurred in the presence of
NH_4Cl-washed ribosomes and the NH_4Cl wash fraction. Washed ribosomes
alone were devoid of activity, indicating that a factor present in
the high-salt wash was essential. The high-salt wash from *rel*[+] ribosomes catalyzed the reaction in the presence of washed ribosomes from
either *rel*[+] or *rel*[−] cells, but no activity was found in the high-salt
wash of *rel*[−] ribosomes. Thus, the genetic specificity of the *rel*[−]
mutation pertains to a ribosome-associated factor that can be extracted in high concentrations of NH_4Cl. Haseltine et al. (1972) have
named this material the "stringent factor."

Recently, Block and Haseltine have detected residual stringent-factor activity in the high-salt wash of ribosomes from a number of independent *rel*⁻ mutants. In most cases, this activity was abnormally thermolabile, and each such *rel*⁻ stringent factor showed a characteristic degree of thermolability (Block and Haseltine, 1973). Thus, different mutant alleles of the *rel* locus lead to the production of a set of stringent-factor proteins with distinctively altered physical properties. This observation constitutes strong evidence that the stringent factor is indeed the product of the *rel* gene.

Purification of the stringent factor and further examination of its properties is proceeding in several different laboratories (Sy et al., 1973; Cochran and Byrne, 1974; Cashel, personal communication; Block, personal communication). The factor sediments through sucrose gradients with a velocity indicative of a molecular weight near 80,000 (Haseltine and Cashel, personal communication). Stringent factor shows a low level of activity even in the absence of ribosomes, an activity that can be stimulated by the addition of methanol (Sy et al., 1973; Cashel and Block, personal communications). The reaction catalyzed by the stringent factor is the transfer of the beta-gamma pyrophosphate from ATP to GDP, to produce ppGpp, and to GTP, to produce pppGpp (Haseltine et al., 1972; Sy and Lipmann, 1973). The only known counterpart of this unusual pyrophosphate transfer reaction is in the production of 5'-phosphoribosylpyrophosphate from ribose-5'-phosphate and ATP.[*]

The ribosome-stringent-factor complex is very much more active than isolated stringent factor, but only under special conditions. Groups at Harvard and the University of Aarhus have shown that bare

---

[*]An ATP: purine nucleotide pyrophosphotransferase from several species of *Streptomyces* has been identified and characterized (Nishino and Murao, 1975; Hamagishi et al., 1975; Oki et al., in press). The enzyme carries out a reaction identical to that of the stringent factor, with adenosine or guanosine mono-, di-, or triphosphates as acceptor. From a *coli*centric point of view, it would appear that *E. coli* has developed the biochemical trick of linking this possibly widespread enzymatic activity to a ribosomal function. (We are indebted to Dr. Akihiro Yoshimoto for bringing the *Streptomyces* work to our attention.)

70S ribosomes (when sufficiently purified) are insufficient to trigger stringent activity. In order to obtain maximum rates of MS nucleotide synthesis, the ribosomes must be programmed with a messenger
and must interact with uncharged tRNA cognate to triplets in the messenger (Haseltine and Block, 1973; Pederson et al., 1973). Thus, for
example, when the ribosomes are programmed with poly(U), the addition
of uncharged tRNA^{Phe} greatly stimulates the reaction, whereas the addition of uncharged tRNA species that do not recognize the UUU trip-
let has no effect. These observations agree very neatly with the
conditions necessary for MS nucleotide biosynthesis in rel^+ cells
(Sect. II-C, discussed previously). In retrospect, it is clear that
the discovery of in vitro stringent-factor activity by Haseltine et
al. depended on the fact that messenger RNA and tRNA are fortunately
present as "contaminants" in ribosome preparations.

It has long been known that uncharged tRNA binds to ribosomes
(Cannon et al., 1963). This binding, which occurs at high magnesium
concentrations and does not require elongation factors (Levin and
Nirenberg, 1968; Levin, 1970) is often referred to as "nonenzymatic
binding." Moreover, it is codon specific and subject to inhibition
by tetracycline. Stringent-factor activity on the ribosomes shows
precisely this constellation of properties (Haseltine and Block, 1973;
Pederson et al., 1973). It therefore appears that the stringent-factor reaction explains the biological significance of nonenzymatic
binding.

The normal codon-anticodon interaction with charged tRNA, which
is also inhibited by tetracycline, occurs in the acceptor site of the
ribosome. It might be anticipated, therefore, that the codon-anticodon interaction with uncharged tRNA that triggers stringent-factor
activity also occurs in the acceptor site. This has been elegantly
demonstrated by Haseltine and Block (1973) through experiments with
phage R17 messenger (see Table 1). First, an initiation complex was
formed between R17 RNA, 30S and 50S ribosome subunits, and formyl-
methionyl-tRNA^{fMet} in the presence of purified initiation factors.
Under these conditions, the ribosomes position themselves at the
starts of the messages for R17 coat protein and synthetase (Steitz,

TABLE 1   Synthesis of ppGpp in vitro[a]

Components	Reaction mixtures[b]				
R17 RNA	+	+	−	+	+
fMet-tRNA	+	+	+	−	+
Uncharged tRNA^{Ala}	+	−	+	+	−
Uncharged tRNA^{Ser}	−	+	−	−	−
Uncharged tRNA^{Val}	−	−	−	−	+
G_{4p} + G_{5p} (nmol) formed	3.2	2.1	<0.1	<0.1	<0.1

[a]The composition of reaction mixtures and their production of ppGpp and pppGpp. From Haseltine and Block (1973).
[b]Plus sign indicates presence, minus sign indicates absence.

1969) with most of the formylmethionyl-tRNA^{fMet} bound in the donor sites, leaving the acceptor sites programmed by the codons for the second amino acid in each of the two proteins, which are alanine and serine, respectively. In the presence of stringent factor, the addition of uncharged tRNA^{Ala} and tRNA^{Ser} specifically stimulated the production of the MS nucleotides. Moreover, when translocation was accomplished through the addition of charged tRNA^{Ala} and tRNA^{Ser} and elongation factors, thus bringing the triplets for the third amino acid in each protein into the acceptor site, then uncharged tRNA^{Ala} and tRNA^{Ser} no longer stimulated MS nucleotide synthesis. [More complete discussions of the in vivo and in vitro synthesis of the MS nucleotides can be found in Cashel and Gallant (1974) and Block and Haseltine (1974).]

E. Aberrant Translation in rel⁻ Cells

Besides triggering the synthesis of the MS nucleotides, the codon-specific binding of uncharged tRNA may well play an important role in maintaining the fidelity of translation. A degree of ambiguity is inherent in the decoding machinery, as shown, for example, in the poly(U)-promoted binding of such illegitimate species as leucyl-tRNA^{Leu} and isoleucyl-tRNA^{Iso}. All else being equal, a cell in which the level of charged tRNA^{Phe} was reduced through phenylalanine star-

vation would necessarily suffer a relative increase in the erroneous
incorporation of leucine and isoleucine in place of phenylalanine.
But the codon-specific binding of uncharged tRNA at the acceptor site
undoubtedly competes with the binding of illegitimate charged species
and may thus act to reduce the frequency of error.

If the *rel* gene product were itself involved in this binding of
uncharged tRNA, then one might expect *rel⁻* mutants to suffer an increase in translational errors during amino acid starvation. This
appears to be precisely the case. The phenomenon was first uncovered
through studies of enzyme synthesis in amino-acid-starved cells. Under these conditions, protein synthesis continues at a small residual
rate, due to the release of small quantities of amino acids through
protein turnover, which is about the same in *rel⁺* and *rel⁻* cells
(Goodman et al., 1969; Sells and Ennis, 1970; Gallant et al., 1970).
Yet the presence of the *rel⁻* allele greatly impairs the synthesis of
a number of different active enzymes: tryptophan synthetase (Lavalle
and DeHauwer, 1968); β-galactosidase (Fiil, 1969); alkaline phosphatase (Foster, 1968; Hall and Gallant, 1972); and ornithine transcarbamylase, acid phosphatase, and d-serine deaminase (Gallant et al.,
1970).

This *rel⁻* defect in active enzyme production is independent of
repression, inasmuch as it still occurs in a *rel⁻* strain in which
alkaline phosphatase synthesis is genetically derepressed through the
presence of a constitutive mutation (Hall and Gallant, 1971). Moreover, several different studies argue that the defect pertains to
translation rather than transcription. Lavalle and DeHauwer (1968)
found normal quantities of hybridizable *trp* messenger RNA in trypto-
phan-starved *rel⁻* cells that failed to produce active tryptophan synthetase. Although Edlin et al. (1968) disputed this conclusion, es-
timations of messenger level, by means of enzyme-forming capacity,
have confirmed it in the cases of β-galactosidase (Morris and Kjeldgaard, 1968), alkaline phosphatase (Foster, 1968), and ornithine
transcarbamylase (Gallant et al., 1970). The presence of normal levels of enzyme-forming capacity (assayed upon restoration of the missing amino acids) indicates that *rel⁻* cells elaborate potentially

functional mRNA for the enzymes in question, but fail to translate it, or translate it incorrectly, so long as a required amino acid is withheld.

Novick and Smith (unpublished observations) first obtained suggestive evidence for the interpretation that erroneous translation was the cause of this phenomenon: They found that the β-galactosidase formed in amino-acid-starved rel⁻ cells was abnormally thermobile. This observation was confirmed by Hall and Gallant (1972), who went on to show that the β-galactosidase monomer produced by amino-acid-starved rel⁻ cells also suffered a diminished specific activity.

Recently, further evidence for this interpretation has been obtained. Goldberg (1972) has shown that aberrations in protein primary structure, brought about through the incorporation of amino acid analogues or the occurrence of translational errors in a ram mutant (Rosset and Gorini, 1969), can be detected through an increase in the sensitivity to trypsin proteolysis of crude protein extracts. The same increased trypsin sensitivity characterizes protein formed by rel⁻ cells during amino acid starvation (Gallant and Harada, unpublished experiments).

There is thus substantial evidence that the presence of a mutant stringent factor increases the frequency of errors in translation. The notion that this phenomenon is due to a decreased codon-specific binding of uncharged tRNA, as we suggested before, is an attractive one, but it has yet to be tested in vitro.

Alföldi et al. (1963) found that rel⁻ mutants suffered a severe growth inhibition when downshifted from broth medium to minimal medium. Needless to say, such a downshift entails amino acid starvation during the period of adaptation to minimal medium, and erroneous translation would hinder this adaptation. Moreover, the notion of rel⁻ ribosomal ambiguity explains a curious pattern of response to individual amino acids, otherwise quite inexplicable, which rel⁻ cells manifest during such a downshift. The growth inhibition is accentuated by leucine and, in some rel⁻ strains, by methionine or phenylalanine as well. Moreover, this inhibition is reversed by the addition of isoleucine and valine (Alföldi and Kerekes, 1964). The metabolic

pathways for the production of these five amino acids are sufficiently distinct to render an explanation based on feedback interrelation-
ships advanced by Alfoldi and Kerekes quite unconvincing.

However, the codons for each of these five amino acids are related to one another by transformations of the first base. Misreading
of the first base in a codon appears to be the basis for the most
frequent types of ambiguity (Szer and Ochoa, 1964; Jones and Nirenberg, 1962; Bretscher and Grunberg-Manago, 1962). We therefore suggest that first-base misreading in *rel⁻* cells results in the incorporation of leucine, methionine, or phenylalanine in place of other
amino acids that are particularly limiting during a downshift from
broth to minimal medium. Reference to the genetic code shows that
there are only a few amino acids that leucine, methionine, and phenylalanine might replace through first-base misreading. Beside the
three inhibitory amino acids themselves, the only others that can be
confounded with them are isoleucine and valine. The striking fact
that isoleucine and valine reverse the inhibitory effects of the other three amino acids fits this interpretation very neatly.

F.   Effects of the MS Nucleotides

Enzymatic studies have shown that ppGpp inhibits a number of enzymes that carry out reactions blocked during the stringent response.
These include IMP dehydrogenase, the first enzyme in the guanylate
pathway, and adenylosuccinate synthetase, the first enzyme in the
adenylate pathway (Gallant et al., 1971); two membrane-bound phos-
phorylases that carry out the uptake and phosphorylation of exogenous
purines and pyrimidines (Hochstadt-Ozer and Cashel, 1972); and sn-
glycerol-3-phosphate-acyltransferase, sn-glycerol-3-phosphate-CMP-
phosphatidyltransferase (Merlie and Pizer, 1973), and acetyl-CoA carboxylase (Polakis et al., 1973), which carry out early steps in lipid
biosynthesis.

Since ppGpp is a structural analog of GDP and GTP, it might be
anticipated that it would inhibit those reactions in protein synthesis that utilize the normal guanine nucleotides as cofactors. This
appears to be the case. The formation of the initiation complex be-

tween formylmethionyl-tRNA$_f$, 70S ribosomes, and messenger RNA depends
on the interaction between IF2 and GTP (Mazumder et al., 1969; Rudland et al., 1971); and the formation of this complex is inhibited
by ppGpp (Yoshida et al., 1972). Elongation factor T_u binds ppGpp
tightly (Blumenthal et al., 1972), and the result appears to be an
inhibition of the exchange of free and T_u-bound GDP (Arai et al.,
1972). The latter authors have also reported that ppGpp inhibits the
GTPase activity of EF-G.

The physiological significance of these effects on protein synthesis in vivo is by no means clear, since ppGpp accumulates under
conditions where peptide chain elongation is restricted, in any case,
by the unavailability of one or another species of charged tRNA. It
seems likely, however, that the way in which protein synthesis responds to amino acid starvation will depend on the intracellular concentration of ppGpp and thus on the function of the rel gene product.

The nucleotide also interacts with elements of the transcription
machinery. Cashel (1970) first reported on the effect of ppGpp on
DNA-dependent RNA synthesis: Inhibition was observed under a number
of conditions and with several different DNA templates. With native
E. coli DNA, the extent of this inhibition was about 50%, in satisfactory agreement with our previous inference that messenger RNA synthesis is inhibited by about this much on the average during the
stringent response. An important control observation was that ppppG,
an isomer of ppGpp that does not occur naturally, produced no such
inhibition.

Zubay and his associates and Pastan and his associates have examined the effects of ppGpp on the transcription of a number of specific genes, using DNA from various special transducing phages in a
coupled transcription-translation system (Zubay et al., 1971a,b;
Yang et al., 1973; de Crombrugghe et al., 1971a,b; Parks et al.,
1972; Aboud and Pastan, 1973). Here the effect of ppGpp varies be-
wilderingly from one gene to another (Table 2): The nucleotide evidently inhibits transcription of the argE gene, and perhaps also the
gal operon (Parks et al., 1972) stimulates transcription of the lac,
ara, and trp operons; and has little or no effect on the formation

TABLE 2   Relative Effect of ppGpp on in vitro Enzyme
           and tRNA Syntheses[a]

	100 μM ppGpp	300 μM ppGpp
β-Galactosidase (tryp)[b,c]	2.4	2.2
β-Galactosidase[c]	2.2	1.6
L-Ribulokinase[c]	2.2	1.7
Acetylornithinase[c]	0.25	0.2
tRNA^{Tyr}[c]	1.1	0.75
Galactokinase[d]	0.43	

[a]Each entry shows the rate of gene expression in vitro relative
to incubation mixtures without any ppGpp.
[b]β-Galactosidase synthesized by a lac gene inserted into the
tryp operon and regulated by tryptophan.
[c]Data from Yang et al. (1974).
[d]Data from Parks et al. (1972).

of suIII-tRNA^{Tyr}. It is clear from these data that ppGpp interacts
with the transcription machinery, probably affecting initiation specificity. A disturbing feature, however, is that there is little
correlation between these in vitro effects and the situation in whole
cells. For example, ppGpp has little effect on the synthesis of
suIII-tRNA^{Tyr}, even though the accumulation of this species of tRNA
in whole cells is clearly subject to stringent control (Primakoff
and Berg, 1970).

The situation is equally murky in regard to ribosomal RNA transcription in vitro. Travers et al. (1970a) have reported that subunits III and IV of phage Qβ replicase, proteins that are determined
by the host E. coli genome (Kamen, 1970; Kondo et al., 1970), confer
on RNA polymerase holoenzyme the ability to synthesize nearly normal
quantities (about 25%) of rRNA in vitro. The addition of ppGpp at
biological concentrations reduced overall RNA synthesis by about 50%
and specifically reduced rRNA synthesis to undetectable levels (Travers et al., 1970b; Travers, 1971). The authors suggest that these
two host-determined proteins constitute a factor, which they term
psi$_r$, whose normal function is to provide RNA polymerase holoenzyme
with recognition specificity for the ribosomal promoter(s), and that

this interaction between holoenzyme and psi_r is specifically inhibited by ppGpp. They go on to suggest that psi_r belongs to a class of activation factors that govern the ability of holoenzyme to transcribe various classes of transcription units, and whose ability to do so is modulated by the level of small molecule effectors, such as regulatory nucleotides. The theory gains greatly in plausibility from the fact that a different factor of precisely this sort has already been well characterized. This is the cAMP receptor protein, which enables holoenzyme to recognize the promoters of the lac operon and several others, and whose activity is dependent on the presence of cAMP (deCrombrugghe et al., 1971b).

The theory is simple, attractive, and has already met with serious difficulties. First, quite appreciable quantities of rRNA can be formed by RNA polymerase holoenzyme in the absence of psi_r (Pettijohn, 1972; Haseltine, 1972; Birnbaum and Kaplan, 1973). Second, Haseltine (1972) was unable to confirm either the stimulation of rRNA transcription by psi_r or the preferential inhibition of rRNA transcription by psi_r. Third, Murooka and Lazzarini (1973) have described a crude DNA:protein complex that produced rRNA in vitro in nearly biological quantities (20-30% of total RNA). As in the case of Haseltine's study of holoenzyme activity, this complex could be inhibited about 50% of ppGpp but there was no preferential restriction of rRNA synthesis (Murooka and Lazzarini, 1973).

In further studies, Travers and his colleagues have reported that a conformational change in DNA, activated by relatively high KCl concentrations and temperatures above 35°C, "opens" the ribosomal promoters for transcription by holoenzyme alone (Travers et al., 1973). They contend that this open conformation is the basis of the reports described above. Continuing this line of investigation, Travers (1973) has reported that it is only when the ribosomal promoters are in the "closed" confirmation that their transcription is markedly stimulated by psi_r and that this stimulation is antagonized by ppGpp.

Travers also reported that psi_r stimulated the synthesis of suIII-tRNA, and that this stimulation was likewise antagonized by ppGpp, in contradiction to the results of Yang et al. (1974). More recently,

Beckman and Daniel (1974) confirmed a thermal activation of suIII-tRNA transcription, but could not confirm any stimulatory effect of psi_r.

It is difficult to assess the pertinance of these studies to the biological situation within whole cells. For one thing, whole cells control the synthesis of rRNA above 35°C, and therefore the proposed conformational change in ribosomal DNA must be something of an in vitro artifact. For another, none of the in vitro studies cited above attain proportions of rRNA synthesis as high as that observed in vivo. In Travers' most recent report, for example, the psi_r-stimulated rRNA synthesis from the "closed" promoter, in the absence of ppGpp, is only 15% (Travers, 1973, Fig. 2).

Small values like these also make for ambiguous data as a matter of simple statistics. Hybridization-competition assay for rRNA synthesis depends on the difference between radioactivity hybridized to DNA in the absence and in the presence of competition by an excess of unlabeled authentic rRNA. The variance of such a difference measurement is necessarily larger than the experimental variance of each of the two hybridization measurements from which the difference is estimated; and the smaller the difference is, the larger the unreliability of the estimate. In more recent studies, this problem has been alleviated through the use of internal standards in each hybridization reaction.

In order to avoid some of these ambiguities, Lazzarini and Johnson (1973) examined a kind of halfway house between the cell and the test tube: Whole cells that, through a particular maltreatment, had been made permeable to triphosphates and to ppGpp. These cells took up triphosphates in a conventional RNA polymerase reaction mixture, and a normal proportion of the product (30-40%) was shown to be ribosomal RNA. At biological concentrations, ppGpp inhibited overall RNA synthesis by the familiar 50%, but failed to produce any preferential reduction in the proportion of rRNA. The authors conclude that, contrary to Travers' proposal, ppGpp does not directly or by itself inhibit any element of the enzymic machinery catalyzing the

synthesis of rRNA. Precisely similar results have been reported by
Atherly (1974) with cells permeabilized in a slightly different way.

These reports suggest, at the very least, that restriction of
rRNA synthesis involves another entity in addition to ppGpp, and one
which is inactivated or lost in permeabilized cells. Whatever the
involvement of open and closed promoters in vitro, it seems to us
that the question of how rRNA synthesis is controlled in live cells
remains open.*

Although the effect of psi_r on rRNA synthesis in vitro is at issue, the existence of these two proteins is not. They are relatively
well characterized components of Qβ replicase and, as they are host
determined, it is reasonable to suppose that they carry out some
function(s) in uninfected cells. Blumenthal et al. (1972) have added
a fascinating new element to the mystery by demonstrating what one of
these functions is: They have shown that the psi_r proteins, subunits
III and IV of Qβ replicase, are almost certainly identical to, or
modifications of, the two protein-synthesis elongation factors T_u and
T_s. The demonstration, which is wholly convincing, rests on immunological cross reaction, the presence of the same seven N-terminal
amino acids in T_s and Qβ replicase III, and the reconstitution of
replicase activity from isolated T_u, T_s, and the other two QB replicase subunits.

Since T_u binds ppGpp (Arai et al., 1972; Blumenthal et al., 1972),
these results are consistent, from an entirely unexpected angle, with

---

*Despite this caveat, we should note that further evidence for
repression of rRNA synthesis by ppGpp in vitro has accumulated. Van
Ooyen et al. (1975) reported that ppGpp preferentially inhibited rRNA
synthesis by purified RNA polymerase, although the proportion of rRNA
synthesis detected in this system was unbiologically low. More recently, Block (in press) and Reiness et al. (1975) have developed
crude cell-free systems in which proportions of rRNA synthesis as
high as 25% could be demonstrated. In both studies, biological concentrations of ppGpp produced a marked preferential inhibition of
rRNA transcription. Block further reported a partial separation from
RNA polymerase of a factor which stimulated the ppGpp sensitive synthesis of rRNA. It remains to be seen what relationship, if any,
Block's factor bears to Travers' psi_r.

the findings reported by Travers et al. in regard to rRNA synthesis
in vitro. The two elongation factors evidently possess double and
perhaps even triple identities: Their aliases are T_u and T_s, QB replicase subunits III and IV, and perhaps, subject to further verifi-
cation, psi_r. The question raised by this bizarre situation is:
What functional or regulatory characteristics do these three activities share in common? Interaction with guanine nucleotides, including ppGpp, is the only obvious common feature.

G. Another Genetic Locus Involved in the
Stringent Response

Most stringent strains of E. coli accumulate both of the MS nucleotides during amino acid starvation (Cashel, 1969), as do other
bacterial species (Gallant and Margason, 1972; Swanton and Edlin,
1972). E. coli strains of the 58-161 lineage, however, show an abnormal behavior termed the "spotless" phenotype: MSII (which is probably pppGpp) fails to accumulate detectably, even though the stringent factor is apparently active, inasmuch as large quantities of
ppGpp do accumulate. Indeed, these strains accumulate much more
ppGpp than do normal rel^+ strains.

Genetic analysis has shown that this peculiarity is due to the
presence of a recessive allele of a locus mapping close to 72 min
(Laffler and Gallant, 1974a). This locus, which has been designated
spoT, is distinct from rel and evidently affects the levels of the MS
nucleotides in a different manner. The presence of the "spotless"
or $spoT^-$ allele has the following consequences: (a) the accumulation
of pppGpp during the stringent response is depressed to undetectable
levels; (b) the accumulation of ppGpp is increased by a factor of two
to three. This enhanced accumulation of ppGpp is apparently due to
a reduction in the rate at which ppGpp is further metabolized: When
ppGpp synthesis is shut off through the restoration of a missing amino acid or through the addition of tetracycline, then the disappearance of ppGpp occurs 10-30 times more slowly than normal in $spoT^-$
strains (Laffler and Gallant, 1974a; Stamminger and Lazzarini, 1974).

The fact that the *spo*T mutation decreases both the consumption
of ppGpp and the production of pppGpp strongly suggests that the
*spo*T gene product is involved in the conversion of ppGpp to pppGpp.
Attempts to detect such a reaction in vitro have so far proven fruit-
less, perhaps because of the occurrence of competing dephosphorylation reactions. Nonetheless, the existence of the *spo*T⁻ phenotype
shows that the in vitro system of Haseltine et al. (1972) provides
an incomplete picture of the metabolism of the MS nucleotides. Even
though *spo*T⁻ cells fail to accumulate pppGpp during amino acid starvation, ribosomes derived from these cells do generate pppGpp in vitro (Laffler and Gallant, unpublished experiments).

Stamminger and Lazzarini (1974) have discovered an interesting
additional consequence of the *spo*T⁻ mutation: The consumption or
further metabolism of ppGpp becomes sensitive to high levels (greater than 100 μg/ml) of tetracycline. This suggests that the *spo*T gene
product, like the *rel* gene product, may function on the ribosome.

In Sect. II-B, we described circumstantial evidence that a high
level of ppGpp is responsible for the inhibition of RNA accumulation
during the stringent response. Nonetheless, it remains conceivable
that some other consequence of amino acid starvation affects both RNA
accumulation and the generation of the MS nucleotides independently.
Studies of *spo*T⁻ cells have made it possible to rule out this alternative. When a missing amino acid is added back to *spo*T⁻ cells, high
levels of ppGpp persist for many minutes even in the presence of all
required amino acids, because of the extraordinarily slow disappearance of ppGpp: In this situation, a normal rate of RNA accumulation
does not resume until the level of ppGpp has returned to near its
basal value (Stamminger and Lazzarini, 1974). A similar but somewhat shorter lag is observed before protein synthesis resumes, suggesting that protein synthesis, even in the presence of all required
amino acids, is also sensitive to ppGpp level (Laffler and Gallant,
1974b). This is, of course, consistent with evidence that ppGpp inhibits protein initiation and chain elongation in vitro (Sect. II-E,
discussed previously).

The existence of a genetically defined system for the degradation
or further metabolism of ppGpp immediately suggests the possibility
of controlling the nucleotide's level through adjustment of this rate,
independently of the rate of ppGpp synthesis by the *rel* product. Regulation at this level would be independent of stringent-factor activity and therefore occur in both *rel*[+] and *rel*[−] strains. Evidence
that energy-source metabolism affects the level of ppGpp in precisely
this manner will be discussed further (Sect. III-D).

III. GROWTH-RATE CONTROL

A. Adjustment of Cellular Composition

It might be suspected that bacteria which are impaired in their
ability to reduce ribosomal RNA synthesis in response to amino acid
deprivation might also be unable to adjust and maintain their RNA
content during balanced growth. However, this does not appear to be
the case. A comparison of several relaxed and stringent strains of
E. coli grown on a variety of media did not reveal any pronounced
difference in RNA content (Neidhardt, 1962, 1963). More recently,
this question has been reinvestigated with the nearly isogenic pair
of relaxed and stringent strains, *E. coli* NF161 and NF162 (Gallant
et al., 1970). After balanced growth on glucose minimal medium, these
strains differ by 15-25% in their RNA content, but also differ by more
than a factor of two in their ppGpp content. On the other hand, this
isogenic pair of strains have nearly identical ppGpp concentrations
under other growth conditions, and this small difference in RNA content is not observed (Lazzarini et al., 1971).

Not only do relaxed and stringent cells have approximately the
same gross composition (RNA/protein/DNA), but they also adjust these
ratios with equal facility in response to most changes in the culture
environment. The exception to this, the inability to efficiently negotiate a change from rich complex media to glucose minimal medium
by the relaxed strains might be anticipated, since this transition
entails an amino acid deprivation until the cells derepress their

amino acid biosynthetic pathways. Taken through such a shiftdown in culture media, the relaxed strains behave much as they do during amino acid deprivation and overaccumulate ribosomal RNA (Neidhardt, 1963; Alföldi et al., 1963; Sokawa et al., 1971). Shiftdowns in the major carbon or nitrogen source that do not involve the direct removal of amino acid from the media are traversed with relative ease by the relaxed strains (Neidhardt, 1963; Sokawa et al., 1970; Lazzarini and Winslow, 1970; Winslow, 1971). Even this is somewhat surprising, since these stepdowns also are likely to lead to a transient deprivation of the endogenously synthesized amino acids. Nonetheless, the relaxed cells show no evidence of overaccumulating RNA. Perhaps the absence of a readily utilizable carbon or nitrogen source itself prevents the RNA accumulation that occurs when the same cells are transferred from complex to glucose minimal medium. In all outward details the relaxed strains traverse a carbon or nitrogen shiftdown exactly as their stringent counterpart traverses any shiftdown.

In the sections that follow we will review some of the recent work concerning the molecular mechanism(s) that operate to restrict the accumulation of RNA and bring about the changes in gross cellular composition during shiftdown transitions. Despite the many obvious outward similarities in response of rel^+ and rel^- strains, we will group the studies along relaxed and stringent lines to emphasize any role that the rel and $spoT$ genes may play in these transitions.

B. RNA Synthesis in Downshifted Stringent Cells

The cessation of RNA accumulation that occurs in stringent cells during a downshift could be explained by any of the mechanisms that have been proposed for the amino acid regulation of RNA synthesis: coordinate control, noncoordinate control, and degradative control. However, the early reports dealing with this question so clearly established the abundant synthesis of at least some messenger RNA species during the shiftdown transition that noncoordiante control soon became the favorite interpretation (Spiegelman, 1961; Hayashi and Spiegelman, 1961; Naono et al., 1965). Nonetheless, these early reports did not completely resolve this issue; they showed only that RNA was synthesized and that ribosomal RNA was not appreciably labeled by uridine or phosphate during the growth-rate transition. Consequently, two of the three mechanisms were still consistent with the available data: Ribosomal RNA could be made and rapidly degraded, or made at a very slow rate. These remaining alternatives are not mutually exclusive and, as we suggested in the introduction, evidence has been obtained suggesting that both are operative.

Nierlich (1972a, 1972b) measured both the rate of synthesis and composition of pulse-labeled RNA in stringent *E. coli* undergoing a shiftdown from complex to glucose minimal medium. His results show that *rel*[+] *E. coli* regulates RNA synthesis during a stepdown from Casamino acids exactly as it does during direct amino acid starvation. The rate of global RNA synthesis falls to 45%, the preshift rate, and there is a preferential restriction on the synthesis of ribosomal RNA species.

The restriction on the rate of synthesis and composition observed during shift from complex to minimal media also is seen during simple shifts from one carbon source to a poorer one. Erlich (1972) has observed a twofold reduction in the relative rate of ribosomal RNA synthesis during a shiftdown from glucose to succinate. More severe reductions in the rate of global RNA synthesis have been observed in *Bacillus subtilis* traversing a carbon source stepdown (Santangelo, 1970) and in *E. coli* traversing a glucose to lactate downshift (Winslow, 1971). These differences suggest that the severity of the restrictions depends, in large measure, on the particular shift employed and the ease with which the strain can adapt to the poorer carbon source.

C. RNA Synthesis in Downshifted Relaxed Cells

If wild-type (*rel*[+]) bacteria negotiate shiftdown and amino acid deprivation by restricting both the rate of global RNA synthesis and the composition of the RNA synthesized, how then do relaxed mutants, which are unable to elicit these responses during amino acid deprivation, traverse a carbon source downshift? The fact that isogenic relaxed and stringent strains alter their RNA and protein content in identical manner when traversing the same carbon source downshift suggests that these macromolecules are regulated in the same way in the two strains. Lazzarini and Winslow (1970) have shown that this similarity is more than superficial. Employing hybridization competition analyses, they demonstrated that the relaxed strain *E. coli* CP79 preferentially restricts its synthesis of ribosomal RNA during a glucose to succinate shiftdown. Thus, relaxed strains undergoing a downshift exhibit the very response that they fail to show during amino acid starvation. However, it should be pointed out that the degree of noncoordinate control exhibited by relaxed strains during a shiftdown is considerably less than that shown by stringent cells during amino acid starvation. Typically, RNA synthesized by stringent cells undergoing amino acid starvation is 5-10% ribosomal RNA, while that synthesized by relaxed strains undergoing downshift is 15-20% ribosomal RNA. These numbers are to be compared with the 30-35% rRNA found in RNA pulse-labeled during balanced growth. Nonetheless, these results do show that relaxed strains possess the ability to reduce rRNA synthesis in response to changes in the environment.

The extent to which relaxed strains exercise their capacity to restrict rRNA synthesis preferentially not only differs between strains but in some cases changes during the course of an experiment. During a glucose to succinate downshift, the relaxed-strain *E. coli* NF162 increasingly restricts the synthesis of rRNA during the diauxic lag. Starting at 35% rRNA before the lag, the composition of pulse-labeled RNA gradually falls to 25% rRNA by mid-lag and finally to 15% rRNA by the end of the lag phase (Lazzarini, unpublished results). This progressive change in the composition of pulse-labeled RNA may account for the small restriction in the relative rate of rRNA synthesis observed by Erlich (1972) in the same strain. The effect, which at best is small, is most apparent during the end of the diauxie.

The progressive restriction observed with *E. coli* NF162 has not been seen with another relaxed strain, *E. coli* CP79 (Lazzarini and Winslow, 1970). With this strain, pulse-labeled RNA contained 15% rRNA sequences whether labeled early or late during the diauxie.

The reason underlying the different responses of the two relaxed strains has not been established as yet. However, differences in the metabolism of ppGpp that have been detected in these strains (see Sect. III-D) may be relevant.

D. Accumulation of ppGpp During Downshift

The ability of relaxed cells to restrict rRNA synthesis preferentially during downshift is paralleled by their ability to accumulate high levels of ppGpp under the same conditions (see Fig. 2). This surprising observation is another piece of circumstantial evidence implicating ppGpp in the control of rRNA synthesis. During outright glucose starvation (Lazzarini et al., 1971) or downshift from glucose to succinate (Lazzarini et al., 1971; Harshman and Yamazaki, 1971) or lactate (Winslow, 1971), ppGpp rises to many times its basal level in two different *rel⁻* mutants. Plasmolysis in hypertonic NaCl (Harshman and Yamazaki, 1971) or sucrose (Cashel, unpublished) has a similar effect. We believe that leakage of intermediary metabolites out of plasmolyzed cells makes this condition physiologically equivalent to downshift. Similarly, when the downshift physiology is induced with cellular poisons (Nazar and Wong, 1970, 1971), such as cyanide or hydroxylamine, the intracellular concentration of ppGpp rapidly increases in *rel⁺* strains (Edlin and Donini, 1971; Lund and Kjeldgaard, 1972).

Typically, downshifted cells (either *rel⁺* or *rel⁻*) accumulate levels of ppGpp 1/4-1/2 of the value attained by *rel⁺* cells during amino acid starvation. One *rel⁻* mutant (CP79) accumulates ppGpp during downshift virtually as fast as its *rel⁺* parent. Another *rel⁻* mutant (NF162) accumulates ppGpp during downshift much less rapidly than its *rel⁺* sibling. It is not known whether this difference pertains to the different *rel⁻* alleles carried by the two relaxed mutants, or the fact that they differ also in their *spoT* genotype. In any case, the different rates of ppGpp accumulation exhibited by the two strains correlate nicely with the different kinetics with which restriction of rRNA synthesis develops (previous Sect. III-C).

Since *rel*⁻ mutants do not accumulate ppGpp in response to amino
acid starvation, it is clear that downshift and amino acid starvation affect the metabolism of ppGpp in different ways. An additional
clue to the difference between the two mechanisms comes from measurements of the less abundant MS nucleotide, MSII or pppGpp. It will be
recalled that *spo*T⁺ strains accumulate both MS nucleotides in parallel
during the stringent response (see previous Sect. II-G). When the
same *spo*T⁺ strains are subjected to downshift, however, only ppGpp
accumulates (see Fig. 2) (Harshman and Yamazaki, 1971; Stamminger
and Lazzarini, unpublished experiments). In certain downshift conditions, in fact, the level of MSII actually decreases while ppGpp accumulates (Harshman and Yamazaki, 1971; Gallant, unpublished experiments).

In short, the downshift physiology seems to mimic the effect of
the *spo*T⁻ mutation. If this surmise is correct, then one would expect downshifted cells to exhibit a decrease in the rate of ppGpp
turnover or further metabolism, just as *spo*T⁻ mutants do (see previous Sect. II-G). This prediction has been verified both in the case
of cells downshifted from glucose to succinate (Gallant et al., 1972)
and in cells plasmolyzed in hypertonic NaCl (Laffler and Gallant,
unpublished).

These considerations suggest that the accumulation of ppGpp during downshift is attributable, at least in part, to a reduction in
the rate of the *spo*T reaction, namely the phosphorylation of ppGpp
to pppGpp preparatory to futher metabolism.[*] Stamminger and Lazza-

---

[*]Mention should be made, in this connection, of the effects of
morphine analogs such as levallorphan and levorphanol. In *E. coli*,
these compounds bring about a strong inhibition of ribosomal RNA synthesis (Green and Magasanik, 1967; Simon and Van Praag, 1964; Roe-
schenthaler et al., 1969). More recent studies have shown that accumulation of ppGpp is probably the explanation (Harshman and Yamazaki, 1972; Boquet et al., 1973). The drugs apparently induce an
artificial downshift physiology, characterized by reduced turnover
of ppGpp, and a simultaneous accumulation of this nucleotide and depletion of pppGpp. Indeed, on this basis Boquet et al. (1973) have
come to precisely the same conclusions about the postulated *spo*T reaction as we have, independent of the further evidence provided by

rini (1974) suggest that this reduction in the *spo*T reaction may result from a diminished intracellular level of the phosphate donor
during downshift or carbon starvation. This hypothesis also accounts
for the accumulation of ppGpp in relaxed mutants during downshift,
given the fact that relaxed mutants do show some residual synthesis
of ppGpp during normal growth: If the nucleotide's further metabolism were restricted, it would of course accumulate. In the case of
one *rel⁻* mutant, kinetic calculations show that the time course of
ppGpp accumulation during downshift is roughly consistent with such
a mechanism (Gallant et al., 1972). In at least one other *rel⁻* mutant, on the other hand, similar calculations show that an increase
in ppGpp synthesis is probably also involved in the response to downshift (Lazzarini and Gallant, unpublished experiments). The mechanism of this latter response is quite unknown. We might point out
that all *rel⁻* mutants that have been studied show some basal level
of ppGpp, demonstrating that all of the mutants retain some enzymatic activity for ppGpp formation. It is conceivable that this residual activity can be stimulated by conditions that occur during
downshift but not during amino acid starvation.

E. Degradation of rRNA During Downshift

During amino acid starvation of stringent strains, degradation
of newly synthesized rRNA was not detected (Lazzarini and Dahlberg,
1971). In these experiments the rate of rRNA synthesis was calculated from the composition of pulse-labeled RNA and the rate of global
RNA synthesis. These measurements agreed with the observed rate of
increase in the RNA content of the culture, measured as orcinol re-
active material. When these same computations are made with data
from relaxed cells undergoing a downshift, an entirely different answer is obtained. For example, during a glucose to lactate down-

the phenotype of the *spo*T⁻ mutant. A more detailed discussion of
the postulated MS nucleotide pathway will be published in the Alfred
Benzon Symposium IX, "Control of Ribosome Biosynthesis," Munksgaard,
Copenhagen.

shift of *E. coli* CP79, rRNA accounts for 15% of all RNA synthesized. At the same time, the average global rate of RNA synthesis, calculated from the rate of uracil incorporation, is 1.9 nmoles of nucleotide/min per 5×10^8 cells. Consequently, the total amount of rRNA synthesized by 5×10^8 cells during the 165-min diauxic lag is 4.7 nmoles (1.9 x 0.15 x 165) or 16.5 µg. If all this rRNA accumulated, the RNA content of the cells would increase by 22%. These calculations are probably underestimates since they ignore any contribution by tRNA synthesis. As pointed out by Lazzarini and Winslow (1970), these expectations are clearly inconsistent with the observed results. Using an orcinol colorimetric assay sensitive enough to detect a 7% increase, they did not observe any increase in the RNA content of the culture during the diauxie.

The situation with *E. coli* NF162 is quite similar. Erlich's data suggest that enough rRNA is synthesized during the diauxie to increase the RNA content of the cell 27%. Again, no significant increase in the RNA content of NF162 cultures was observed (Erlich, 1972). Thus, degradation of rRNA must occur during stepdown transitions.

The picture that emerges is that noncoordinate synthesis of ribosomal RNA and degradative control, far from being two mutually exclusive modes of regulation, rather are two systems that can act in concert. Under conditions where the restriction of ribosomal RNA synthesis is inadequate (as during a downshift or growth at very slow rates), degradative reactions take up the slack. In this regard, shiftdown transitions resemble balanced growth on poor media. Both of the conclusions drawn from the work on diauxie, i.e., a restriction on the rate of ribosomal RNA synthesis and a degradation of excess rRNA, are the same as those drawn by Norris and Koch (1972) for very slowly growing cells. These authors observe that the composition of newly synthesized RNA varied regularly with growth rate of the cells, showing that cells exercise control over the synthesis of rRNA. Despite this control, the authors observe that cells growing at very slow rates (10-hr doubling time) produced rRNA in excess of that actually accumulated, thus demonstrating that these cells also exercise degradative control of rRNA accumulation.

We have already discussed the evidence that ppGpp is involved in
regulating the synthesis of rRNA. In order to isolate the mechanism
governing rRNA stability, it is instructive to consider those physiological conditions where RNA accumulation is blocked but ppGpp does
not accumulate. Two such conditions have been analyzed in detail:
uracil starvation (Lazzarini et al., 1969) and guanine starvation
(Erlich et al., 1975). In both cases, there is virtually no RNA accumulation, while the global instantaneous rate of RNA synthesis
shows a surprisingly moderate reduction, about a factor of three.
In contrast to the stringent response, however, this result is not
because of a preferential reduction in the synthesis of rRNA. On
the contrary, hybridization-competition analysis shows that the proportion of rRNA (mainly nascent) labeled in short pulses under these
conditions is nearly normal. Therefore, cells starved for a pyrimidine or purine precursor produce rRNA but then turn it over, instead
of accumulating it in stable form as growing cells do.

It is easy to see the selective advantage of preventing RNA accumulation when a pyrimidine or purine precursor is not available.
Were RNA accumulation to continue in this circumstance, then the ri-
boside triphosphate pools would be quickly drained to a point where
messenger RNA synthesis would be greatly retarded, and adaptive responses (such as the elaboration of new enzymes made necessary by the
starvation condition) would be hindered.

Since there is no accumulation of ppGpp in uracil- or guanine-
starved cells, ppGpp is evidently not involved in triggering rRNA degradation. On the contrary, the triggering stimulus must be some
event that is common to uracil starvation, guanine starvation, and
downshift.

One obvious possibility is the synthesis of ribosomal proteins.
One could postulate that naked rRNA molecules are vulnerable to hydrolysis unless they are protected by association with newly made
ribosomal proteins. Were this the case, then the stability of rRNA
molecules would be determined by the rate of synthesis of ribosomal
proteins, and would presumably be correlated with the overal rate of
protein synthesis. However, crude correlations of this sort do not

work out. During amino acid starvation, where the rate of protein synthesis is reduced to some 5% of normal, rRNA accumulates with little degradation in relaxed mutants. Similarly, concentrations of chloramphenicol that inhibit protein synthesis even further allow continued rRNA accumulation in both rel^+ and rel^- cells. Yet, in cells starved for guanine or uracil, a much higher rate of protein synthesis, some 20-30% of normal (Gallant and Harada, 1969; Lazzarini et al., 1969), accompanies virtually complete degradation of newly made rRNA. In short, the signal for RNA degradation must not only be common to the three situations listed above, but it must also be absent in chloramphenicol-inhibited or amino-acid-starved cells.

The most attractive possibility is that the activity of one or more scavenger enzymes that degrade rRNA is directly controlled by small molecules, possibly nucleotides. In this way, the activity of the scavenger system would respond to the cell's overall capacity for phosphorylation and nucleotide generation. Downshift undoubtedly reduces the cell's phosphorylating capacity, and decreases in the nucleotide pools, albeit rather small ones, have in fact been observed to occur during downshift (see Fig. 3) (Winslow, 1971; Lazzarini, Cashel, and Gallant, assorted unpublished downshifts). On the other hand, neither chloramphenicol inhibition nor amino acid starvation has such an effect. On the contrary, nucleotides accumulate under both conditions (Edlin and Neuhardt, 1967; Gallant, unpublished experiments). Thus, neither condition would be expected to activate the hypothetical scavenger system.

We propose, therefore, that there exist specific rRNA nucleases subject to control by small molecules, possibly nucleotides, whose levels may vary with metabolic activity. The identification of such enzymes has yet to be attempted, and it is not possible to specify which nucleotides might control their activity. In fact, they may be nucleotides that have yet to be discovered.*

---

*Very recently, Gallant, Shell, and Bittner (in press) have detected another unusual nucleotide whose level responds dramatically to downshift.

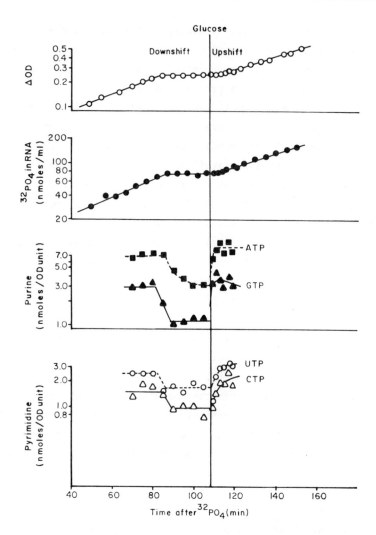

FIG. 3. Ribonucleoside triphosphate pools during growth-rate
shifts. Strain NF162 (arg⁻ met⁻ spoT⁻ rel⁻) was cultivated in minimal medium in the presence of a limiting concentration of glucose
(0.03%) and an excess of succinate (0.3%). At the point indicated
by the optical density curve break, the glucose was exhausted from
the medium and a downshift began. At the point indicated by the
vertical line, excess glucose (0.2%) was added back to the culture
to produce an upshift. RNA and ribonucleoside triphosphate pool
sizes were measured as described by Gallant and Harada (1969).

REFERENCES

Aboud, M., and I. Pastan. 1972. Stimulation of LAC transcription by guanosine 5'-diphosphate 2' (or 3')-diphosphate and transfer ribonucleic acid. *J. Biol. Chem. 248:*3356-3358.

Alföldi, L., and E. Kerekes. 1964. Neutralization of the amino acid sensitivity of RC^{rel} *Escherichia coli. Biochim. Biophys. Acta 91:* 155-157.

Alföldi, L., G. S. Stent, and R. C. Clowes. 1962. The chromosomal site of the RNA control (RC) locus in *E. coli. J. Mol. Biol. 5:* 348-355.

Alföldi, L., G. S. Stent, M. Hooks, and R. Hill. 1963. Physiological effect of the RNA control (RC) gene in *E. coli. Z. Vererbungslehre 94:*285-302.

Arai, K., N. Arai, M. Kawakita, and Y. Kaziro. 1972. Interaction of guanosine 5'-diphosphate 2'-(or 3') diphosphate (ppGpp) with elongation factors from *E. coli. Biochem. Biophys. Res. Commun. 48:*190-196.

Beckmann, J. S., and V. Daniel. 1974. Transcriptional control of in vitro tRNA^{Tyr} synthesis. *Biochemistry 13:*4058-4062.

Birnbaum, L. S., and S. Kaplan. 1973. *In vitro* synthesis of *Escherichia coli* ribosomal RNA. *J. Mol. Biol. 75:*73-81.

Block, R. 1976. Synthesis of ribosomal RNA in a partially purified extract from *Escherichia coli.* in: *Control of Ribosome Synthesis,* Alfred Benzon Symposium IX (N. O. Kjelgaard and O. Maaloe, eds.). Copenhagen: Munksgaard. In press.

Block, R., and W. A. Haseltine. 1973. Thermolability of the stringent factor in *rel* mutants of *Escherichia coli. J. Mol. Biol. 77:* 625-629.

Block, R., and W. A. Haseltine. 1974. In vitro synthesis of ppGpp and pppGpp. in: *Ribosomes* (M. Nomura, A. Tissieres, and P. Lengyel, eds.). Cold Spring Harbor Laboratory. pp. 747-761.

Blumenthal, T., T. A. Landers, and K. Weber. 1972. Bacteriophage Qβ replicase contains the protein biosynthesis elongation factors EF Tu and EF Ts. *Proc. Nat. Acad. Sci. US 69:*1313-1317.

Boquet, P. L., M.-A. Devynck, C. Monnier, and P. Fromageot. 1973. Inhibition of stable RNA synthesis by levallorphan in *Escherichia coli. Eur. J. Biochem. 40:*31-42.

Borek, E., A. Ryan, and J. Rockenbach. 1955. Studies on a mutant of *Escherichia coli* with unbalanced ribonucleic acid synthesis. *J. Bacteriol. 71:*318-323.

Bretscher, M. S., and M. Grunberg-Manago. 1962. Polyribonucleotide directed protein synthesis using an *E. coli* cell-free system. *Nature 195:*283-284.

Cannon, M., R. Krug, and W. Gilbert. 1963. The binding of s-RNA by *Escherichia coli* ribosomes. *J. Mol. Biol. 7:*360-378.

Cashel, M. 1969. The control of ribonucleic acid synthesis in *Escherichia coli.* IV. Relevance of unusual phosphorylated compounds from amino acid starved stringent strains. *J. Biol. Chem. 244:*3133-3143.

Cashel, M. 1970. Inhibition of RNA polymerase by ppGpp, a nucleotide accumulated during the stringent response to amino acid starvation in *E. coli. Cold Spring Harbor Symp. Quant. Biol. 35:*407.

Cashel, M., and J. Gallant. 1968. Control of RNA synthesis in *Escherichia coli.* I. Amino acid dependence of the synthesis of the substrates of RNA polymerase. *J. Mol. Biol. 34:*317-330.

Cashel, M., and J. Gallant. 1969. Two compounds implicated in the function of the RC gene of *Escherichia coli. Nature 221:*838-841.

Cashel, M., and J. Gallant. 1974. Cellular regulation of guanosine tetraphosphate and guanosine pentaphosphate. in: *Ribosomes* (M. Nomura, A. Tissieres, and P. Lengyel, eds.). Cold Spring Harbor Laboratory. pp. 733-745.

Cashel, M., and B. Kalbacher. 1970. The control of ribonucleic acid synthesis in *Escherichia coli.* V. Characterization of a nucleotide associated with the stringent response. *J. Biol. Chem. 245:* 2309-2318.

Chen, J. H., S. M. Weissman, and P. Lengyel. 1972. RC[+] gene-dependent inhibition of RNA synthesis without ppGpp accumulation. *Biochem. Biophys. Res. Commun. 46:*785-789.

Cochran, J. W., and R. W. Byrne. 1974. Isolation and properties of a ribosome-bound factor required for ppGpp and pppGpp synthesis in *Escherichia coli. J. Biol. Chem. 249:*353-360.

De Boer, H. A., H. A. Raue, G. Ab, and M. Gruber. 1971. Role of the ribosome in stringent control of bacterial RNA synthesis. *Biochim. Biophys. Acta 246:*157-160.

de Crombrugghe, B., B. Chen, M. Gottesman, I. Pastan, H. E. Varmus, M. Emmer, and R. L. Perlman. 1971a. Regulation of LAC mRNA synthesis in a soluble cell-free system. *Nature New Biol. 230:*37-40.

de Crombrugghe, B., B. Chen, W. Anderson, P. Nissley, M. Gottesman, I. Pastan, and R. Perlman. 1971b. LAC DNA, RNA polymerase and cyclic AMP, LAC repressor and inducer are the essential elements for controlled LAC transcription. *Nature New Biol. 231:*139-142.

Dimmitt, K., S. Bradford, and M. Simon. 1968. Synthesis of bacterial flagella. I. Requirement for protein and ribonucleic acid synthesis during flagella regeneration in *Bacillus subtilis. J. Bacteriol. 95:*801-810.

Donini, P. 1970. Amino acid control over deoxyribonucleic acid synthesis in *Escherichia coli* infected with T-even bacteriophage. *J. Bacteriol. 102:*616-627.

Donini, P. 1972. Turnover of ribosomal RNA during the stringent response in *Escherichia coli*. *J. Mol. Biol.* 72:553-569.

Edlin, G., and P. Broda. 1968. Physiology and genetics of the "Ribonucleic Acid Control" locus in *Escherichia coli*. *Bacteriol. Rev.* 32:206-226.

Edlin, G., and P. Donini. 1971. Synthesis of guanosine 5' diphosphate, 2'(or 3'-)diphosphate and related nucleotides in a variety of physiological conditions. *J. Biol. Chem.* 246:4371-4373.

Edlin, G., and J. Neuhard. 1967. Regulation of nucleoside triphosphate pools in *Escherichia coli*. *J. Mol. Biol.* 24:225-230.

Edlin, G., G. S. Stent, R. F. Baker, and C. Yanofsky. 1968. Synthesis of a specific messenger RNA during amino acid starvation of *Escherichia coli*. *J. Mol. Biol.* 37:257-268.

Erlich, H. 1972. Ph.D. thesis. University of Washington. The regulation of RNA synthesis and stability by nucleotide pools in *Escherichia coli*.

Erlich, H., T. Laffler, and J. Gallant. 1971. ppGpp formation in *Escherichia coli* treated with rifampicin. *J. Biol. Chem.* 246: 6121-6125.

Erlich, H., R. A. Lazzarini, and J. Gallant. 1975. Synthesis and turnover of ribosomal RNA in guanine-starved cells of *E. coli*. *J. Biol. Chem.* 250:3057-3061.

Fiil, N. 1969. A functional analysis of the rel gene in *Escherichia coli*. *J. Mol. Biol.* 45:195-203.

Fiil, N., and J. D. Friesen. 1968. Isolation of relaxed mutants of *Escherichia coli*. *J. Bacteriol.* 95:729-731.

Fiil, N. P., K. von Meyenburg, and J. D. Friesen. 1972. Accumulation and turnover of guanosine tetraphosphate in *Escherichia coli*. *J. Mol. Biol.* 71:769-783.

Foster, G. G. 1968. Master's thesis, University of Washington. Studies on the amino acid regulation of messenger RNA synthesis in *Escherichia coli*.

Friesen, J. D. 1966. Control of messenger RNA synthesis and decay in *Escherichia coli*. *J. Mol. Biol.* 20:559-573.

Gallant, J., and B. Harada. 1969. The control of RNA synthesis in *Escherichia coli*. III. The functional relationship between purine ribonucleoside triphosphate pool sizes and the rate of ribonucleic acid accumulation. *J. Biol. Chem.* 244:3125-3132.

Gallant, J., and G. Margason. 1972. Amino acid control of messenger ribonucleic acid synthesis in *Bacillus subtilis*. *J. Biol. Chem.* 247:2280-2294.

Gallant, J., H. Erlich, B. Hall, and T. Laffler. 1970. Analysis of the RC function. *Cold Spring Harbor Symp. Quant. Biol.* 35:397-405.

Gallant, J., J. Irr, and M. Cashel. 1971. The mechanism of amino acid control of guanylate and adenylate biosynthesis. *J. Biol. Chem.* 246:5812-5816.

Gallant, J., G. Margason, and B. Finch. 1972. On the turnover of ppGpp in *E. coli*. *J. Biol. Chem. 247*:6055-6058.

Gallant, J., L. Shell, and R. Bittner. 1976. A novel nucleotide implicated in the response of *Escherichia coli* to energy source downshift. *Cell:*in press.

Goldberg, A. L. 1972. Degradation of abnormal proteins in *Escherichia coli*. *Proc. Nat. Acad. Sci. US 69*:422-426.

Golden, N. G., and G. L. Powell. 1972. Stringent and relaxed control of phospholipid metabolism in *Escherichia coli*. *J. Biol. Chem. 247*:6651-6658.

Goodman, D., H. Manor, and W. Rombauts. 1969. Ribosomal protein synthesis during and after amino acid starvation in relaxed and stringent bacteria. *J. Mol. Biol. 40*:247-260.

Greene, R., and B. Magasanik. 1967. The mode of action of levallorphan as an inhibitor of cell growth. *Mol. Pharmacol. 3*:453-472.

Gros, F., J. M. Dubert, A. Tissieres, S. Bourgeois, M. Michelson, R. Soffer, and L. Legault. 1963. Regulation of metabolic breakdown and synthesis of messenger RNA in bacteria. *Cold Spring Harbor Symp. Quant. Biol. 28*:299-313.

Hall, B. G. 1971. Ph.D. thesis. University of Washington. The role of the RC gene product in translation in *Escherichia coli*.

Hall, B. G., and J. Gallant. 1971. The effect of the RC gene product on constitutive enzyme synthesis. *J. Mol. Biol. 61*:271-273.

Hall, B. G., and J. Gallant. 1972. Defective translation in RC⁻ cells. *Nature New Biol. 237*:131-135.

Hamagishi, Y., T. Nishino, and S. Murao. 1975. Structure and some properties of the new nucleotides (adenosine pentaphosphate, guanosine tetraphosphate and inosine pentaphosphate) synthesized by nucleotide pyrophosphotransferase. *Agr. Biol. Chem. 39*:1015-1023.

Harshman, R. B., and H. Yamazaki. 1971. Formation of ppGpp in the relaxed and stringent strains of *Escherichia coli* during diauxie lag. *Biochemistry 10*:3980-3982.

Harshman, R. B., and H. Yamazaki. 1972. MS I accumulation induced by sodium chloride. *Biochemistry 11*:615-617.

Hartwell, L. H., and B. Magasanik. 1963. The molecular basis of histidase induction in *Bacillus subtilis*. *J. Mol. Biol. 7*:401-420.

Haseltine, W. A. 1972. *In vitro* transcription of *Escherichia coli* ribosomal RNA genes. *Nature 235*:329-333.

Haseltine, W. A., and R. Block. 1973. Synthesis of guanosine tetra- and pentaphosphate requires the presence of a codon-specific uncharged transfer ribonucleic acid in the acceptor site of ribosomes. *Proc. Nat. Acad. Sci. US 70*:1564-1568.

Haseltine, W. A., R. Block, W. Gilbert, and K. Weber. 1972. MS I and MS II made on ribosome in idling step of protein synthesis. *Nature* 238:381-384.

Hayashi, M., and Spiegelman, S. 1961. The selective synthesis of informational RNA in bacteria. *Proc. Nat. Acad. Sci. US* 47:1564-1580.

Hochstadt-Ozer, J., and M. Cashel. 1972. The regulation of purine utilization in bacteria. V. Inhibition of purine phosphoribosyl-transferase activities and purine uptake in isolated membrane vesicles by guanosine tetraphosphate. *J. Biol. Chem.* 247:7067-7072.

Imamoto, F., N. Morikawa, K. Sato, S. Mishima, and T. Nishimura. 1965. On the transcription of the tryptophan operon in *Escherichia coli*. II. Production of the specific messenger RNA. *J. Mol. Biol.* 13:157-168.

Irr, J., and J. Gallant. 1969. The control of ribonucleic acid synthesis in *Escherichia coli*. II. Stringent control of energy metabolism. *J. Biol. Chem.* 244:2233-2239.

Jones, O. W., and M. W. Nirenberg. 1962. Qualitative survey of RNA codewords. *Proc. Nat. Acad. Sci. US* 48:2115-2123.

Kamen, R. 1970. Characterization of the subunits of Qβ replicase. *Nature* 228:527-533.

Kaplan, S., A. G. Atherly, and A. Barrett. 1973. Synthesis of stable RNA in stringent *Escherichia coli* cells in the absence of charged transfer RNA. *Proc. Nat. Acad. Sci. US* 70:689-692.

Kellenberger, E., K. G. Lark, and A. Bolle. 1962. Amino acid dependent control of DNA synthesis in bacteria and vegetative phage. *Proc. Nat. Acad. Sci. US* 48:1860-1868.

Kennel, D. 1968. Titration of the gene sites on DNA by DNA-RNA hybridization. II. The *Escherichia coli* chromosome. *J. Mol. Biol.* 34:85-103.

Kepes, A. 1963. Kinetics of induced enzyme synthesis. Determination of the mean life of β-galactosidase-specific messenger RNA. *Biochim. Biophys. Acta* 76:293-309.

Khan, S. R., and H. Yamazaki. 1972. Trimethoprim-induced accumulation of guanosine tetraphosphate (ppGpp) in *Escherichia coli*. *Biochem. Biophys. Res. Commun.* 48:169-174.

Kondo, M., R. Gallerani, and C. Weissmann. 1970. Subunit structure of Qβ replicase. *Nature* 228:525-527.

Kossman, C. R., T. D. Stamato, and D. E. Pettijohn. 1971. Tandem synthesis of the 16S and 23S ribosomal RNA sequences of *Escherichia coli*. *Nature New Biol.* 234:102-104.

Kurland, C. G., and O. Maaloe. 1962. Regulation of ribosomal and transfer RNA synthesis. *J. Mol. Biol.* 4:193-210.

Laffler, T., and J. Gallant. 1974a. SpoT, a new genetic locus involved in the stringent response in *E. coli*. *Cell* 1:27-30.

Laffler, T., and J. Gallant. 1974b. Stringent control of protein synthesis in *Escherichia coli*. *Cell* 3:47-49.

Lavalle, R., and G. De Hauwer. 1968. Messenger RNA synthesis during amino acid starvation in *Escherichia coli*. *J. Mol. Biol.* 37:269-288.

Lazzarini, R. A., and A. E. Dahlberg. 1971. The control of ribonucleic acid synthesis during amino acid deprivation in *Escherichia coli*. *J. Biol. Chem.* 246:420-429.

Lazzarini, R. A., and L. D. Johnson. 1973. On the regulation of ribosomal RNA synthesis. *Nature New Biol.* 243:17-19.

Lazzarini, R. A., and R. M. Winslow. 1970. The regulation of RNA synthesis during growth rate transitions and amino acid deprivation in *E. coli*. *Cold Spring Harbor Symp. Quant. Biol.* 35:383-390.

Lazzarini, R. A., K. Nakata, and R. M. Winslow. 1969. Coordinate control of ribonucleic acid synthesis during uracil deprivation. *J. Biol. Chem.* 244:3092-3100.

Lazzarini, R. A., M. Cashel, and J. Gallant. 1971. On the regulation of guanosine tetraphosphate levels in stringent and relaxed strains of *Escherichia coli*. *J. Biol. Chem.* 246:4381-4385.

Levin, J. 1970. Codon specific binding of deacylated tRNA to ribosomes. *J. Biol. Chem.* 245:3195-3202.

Levin, J., and M. Nirenberg. 1968. Ribonucleic acid codons and protein synthesis. *J. Mol. Biol.* 34:467-480.

Lund, E., and N. O. Kjeldgaard. 1972. Metabolism of guanosine tetraphosphate in *Escherichia coli*. *Eur. J. Biochem.* 28:316-326.

Maaloe, O., and N. O. Kjeldgaard. 1966. *Control of Macromolecular Synthesis*. New York: Benjamin.

Mazumder, R., Y. Chae, and S. Ochoa. 1969. Polypeptide chain initiation in *E. coli*: sulfhydryl groups and the function of initiation factor FZ. *Proc. Nat. Acad. Sci. US* 63:98.

Merlie, J., and L. Pizer. 1973. Regulation of phospholipid synthesis in *Escherichia coli* by guanosine tetraphosphate. *J. Bacteriol.* 116:355-366.

Midgeley, J. E. M. 1974. Imbalance in ribosomal protein and ribonucleic acid synthesis in amino acid starved rel^+ and rel^- mutants of *Escherichia coli*. *Biochem. Biophys. Acta* 335:349-364.

Morris, D. W., and N. O. Kjeldgaard. 1968. Evidence for the noncoordinate regulation of ribonucleic acid synthesis in stringent strains of *Escherichia coli*. *J. Mol. Biol.* 31:145-148.

Murooka, Y., and R. A. Lazzarini. 1973. *In vitro* synthesis of ribosomal RNA by a DNA-protein complex isolated from *Escherichia coli*. *J. Biol. Chem.* 248:6248-6250.

Naono, S., J. Rouviere, and F. Gros. 1965. Preferential transcription of the lactose operon during the diauxic growth of *Escherichia coli*. *Biochem. Biophys. Res. Commun. 18*:664-674.

Nazar, R. N., and J. T. Wong. 1970. Inhibitor-induced shift-downs in *Escherichia coli*. *J. Bacteriol. 100*:956-961.

Nazar, R. N., and J. T. Wong. 1972. Nucleotide changes and the regulation of ribonucleic acid accumulation during growth rate shifts in *Escherichia coli*. *J. Biol. Chem. 247*:790-797.

Neidhardt, F. C. 1962. Role of amino acids in regulating RNA formation. *Biochem. Biophys. Res. Commun. 7*:361-365.

Neidhardt, F. C. 1963. Properties of a bacterial mutant lacking amino acid control of RNA synthesis. *Biochim. Biophys. Acta 68*: 365-379.

Nierlich, D. P. 1968. Amino acid control over RNA synthesis: A reevaluation. *Proc. Nat. Acad. Sci. US 60*:1345-1352.

Nierlich, D. P. 1972a. Regulation of ribonucleic acid synthesis in growing bacterial cells. Control over the total rate of RNA synthesis. *J. Mol. Biol. 72*:751-764.

Nierlich, D. P. 1972b. Regulation of ribonucleic acid synthesis in growing bacterial cells. II. Control over the composition of the newly made RNA. *J. Mol. Biol. 72*:765-777.

Nishino, T., and S. Murao. 1975. Characterization of pyrophosphoryl transfer reaction of ATP: nucleotide pyrophosphotransferase. *Agr. Biol. Chem. 39*:1007-1014.

Nomura, M. 1970. Bacterial ribosome. *Bacteriol. Rev. 34*:228.

Norris, T. E., and A. L. Koch. 1972. Effect of growth rate on the relative rates of synthesis of messenger, ribosomal and transfer RNA in *Escherichia coli*. *J. Mol. Biol. 64*:633-649.

Oki, T., A. Yoshimoto, S. Sato, and A. Takamatsu. 1975. Purine nucleotide pyrophosphotransferase from *Streptomyces morookanesis*, capable of synthesizing pppApp and pppGpp. *Biochim. Biophys. Acta*: in press.

Olsnes, S., and J. G. Hauge. 1968. Amino acid control of RNA synthesis in T₄ infected *Escherichia coli*. *Eur. J. Biochem. 7*:128-136.

Parks, J. S., M. Gottesman, R. L. Perlman, and I. Pastan. 1972. Regulation of galactokinase synthesis by cyclic adenosine 3',5'-monophosphate in cell-free extracts of *Escherichia coli*. *J. Biol. Chem. 246*:2419-2424.

Pederson, F. S., E. Lund, and N. O. Kjeldgaard. 1973. Codon specific, tRNA dependent in vitro synthesis of ppGpp and pppGpp. *Nature New Biol. 243*:13-15.

Polakis, S. E., R. B. Guchhait, and M. D. Lane. 1973. Stringent control of fatty acid synthesis in *Escherichia coli*: possible regulation of acetyl CoA carboxylase by ppGpp. *J. Biol. Chem. 248:* 7957-7966.

Primakoff, P., and P. Berg. 1970. Stringent control of transcription of phage φ80psu₃. *Cold Spring Harbor Symp. Quant. Biol. 35:* 391-396.

Que, L., G. R. Willie, M. Cashel, J. W. Bodley, and G. R. Gray. 1973. Guanosine 5'-diphosphate, 3'-diphosphate; assignment of structure by ^{13}C nuclear magnetic resonance spectroscopy. *Proc. Nat. Acad. Sci. US 70:*2563-2566.

Reiness, G., H.-L. Yang, G. Zubay, and M. Cashel. 1975. Effects of guanosine tetraphosphate on cell-free synthesis of *Escherichia coli* ribosomal RNA and other gene products. *Proc. Nat. Acad. Sci. US 72:*2881-2885.

Roschenthaler, R., M.-A. Devynck, P. Fromageot, and E. J. Simon. 1969. Inhibition of the synthesis of 5S ribosomal RNA in *Escherichia coli* by levallorphan. *Biochim. Biophys. Acta 182:*481-490.

Rose, J. K., and C. Yanofsky. 1972. Metabolic regulation of the tryptophan operon of *Escherichia coli:* Repressor-independent regulation of transcription initiation frequency. *J. Mol. Biol. 69:* 103-118.

Rosset, R., and L. Gorini. 1969. A ribosomal ambiguity mutation. *J. Mol. Biol. 39:*95-112.

Rudland, P. S., W. A. Whybrow, and B. F. C. Clark. 1971. Recognition of bacterial initiator tRNA by an initiation factor. *Nature New Biol. 231:*76.

Ryan, A. M., and E. Borek. 1971. The relaxed control phenomenon. in: *Progress in Nucleic Acid Research and Molecular Biology* (J. N. Davidson and W. E. Cohn, eds.). Vol. 11, pp. 193-228. New York: Academic Press.

Salser, W., J. Janin, and C. Levinthal. 1968. Measurement of the unstable RNA in exponentially growing cultures of *Bacillus subtilis* and *Escherichia coli*. *J. Mol. Biol. 31:*237-266.

Santangelo, E. M. 1970. Master's thesis. Catholic University of America. RNA synthesis in *Bacillus subtilis* cells during step-down growth.

Sells, B. H., and H. L. Ennis. 1970. Polysome stability in relaxed and stringent strains of *Escherichia coli* during amino acid starvation. *J. Bacteriol. 102:*666-671.

Simon, E. J., and D. Van Praag. 1964. Selective inhibition of synthesis of ribosomal RNA in *Escherichia coli* by levorphanol. *Proc. Nat. Acad. Sci. US 51:*1151-1158.

Smith, R. J., and J. E. M. Midgley. 1973a. The effect of trimethoprim on macromolecular synthesis in *Escherichia coli:* general effects on RNA and protein synthesis. *Biochem. J. 136:*225-234.

Smith, R. J., and J. E. M. Midgley. 1973b. The effect of trimethoprim on macromolecular synthesis in *Escherichia coli*: regulation
of RNA synthesis by "magic spot" nucleotides. *Biochem. J. 136*:
249-257.

Sokawa, Y., and Y. Kaziro. 1969. Amino acid-dependent control of
the transport of α-methyl glucoside in *E. coli*. *Biochem. Biophys.
Res. Commun. 34*:99-103.

Sokawa, Y., E. Nkao, and Y. Kaziro. 1968. On the nature of the control by RC gene in *E. coli*: Amino acid dependent control of lipid
synthesis. *Biochem. Biophys. Res. Commun. 33*:108-112.

Sokawa, Y., E. Nakao-Sato, and Y. Kaziro. 1970. RC gene control in
Escherichia coli is not restricted to RNA synthesis. *Biochim.
Biophys. Acta 199*:256-264.

Sokawa, Y., J. Sokawa, and Y. Kaziro. 1971. Function of the rel
gene in *Escherichia coli*. *Nature New Biol. 234*:7-10.

Spiegelman, S. 1961. The relation of informational RNA to DNA.
Cold Spring Harbor Symp. Quant. Biol. 26:75-90.

Stamato, T. D., and D. E. Pettijohn. 1971. Regulation of ribosomal
RNA synthesis in stringent bacteria. *Nature New Biol. 234*:99-102.

Stamminger, G., and R. A. Lazzarini. 1974. Altered metabolism of
the guanosine tetraphosphate, ppGpp, in mutants of *E. coli*. *Cell
1*:85-90.

Steitz, J. 1969. Polypeptide chain initiation: Nucleotide sequences of the three ribosomal binding sites in bacteriophage R17 RNA.
Nature 224:957-964.

Stent, G. S., and S. Brenner. 1961. A genetic locus for the regulation of ribonucleic acid synthesis. *Proc. Nat. Acad. Sci. US 47*:
2005-2014.

Stubbs, J. D., and B. D. Hall. 1968. Level of tryptophan messenger
RNA in *Escherichia coli*. *J. Mol. Biol. 37*:289-302.

Stubbs, J. D., and E. A. Stubbs. 1970. Effects of amino acid starvation on constitutive tryptophan messenger ribonucleic acid synthesis. *J. Mol. Biol. 51*:717-725.

Sussman, A. J., and C. Gilvarg. 1969. Protein turnover in amino
acid-starved strains of *Escherichia coli* K12 differing in their
ribonucleic acid control. *J. Biol. Chem. 244*:6304-6308.

Swanton, M., and G. Edlin. 1972. Isolation and characterization of
an RNA relaxed mutant of *B. subtilis*. *Biochem. Biophys. Res. Commun. 46*:583-588.

Sy, Y., and F. Lipmann. 1973. Identification of the synthesis of
guanosine tetraphosphate (MSI) as insertion of a pyrophosphoryl
group into the 3'-position in guanosine 5'-diphosphate. *Proc.
Nat. Acad. Sci. US 70*:306-309.

Sy, J., Y. Ogawa, and F. Lipmann. 1973. Nonribosomal synthesis of guanosine 5',3'-polyphosphates by the ribosomal wash of stringent *Escherichia coli*. *Proc. Nat. Acad. Sci. US 70*:2145-2148.

Szer, W., and S. Ochoa. 1964. Complexing ability and coding properties of synthetic polynucleotides. *J. Mol. Biol. 8*:823-834.

Then, R., and P. Angehrn. 1972. Effects of trimethoprim and its antagonists on RNA synthesis in *Escherichia coli*. *Biochim. Biophys. Acta 287*:98-105.

Travers, A. 1971. Control of transcription in bacteria. *Nature New Biol. 229*:69-74.

Travers, A. 1973. Control of ribosomal RNA synthesis *in vitro*. *Nature 244*:15-18.

Travers, A. A., R. I. Kamen, and R. F. Schleif. 1970a. Factor necessary for ribosomal RNA synthesis. *Nature 228*:748-751.

Travers, A., R. Kamen, and M. Cashel. 1970b. The in vitro synthesis of ribosomal RNA. *Cold Spring Harbor Symp. Quant. Biol. 35*:415-418.

Travers, A., D. L. Baillie, and S. Pedersen. 1973. Effect of DNA conformation on ribosomal RNA synthesis *in vitro*. *Nature New Biol. 243*:161-163.

Van Ooyen, A. J. J., H. A. de Boer, G. Ab, and M. Gruber. 1975. Specific inhibition of ribosomal RNA synthesis *in vitro* by guanosine 3' diphosphate, 5' diphosphate. *Nature 254*:530-531.

Venetianer, P. 1968. Preferential synthesis of the messenger RNA of the histidine operon during histidine starvation. *Biochem. Biophys. Res. Commun. 33*:959-968.

Venetianer, P. 1969. Level of messenger RNA transcribed from the histidine operon in repressed, derepressed and histidine-starved *Salmonella typhimurium*. *J. Mol. Biol. 45*:375-384.

Winslow, R. M. 1971. A consequence of the rel gene during a glucose to lactate down shift in *Escherichia coli*. *J. Biol. Chem. 246*:4872-4877.

Winslow, R. M., and R. A. Lazzarini. 1969a. The rate of synthesis and chain elongation of ribonucleic acid in *Escherichia coli*. *J. Biol. Chem. 244*:1128.

Winslow, R. M., and R. A. Lazzarini. 1969b. Amino acid regulation of the rates of synthesis and chain elongation of ribonucleic acid in *Escherichia coli*. *J. Biol. Chem. 244*:3387-3392.

Wong, T. J., and R. N. Nazar. 1970. Relationship of the MS nucleotides to the regulation of ribonucleic acid synthesis in *Escherichia coli*. *J. Biol. Chem. 245*:4591-4595.

Yang, H., G. Zubay, E. Urm, G. Reiness, and M. Cashel. 1974. The effect of guanosine tetraphosphate (ppGpp) and β-γ-methylenyl guanosine pentaphosphate pcppGpp on *Escherichia coli* gene exprssion *in vitro*. *Proc. Nat. Acad. Sci. US 71*:63-67.

Yoshida, M., A. Travers, and B. F. C. Clark. 1972. Inhibition of translation initiation complex formation by MSI. *Fed. Eur. Biochem. Soc. Lett. 23*:163-166.

Zagorska, L., L. J. Dondon, J. C. Lelong, F. Gros, and M. Grunberg-Manago. 1971. The effect of tetracycline on the binding of the aminoacyl tRNA-Tu complex to ribosomes. *Biochimie 53*:63-72.

Zubay, G., L. Cheong, and M. Gefter. 1971a. DNA-directed cell-free synthesis of biologically active transfer RNA: su$^+$III tyrosyl-tRNA. *Proc. Nat. Acad. Sci. US 68*:2194-2197.

Zubay, G., L. Gielow, and E. Englesberg. 1971b. Cell-free studies on the regulation of the arabinose operon. *Nature New Biol. 233*: 164-165.

This review was completed early in 1974. It was subsequently updated in the fall of 1975, but only in regard to certain selected topics. Therefore, work published in 1974 and 1975 is covered in some areas but not in others.

AUTHOR INDEX

A

Aaij, C., 264, 267, 268, *290, 295*

Ab, G., 322, 335, *350, 358*

Abelson, H.T., 26, 52, *55, 61, 64*

Aboud, M., 331, *349*

Adams, J.M., 47, *55*

Adelman, M.R., 182, *227*

Adesnik, M., 5, 6, 12, 25, 26, 27, *55, 57, 60,* 108, *114*

Adler, K., 205, *240*

Agroudelis, A.D., 134, *167*

Agsteribbe, E., 251, 273, 274, 285, 287, *290, 293, 298*

Akai, A., 257, *305*

Akazawa, T., 196, *246*

Alberte, R.S., 205, *228*

Alfert, M., 73, 91, 92, 93, 95, *114, 115, 116*

Alföldi, L., 313, 314, 329, 339, *349*

Aliev, K.V., 186, *228*

Aliyev, K.A., 184, *245*

Allen, M.B., 171, *228*

Allen, N.E., 288, *290*

Allfrey, V.G., 23, 24, *65,* 86-89, *115*

Aloni, Y., 269, *303*

Amenson, C.S., 33, 34, 38, *57*

Andersen, W.R., 223, *228*

Anderson, J.M., 205, *228*

Anderson, L.A., 182, *228*

Anderson, W.F., 127, 130, *161,* 331, 333, *350,*

Andre, J., 267, 273, 275, 276, 277, 280, 281, *290, 293, 304, 305*

Angehrn, P., *358*

Apel, K., 222, *228*

App, A.A., 173, *228*

Arai, K., 331, 335, *349*

Arai, N., 331, 335, *349*

Arcos, J.C., 255, *290*

Arendzen, A.J., 285, 286, *293, 297*

Argus, M.F., 255, *290*

Armstrong, J.J., 295, 297, 201, 202, 205, 206, 207, 224, *228, 239*

Arnon, D.I., 171, 175, *228, 243*

Aronoff, S., 171, *228*

Ashwell, M., 251, 257, 264, 265, *290*

Atherly, A.G., 322, 323, *353*

Atkinson, M.R., 151, *161*

Attardi, B., 4, 5, 10, *55,* 263, *290*

Attardi, G., 4, 5, 10, 12, 13, 26, 36, 37, *55, 60, 61, 63,* 109, *120,* 254, 257, 258, 263-265, 269, 270, 286, *290, 293-295, 299, 302, 303*

Aurich, O., 204, 205, 221, *240*

Autilio-Gambetti, L.A., 262, *295*

Avadhani, N.G., 177, 181, 182, 183, *228,* 265, 280-282, 288, *291*

Avakeyan, E.R., 31, 40, 41, *59*

Avers, C.J., 273, 275, *292, 304*

Aviv, H., 8, 9, *47, 62*

Ayuso, M., 135, 136, 143, 145, *162*

Ayuso-Parilla, M., 132, *159*

B

Bachenheimer, S.L., 29, *66*

Bachmayer, H., 184, *228*

Baglioni, C., 35, 36, *56, 58,* 135, 143, 158, 159, *161, 163*

Baierlein, R., 178, *237*

SUBJECT INDEX

A

Abrin, 155
Acetabularia, 175, 184, 222
Acetoxycycloheximide, 150
Actinomycin D
  effect on histone mRNA, 109, 112
  effect on histone synthesis, 81
  effect on oligo(A) synthesis, 32
  effect on protein synthesis by chloroplasts, 175
  effect on rRNA synthesis, 4
Adenovirus hnRNA, 17-19
Adenylosuccinate synthetase, inhibited by ppGpp, 330
Amicetin, 145-147
Amino acid pool in mitochondria, 253
Amino acids, attachment to preexisting proteins, 254
Amino acyl-tRNA synthetases of chloroplasts, 184-188, 192-193
Aminosugar nucleoside antibiotics, 145-147
Anacystis, 193
Anguidine, 139
Anisomycin, 139, 141-143, 147, 279
Antibiotic susceptibility:
  of chloroplast ribosomes, 182-183
  of mitoribosomes, 285-286
Antirrhinum, 220

Artemia, ribosome hybrid with *E. coli*, 287
ATP
  in amino acid synthesis, 313
  in mitochondria, 253
  requirement for protein synthesis by isolated chloroplasts, 173-174
ATPase
  of chloroplasts, 190
  of mitochondria, 255
Aurintricarboxylic acid (ATA), 129-134

B

Bacillus subtilis, mRNA synthesis during amino acid starvation, 317
Bacteria
  as contaminants of isolated chloroplasts, 173-174
  as contaminants of isolated mitochondria, 260-261
Bacterial ribosomes
  comparison with chloroplast ribosomes, 176-179
  comparison with mitoribosomes, 284-289
Balbiani ring two, of *Chironomous*, 15
Blasticidin, 145-147
Bruceantin, 136-137
Buoyant density
  of animal mitoribosomes, 266
  of chloroplast DNA, 213
  of fungal mitoribosomes, 276
  of mitochondrial DNA, 213
  of plant mitoribosomes, 284